CONTENTS

MODERN FLUID MECHANICS

MODERN FLUID MECHANICS

SHIH-I. PAI

SCIENCE PRESS

Beijing, 1981

Distributed by

VAN NOSTRAND REINHOLD COMPANY

New York, Cincinnati, Toronto, London, Melbourne

7/20-3588

CHEMISTRY

To My Wife

Alice (Jen-Lan Wang)

To My Wife

Alice (Yeh-Lan Wang)

PREFACE

Even though the basic concepts of fluid mechanics had been well developed in the nineteenth century, fluid mechanics is essentially a science of the twentieth century. In the nineteenth century, there were two unrelated sciences concerned with fluid flow: one was hydrodynamics and the other was hydraulics. The old hydrodynamics was mainly the mathematical theory of fluid flow and its results did not agree with experimental facts investigated in the nineteenth century. It was considered as a hunting ground for mathematicians. The old hydraulics was an engineering science and its results were expressed in terms of various coefficients. Since these coefficients depend on many parameters, the old hydraulics was referred to as a science of coefficients. These two sciences of fluid flow were combined into a new science of fluid mechanics in the twentieth century by such pioneers as L. Prandtl, Th. von Kármán, G.I. Taylor, J.M. Burgers, S. Goldstein, and many others.

The development of fluid mechanics in the twentieth century may be divided into four periods.

(1) *Low Speed Aerodynamics.* *1900 — 1935*

The first development of fluid mechanics was closely associated with aeronautical science. Because of the stringent requirement on weight, one needs very accurate theoretical prediction to practical problems. As a result, one has to combine the essential features of old hydrodynamics and hydraulics into one rational science of fluid mechanics. Some of the important developments in these periods are: (a) Prandtl's boundary layer theory; (b) Prandtl's finite wing theory to explain the phenomenon of induced drag; (c) the theory of turbulent flow by von Kármán and G.I. Taylor and others. In this period, the velocity of the fluid flow is low and the temperature of the fluid is also low. As a result, we may neglect the compressibility effect of the fluid. Both the gas and the liquid may be treated by the same method of analysis. There is practically no difference in principle for hydrodynamics and aerodynamics.

(2) *Aerothermodynamics.*   *1935 — 1950*

The speed of the gas flow was gradually increased from subsonic to supersonic speed. The compressibility effect of the gas is no longer negligible. We have to treat gas and liquid separately. For gasdynamics, we have to consider the mechanics of the flow simultaneously with the thermodynamics of the gas. Hence the term of Aerothermodynamics was suggested for this new branch of fluid mechanics. The most important parameter is the Mach number. However, the temperature range of the gas or air was still below 2000°K and the air may be considered as an ideal gas with constant specific heat. The molecular structure has very little influence on the gas flow and we may use the same formula to deal with monatomic gas and polyatomic gas. Many new phenomena, however, such as shock wave, supersonic flow, etc., had been analyzed in this period.

(3) *Physics of Fluid.*   *1950 — 1960*

This is the start of the space age. The speed of the flow and the temperature of the fluid are high enough so that we have to consider the interaction of mechanics of fluid with other branches of physics and that the molecular structure of the gas has a large influence on the fluid flow. We have to consider the influence on dissociation, ionization, and thermal radiation. New subjects such as aerothermochemistry, magnetogasdynamics, plasma dynamics, and radiation gasdynamics have been extensively studied. We have to deal with the whole physics of fluids.

(4) *New Era of Fluid Mechanics.*   *1960 and on*

In the above three periods, our main interests are still the flow of fluids which consists of liquid, gas, or plasma only. During the recent years, the interest of many technical developments is so broad that we have to deal with flow problems beyond those of fluid alone. For instance, we have to deal with the mixture of solid and fluid, the so-called two-phase flow. In many rheological problems, the fluids behave partly as ordinary fluid and partly as solid. In the above three periods, we treat the fluid flow problems mainly according to the principles of classical physics. In many new problems of fluid flow, we have to consider the principles beyond those of classical physics such as superfluid for which the quantum effects are important even for macroscopic properties (quantum fluid mechanics); relativistic fluid mechanics in which the relativistic mechanics should be used because the velocity of the flow is no longer negligible in comparison with the speed of light. We are also interested in bio-fluid mechanics in which we study the interaction between the physical science of fluid flow and biological science.

It should be noted that even though we divide the development of modern fluid mechanics into the above four periods, there are overlappings in time for these periods as far as the study of various subjects are concerned. For instance, the study of turbulent flow of low speed fluid flow which was one of the major subjects in the first period is still a very active research subject at the present time and many basic problems are far from solved yet. Such new subjects as quantum fluid mechanics and relativistic fluid mechanics have been extensively studied in the 1940s. However, many of the textbooks of fluid mechanics are still concerned with only those subjects for the first two periods, that is, low speed aerodynamics and simple aerothermodynamics. It is the author's aim to write a book which gives a simple general review of the fundamental principles for all the subjects in these four periods, particularly those in the last two periods.

In chapter I, we discuss the properties of fluids: liquid, gas, and plasma. These are the fundamental facts of all fluids which are useful in the study of fluid flow. In chapter II, we discuss the statics of fluids including those new topics such as plasma and rarefied gas. In chapter III, we review some fundamental principles of fluid dynamics, while in chapter IV, we discuss the dimensional analysis and dynamic similarity and their applications to fluid mechanics.

In chapters V to IX, we are concerned with the modern aspects of fluid mechanics. In chapter V, we deal with the aerothermochemistry, that is, the flow problem with chemical reactions. In chapter VI, we deal with magnetofluid dynamics and plasma dynamics, that is, the flow problems of electrically conducting fluid. In chapter VII, we deal with radiation gasdynamics, that is, the flow problems including the effects of thermal radiation.

In chapters II to VII, we consider the fluid as a continuum and use the classical continuum theory such as the Navier-Stokes equations or their modifications to study the fluid flow. In chapter VIII, we consider the flow of gas from the microscopic point of view and the relations between the results of the kinetic theory of gases and the continuum theory of fluid mechanics. Finally in chapter IX, some special topics of modern fluid mechanics, including rheology, two–phase flow, multifluid theory, superfluids, relativistic fluid mechanics, and bio-fluid mechanics are discussed.

During the month of June, 1979, I was invited by the Northwestern Polytechnical University (NWPU),Xi'an, Shanxi Province, the People's Republic of China, to deliver a series of lectures on Modern Fluid Mechanics. My lectures were based essentially on the materials of this book. Because of the hospitality of NWPU to me and my wife Alice (Jen-Lan Wang), and their extreme interest in my lectures, I have presented this book to NWPU as one of their publications.

The author takes this opportunity to thank his wife, Alice, for her constant interest and encouragement during the preparation of the manuscript and his son, Robert Yang Pai, for his proof-reading of part of the manuscript.

Shih-I. Pai

College Park, Maryland
July, 1980

LIST OF FIGURES

THE PROPERTIES OF FLUIDS

1. Fluids and Modern Fluid Mechanics

There are four states of matter: solid, liquid, gas, and plasma (ionized gas). Excepting the solid state, matter in the other three states may be deformed without applying any force, provided the change of shape takes place over a sufficiently long time. The term Fluid has been used as a general name for the three states of matter: liquid, gas, and plasma. There are many similar properties of all fluids which may be treated by the method known as Fluid Mechanics. One of the objects of this book is to discuss the essential results of fluid mechanics. However, there are many special properties of each of these three states which differ significantly from those of the other two states. We shall also point out these special properties.

In our present concept of matter, matter consists of a large number of discrete particles called molecules which are in continual motion[1]. The term, molecule, in a general sense includes ordinary molecules as well as atoms, ions, and electrons. For many practical problems, because of the large number of molecules in a very small volume, the discrete character of the molecules may be replaced by some average properties of the fluid as a whole. As a result, the fluid may be considered as a continuum as a first approximation. The properties of the continuum as a whole are the average properties of the molecules of the matter. Another object of this book is to discuss the relations between the properties of the molecules of a fluid and the properties of the fluid as a whole. At very high speed and high temperature flows, the influence of the individual properties of the molecules on the flow of the fluid as a whole becomes very significant. These influences have not been discussed in the classical fluid mechanics. They may be regarded as a modern development of fluid mechanics which is very important for engineers and physicists in this "space age".

Strictly speaking, fluid mechanics in the sense of classical physics, deals with the flow of the fluid (kinematics) and the forces acting on the fluid and produced by the fluid on solid bodies (statics and dynamics). However, in many physical conditions which are very important in many practical problems in this space age, we have to study the mechanics of fluids simultaneously with other branches of classical physics. For example:

(1) For high speed and high temperature flow of gases, we have to study the mechanics of gases simultaneously with thermodynamics. It is sometimes known as aerothermodynamics (chapter III).

(2) In combustion problems and very high temperature flow, we have to study the mechanics of fluids simultaneously with chemical reaction and thermodynamics. It is sometimes known as aerothermochemistry (chapter V).

(3) For the flow of a plasma, we have to study the mechanics of plasma simultaneously with electromagnetism. It is known as plasma dynamics of which the well known magnetofluid dynamics is an important branch (chapter VI).

(4) For extremely high temperature, the thermal radiation has significant effect on the fluid flow. We may represent the thermal radiation by heat rays. Hence we have to study the interaction of fluid mechanics with geometric optics which is known as radiation gasdynamics (chapter VII).

(5) At very low temperature, the fluid has some very peculiar properties. Such a fluid is known as superfluid. In the study of the flow of superfluid, we have to study the mechanics of fluid simultaneously with quantum mechanics. It is known as quantum fluid mechanics (chapter IX).

(6) When the velocity of the fluid, whether it is the microscopic motion of the fluid particles or the macroscopic motion, is not negligibly small in comparison with the velocity of light, we have to consider the relativistic effects on the fluid motion. It is known as relativistic fluid mechanics. Except in section 6 of chapter IX, we

shall neglect the relativistic effects in this book.

(7) When the mean free path of the gas is not small, we have to consider the discrete character of the molecule. The treatment of flow problems is then significantly different from the classical fluid mechanics of continuum theory. It is known as rarefied gas-dynamics (chapter VIII).

The third object of this book is to give some essential results on the interaction of mechanics of fluid with other physical phenomena.

2. Properties of Liquids

The molecules of a liquid are quite close together. One of the properties of the liquid is its great resistance to the change of volume. Under extremely high pressure, the volume of liquid may be changed slightly. For water, if we apply a pressure of 1000 atmospheres, its volume may be reduced by 5%. Because of the enormously large resistance to the change of volume, it is a very good approximation that the liquid may be considered as an incompressible fluid in the treatment of fluid mechanics.

From the kinetic theory of matter, the particles of a solid oscillate about a fixed position while those of a liquid change position frequently. This change of position is the reason that the shape of liquid may be changed without force because the particles yield in the direction of stress and there will be no stress when the liquid is at rest.

Some of the particles of a liquid may leave its free surface on which the liquid is in contact with a gas or a vacuum and reach the gaseous state. Such a gas is usually known as vapor. Under a given physical condition, there is an equilibrium condition between the vapor and the liquid (chapter II section 2). When the temperature of a liquid increases, the number of particles of the liquid reaching the vapor state increases too. At very high temperature, the liquid will change completely into gaseous state.

The free surface of a liquid exhibits a tendency to contract. We may say that there is a surface tension for liquids (chapter II section 2). Surface tension is a special property for liquid which does not occur in a gas or a plasma and which appears in liquid when the surface is curved.

Some liquids, such as mercury, are electrically conductors. Electrons may flow in these liquids under the influence of electric and magnetic fields. Such liquids behave in a manner similar to a plasma. We shall discuss these phenomena in chapter VI.

3. Properties of Gases

The main difference between the properties of gases and those of liquids is that the volume of a gas can easily be changed. A gas will completely fill the region which encloses it no matter what amount of gas is enclosed. If we apply suitable pressure to a given amount of gas, it can be compressed into a very small space. When the gas flows, its volume will change with the velocity of the flow. There are, however, also many similar properties between liquid and gas. For instance, if the Mach number of a gas (which is the ratio of the velocity of the flow to the sound speed of the gas) is small, the change of volume of the gas due to the change of velocity is negligibly small. Then the gas may be considered as an incompressible fluid which behaves exactly like a liquid. The variation of density is proportional to the square of the Mach number when the variation of the velocity is small. The sound speed of air at sea level under standard conditions is 1120 feet per second. If the velocity of the air is less than 110 feet per second, the Mach number will be less than 0.1. Then the variation of the density of the air will be less than 1% of the variation of the velocity. Under such a condition, it is a good approximation to consider the air as an incompressible fluid. Many new phenomena will occur due to the compressibility effect of a gas when the Mach number is large and the gas can no longer be considered as an incompressible fluid. For instance, a shock wave may occur. When the Mach number and the temperature of a gas are very high, the molecular structure of the gas will have significant influence on the flow of the gas. Such influences have been extensively investigated during the last twenty years and they constitute a major portion of modern fluid mechanics.

Much matter is in the gaseous state under room temperature

and pressure such as oxygen, nitrogen, etc. The most common gas is air which is a mixture of oxygen, nitrogen, and a small amount of other gases. Even though air is a mixture of gases, its composition under normal conditions is so constant that it may be considered as a single gas in the treatment of flow problems. However, at very high temperature, the composition of air is no longer constant. Then the effects of the mixture of gases in the air, such as diffusion and chemical reaction, will have a large influence on the flow field. These are important problems in modern fluid mechanics.

4. Properties of a Plasma

Even though each molecule of a gas consists of negatively charged electrons and a positively charged nucleus, it is electrically neutral under normal conditions. Hence in the study of ordinary fluid mechanics, the electromagnetic forces have no influence on the gas flow. However, at high temperature, say above $10^{4}°K$, the gas will be ionized. In other words, at very high temperature, some of the electrons may move a large distance away from their nuclei, and they can not be considered as integral parts of the molecule. As a result, we should consider the gas as a mixture of electrons and positively charged molecules, known as ions. The term *plasma* was first used to represent a fully ionized gas which consists of equal numbers of electrons and singly charged ions. However when the temperature of the gas is not very high, not all molecules decompose into ions and electrons. At such moderate temperature, the gas will consist of electrons, ions, and neutral molecules. Such a gas is said to be partially ionized. The current practice is to use the term *plasma* for ionized gas whether it is fully ionized or partially ionized. The main difference between a plasma and an ordinary neutral gas is that the electromagnetic forces play important roles in the dynamics of plasma. Otherwise the plasma behaves very similarly to a gas in many flow problems. The science which studies the properties of plasma is known as plasma physics, of which the branch dealing mainly with the forces and motion of a plasma is known as magneto-fluid dynamics which will be discussed in chapter VI.

5. Simple Kinetic Theory of Fluids. Mean Free Path

Since the fluid·is composed of a large number of particles, the most accurate, although a very complicated method of description of the flow of a fluid, is the molecular theory of a fluid. The kinetic theory of a liquid has not been satisfactorily developed but the kinetic theory of gases has been satisfactorily developed. It is very instructive if we consider many properties of a fluid from the kinetic theory of gas point of view and compare them with the classical theory of fluid mechanics which is based on the theory of continuum. To illustrate the principle, we consider first the simple kinetic theory of gases.

The simplest assumed picture of a gas is as follows:

The gas is assumed to be an aggregate of rapidly moving particles which are rigid bodies of zero diameter but of finite mass and which are constantly colliding with one another to exchange energy. The influence of the particles on each other can be conveniently neglected until they are so close together that a *collision* takes place. The coarseness of the structure of the gaseous medium can be expressed by its free path which is the distance that particles travel between collisions. Since the instantaneous velocity and density distributions of the particles are far from uniform, we can only conveniently use the statistical average of a quantity instead of the instantaneous value of a quantity. The statistical average of the free path is known as the mean free path, which is one of the most important parameters in the kinetic theory of gases. If the mean free path is much smaller than the dimension of the flow field or the dimension of the bodies in the flow field, we may consider the gas as a continuum. If the mean free path is not negligible in comparison with the dimension of the body in the flow field, the effects of the discrete character of the gas must be taken into account in the study of flow problems (see chapter VIII).

Let ξ, η, and ζ be the instantaneous x-, y-, and z-component of the velocity of a particle of gas respectively and u, v, and w be the x-, y-, and z-component of the velocity of the gas flow as a whole respectively. The average values of ξ, η, and ζ are respectively:

$$\bar{\xi}=u, \quad \bar{\eta}=v, \quad \bar{\zeta}=w \tag{1.1}$$

where the bar represents the statistical average values.

We define a peculiar velocity of a particle of the gas as c_a with components c_x, c_y, and c_z such that

$$c_x=\xi-u, \quad c_y=\eta-v, \quad c_z=\zeta-w \tag{1.2}$$

By definition, the average of the peculiar velocity is zero, i.e.,

$$\bar{c_x}=\bar{c_y}=\bar{c_z}=0 \tag{1.3}$$

But the average of powers of the peculiar velocity components are, in general, not zero. For instance:

$$\bar{c_x^2}=\bar{\xi^2}-u^2, \quad \bar{c_y^2}=\bar{\eta^2}-v^2, \quad \bar{c_z^2}=\bar{\zeta^2}-w^2 \tag{1.4}$$

The peculiar velocity represents the velocity fluctuation over the mean velocity of the flow. If we assume that the fluctuations are perfectly random, there is no preferential direction and then we have

$$\bar{c_x^2}=\bar{c_y^2}=\bar{c_z^2}=\frac{1}{3}\bar{c_a^2} \tag{1.5}$$

We shall show later that many properties of the gas as a whole such as pressure, temperature, viscosity, etc., can be expressed in terms of average values of some powers of peculiar velocity components.

In order to get better results from the kinetic theory, we have to improve the kinetic picture of the simple theory. For a next approximation, we may assume that the molecules are still rigid bodies but are surrounded by a field of forces. When two or more molecules approach closely, their paths will be influenced by the presence of the others. We call this an encounter. An encounter in this case takes the place of collision in the simple kinetic theory. For a more refined theory, we may assume that the molecules are nonrigid bodies surrounded by field of forces. Hence the concept of encounter may still be used.

In the case of a plasma which consists of both neutral and charged particles, besides the collision and encounter within a short range just discussed, there is a kind of distant encounter which is due to the electromagnetic interaction between charged particles. This involves taking account of very small angle scattering and thus of a great number of acts of very small momentum transfer. Furthermore the interaction between the electromagnetic fields and the charged particles should also be taken into consideration.

If we could investigate the detailed motion of all molecules in a gas flow, a complete description of the gas flow may be obtained. However it is not possible to do so because, in the first place, we do not know the detailed conditions of all the gas molecules at a prescribed initial instant and, in the second place, the number of molecules of a gas is so large that it is not possible at present to calculate the motions of all the molecules even if we know the initial conditions of all the molecules. Hence we can only conveniently consider the statistical average of the motion of the molecules of a gas.

One way to study the statistical motion of a gas is to use the one particle distribution function (see section 9) which is governed by a nonlinear partial differentio-integral equation known as the Boltzmann equation (see chapter VIII). At present, it is not possible to solve the Boltzmann equation even for some simple practical flow problem. However, the Boltzmann equation and the one particle distribution function serve two important aspects of fluid mechanics. In the first place, the fundamental equations of fluid dynamics may be derived from the Boltzmann equation as a first approximation. Thus we may have some guides about the validity of the fundamental equations for a macroscopic description from the analysis of the Boltzmann equation. In the second place, the Boltzmann equation may give us valuable information about the transport coefficients, such as the coefficient of viscosity, heat conductivity, etc. In the macroscopic analysis, these transport coefficients are simply introduced as known functions of physical quantities of fluid dynamics. They should be either determined experimentally or calculated from the kinetic theory.

6. Pressure

For many practical problems, the description of the motion of a gas in terms of a distribution function is too detailed to be useful. We do not care about the motion of individual particles in the gas and are interested only in the resultant effects due to the motion of a large number of molecules. In other words, we are interested in the macroscopic quantities only, such as pressure, density, temperature, mean flow velocity, etc. These macroscopic properties are the mean values for a large number of molecules. For instance, air at room temperature and pressure contains 2.7×10^{19} molecules per cubic centimeter. If we consider a *point* in space for ordinary flow problem as a cube of side of 1/1000 millimeter, this volume will contain 2.7×10^7 molecules, which is a number generally quite adequate for taking a mean value. These macroscopic properties should be also derived from the kinetic theory. It is very useful to know the relations of these macroscopic properties from both the kinetic theory point of view and the ordinary theory of a continuum (see chapter VIII).

One of the most important properties of a fluid in the analysis of fluid mechanics is the concept of hydrostatic pressure or simply pressure. First we consider the pressure from a macroscopic point of view such that the mean free path is so small that the fluid may be considered as a continuum. It is evident that every portion of the fluid must be subjected to forces because of the surrounding fluid, otherwise the fluid will never reach an equilibrium state under the influence of gravity. For simplicity, let us consider an elementary volume (Fig. 1.1) of a fluid at rest in which the average flow velocity is zero. We may make the element so small that the body forces of the fluid, which is proportional to the cube of the linear dimension of the volume, are negligibly small in comparison with the surface force, the stresses. For fluid at rest, pressure is the only stress which is acting normal to the surface on which the force is considered. The pressure p at a given point in the fluid is defined as

$$p = \lim_{\Delta A \to 0} \frac{\Delta F}{\Delta A} = \frac{dF}{dA} \qquad (1.6)$$

where ΔF is the force on the elementary area ΔA. The elementary area ΔA should be so chosen that it is very small in comparison with the dimension of the flow field but very large in comparison with the dimension of the molecule of the fluid. For instance, the cube of the side of 1/1000 millimeter mentioned above satisfies such a condition in an ordinary flow problem.

From Fig. 1.1, let p, p_x, and p_y be the pressure on the side BC, CA, and AB respectively. Since the fluid is at rest, it must be in static equilibrium. The summation of force in the x-direction as well as that in the y-direction should be zero respectively, i.e.,

$$p(BC \cdot dz)\cos\theta = p_y dx \cdot dz; \quad p(BC \cdot dz)\sin\theta = p_x dz \cdot dy \quad (1.7)$$

But $BC \cdot \cos\theta = dx$ and $BC \cdot \sin\theta = dy$, where θ is the angle between sides BC and AB, Eq. (1.7) gives

$$p = p_x = p_y \quad (1.8)$$

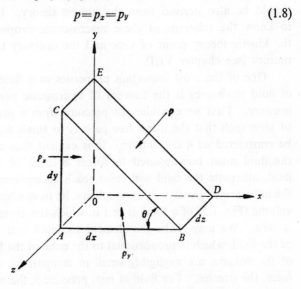

Figure 1.1   An elementary volume of a fluid

Since the volume of the prism shown in Fig. 1.1 may be chosen so small that it may represent a point in space from a fluid mechanics point of view, Eq. (1.8) holds for a point in space. Furthermore, the orientation θ is arbitrary. Hence we conclude that the pressure

p at any point in the fluid is the same in all directions. This is one of the main properties of the pressure of a fluid.

The pressure p may also be derived from kinetic theory. For simplicity, we image a plane BC (Fig. 1.2) in the fluid passing through a point P and perpendicular to the x-axis. The pressure force in the positive x-direction should be equal to the net rate of change of x-momentum of all fluid particles passing through the point P. We consider a unit area in the plane BC around the point P, the x-momentum change due to particles of x-component of velocity ξ is $mn_\xi(\xi^2 - u^2)$, where m is the mass of a particle and n_ξ is the

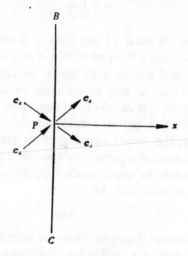

Figure 1.2   Molecules moving through a point P

number of particles of x-component of velocity ξ. The net pressure should be the sum of all the momentum of all particles at the point P, i.e.,

$$p = \overline{mn_\xi(\xi^2 - u^2)} = mn\,(\overline{\xi^2} - u^2) = mn\,\overline{c_x^2} = \frac{1}{3} mn\,\overline{c_a^2} \qquad (1.9)$$

where n is the number density of the gas at the point P and the assumption (1.5) is used. Eq. (1.9) shows again that the pressure is the same for all directions because the x-direction is arbitrary. Furthermore, the pressure of the gas is proportional to the density of the gas $\rho = mn$ and the mean value of the molecular velocity $\overline{c_a^2}$ which is closely associated with the temperature of the gas. Hence we may regard the pressure as one of the fundamental properties for the state of the fluid.

7. Density

Another important property of a fluid in the study of fluid mechanics is the density ρ of the fluid which is defined as

$$\rho = \lim_{\Delta V \to 0} \frac{\Delta m}{\Delta V}. \tag{1.10}$$

where Δm is the mass of the fluid in a volume ΔV. We should choose the volume ΔV so small that it may be considered as a point in the flow field and so large that it contains sufficient number of molecules so that we may take a mean value in this volume.

For liquids, it is usually good enough to assume that the density ρ is a constant in ordinary fluid mechanics problems. For a gas or a plasma, the density depends on its pressure and temperature. If the mean mass of a particle in a gas or a plasma is m, it is sometimes convenient to use a number density n of the gas or plasma in fluid mechanics such that

$$\rho = mn \tag{1.11}$$

In the kinetic theory of gases, we describe the motion of molecules by means of a molecular distribution function $F(r, q_m, t)$ which represents the expectation of the number of molecules per unit volume at the position r and time t within the molecular velocity range q_m and $q_m + dq_m$, i. e.,

$$dn = F(r, q_m, t)dq_m \tag{1.12}$$

The average number density of the gas at r and t is then

$$n = \int F(r, q_m, t)dq_m \tag{1.12a}$$

where dq_m means the integration over the whole velocity plane, i.e., $dq_m = d\xi d\eta d\zeta$ where ξ, η, and ζ are the x-, y-, and z- component of the molecular velocity q_m and the range of ξ, η, or ζ is from minus infinity to positive infinity.

The average of any physical quantity Q of the gas at any given point r in space and time t is

$$\overline{Q} = \frac{1}{n}\iiint QF \, d\xi d\eta d\zeta \tag{1.13}$$

For instance, the average x-component of velocity of the gas flow is

$$u = \overline{\xi} = \frac{1}{n}\iiint F\xi d\xi d\eta d\zeta \tag{1.14}$$

One of the main objects of the kinetic theory of fluid is to find the distribution function under given conditions. We shall discuss it later (see chapter VIII).

8. Temperature

Temperature is an important property of a fluid in the study of energy transfer from one system to another. The temperature is closely associated with the mean kinetic energy of the fluid. The absolute gas-temperature T is so defined that the mean kinetic energy of the molecule of the gas is equal to $kT/2$ per degree of freedom. In the simple kinetic picture of section 5, we have

$$\frac{1}{2}kT = \frac{1}{2}m\overline{c_x^2} = \frac{1}{2}m\frac{1}{3}\overline{c_a^2} \tag{1.15}$$

where k is known as the Boltzmann constant and is

$$k = 1.379 \times 10^{-16} \text{ erg}/^\circ\text{K} \tag{1.16}$$

Eq. (1.15) may be written as

$$T = \frac{1}{3}\frac{m}{k}\overline{c_a^2} \tag{1.17}$$

From Eqs. (1.9), (1.11), and (1.17), we have

$$p = \rho\frac{k}{m}T = \rho RT = nkT \tag{1.18}$$

where R is the gas constant. Eq. (1.18) is known as the perfect gas law which is a very good approximation for all gases and plasma which are far away from the condensation region.

From Eq. (1.18), we see that at a given temperature and a given pressure, the number density n is the same for all gases. Let m_W be the molecular weight of a substance. We assume that for atomic oxygen O, the molecular weight is 16. The molecular weight of all other substances are measured with reference to atomic oxygen $O=16$. The quantity of a substance of weight m_W grams is called a mole. The number of particles of a mole of any substance is known as Avagadro number n_A:

$$n_A = 6.025 \times 10^{23}/\text{mole} \qquad (1.19)$$

At $T=273°K$ and $p=760$ mm Hg, the volume of a mole of any gas is 22.4 liters.

If we consider a more accurate kinetic picture of gases, the equation of state is more complicated than that of Eq. (1.18). It is an empirical fact that there is a functional relation between the pressure p, density ρ, and temperature T of a fluid. For gases at reasonably high temperature and low pressure, this functional relation may be written in the following form:

$$\frac{p}{\rho RT} = 1 + B(T)\rho + C(T)\rho^2 + \cdots \qquad (1.20)$$

where $B(T)$, $C(T)$, etc., are functions of temperature T only but not of density nor pressure. The coefficients $B(T)$ etc., are known as virial coefficients. Eq. (1.18) is the first approximation of Eq. (1.20). In the second approximation, we may take the first two terms of Eq. (1.20) with

$$B(T) = b_1 - \frac{b_2}{RT} \qquad (1.21)$$

where b_1 and b_2 are constants for a given gas. The second approximation gives the van der Waals equation of gas which gives a fairly satisfactory description of fluid behaviour including both gas and liquid phases.

For engineering problems, the general equation of state of gas at higher pressure and low temperature near the liquid phase may be written in the following form:

$$p = Z\rho RT \qquad (1.22)$$

where the correction factor or the compressibility factor Z is a function of the temperature and the pressure of the gas which is usually given in the form of charts or tables (see Appendix III). The formula (1.22) may be used for a mixture of gases too, such as air. The factor Z then depends not only on temperature and pressure of the mixture but also on composition of the mixture (see chapter V).

The compressibility of any fluid is inversely proportional to its volume modulus of elasticity M_{ev} which is defined as

$$M_{ev} = \rho \frac{dp}{d\rho} \qquad (1.23)$$

For water at $p = 15$ psi and $T = 32°F$, M_{ev} is 292,000 psi. Hence the compressibility of water is negligibly small in ordinary conditions. For gas at constant temperature, $M_{ev} = p$.

9. Boltzmann-Maxwellian Laws of Distribution

The main object of the kinetic theory of gases is to find the distribution function $F(r, q_m, t)$ which should be calculated from the statistical theory. In the equilibrium condition, the distribution function is the well known Boltzmann-Maxwellian distribution which is the most probable distribution and which is obtained from the calculus of probabilities based on the assumption of a perfectly chaotic state, i.e., there is no coupling between the states at different instants. Let ϵ be the energy of a molecule. The probability of this molecule having ϵ energy according to the Boltzmann law of distribution in an equilibrium state is

$$F_0 = A \exp[-\epsilon/(kT)] \qquad (1.24)$$

where A is a constant. For a monatomic gas, the kinetic energy of a particle over its mean value of the flow is given by

$$\epsilon = \frac{m}{2}(q_m - q)^2 = \frac{m}{2}(c_x^2 + c_y^2 + c_z^2) \qquad (1.25)$$

Substituting Eq. (1.25) into (1.24) and determining the constant A from Eq. (1.12), we have the well known Maxwellian distribution function for the monatomic gas at a uniform temperature T and uniform flow velocity q as follows:

$$F_0 = n \left(\frac{m}{2\pi kT}\right)^{3/2} \exp\left[-\frac{m(q_m - q)^2}{2kT}\right] \qquad (1.26)$$

Since the Maxwellian distribution function F_0 of Eq. (1.26) depends only on the magnitude of the peculiar velocity $c'_a = q_m - q$, i.e., c_a, the integral (1.12a) may be written as follows:

$$\iiint F_0 \, d\xi d\eta d\zeta = 4\pi \int_0^\infty F_0 c_a^2 \, dc_a \qquad (1.27)$$

Sometimes, we may refer to $c_a^2 F_0$ as the Maxwellian distribution function. In Fig. 1.3, we plot both $(F_0/B_1) = \exp(-x^2)$ and $(F_0 c_a^2/B_2) = 2x^2 \exp(-x^2)$ vs $c_a = x$ which show the distribution of molecules for various peculiar velocities, where B_1 and B_2 are constants.

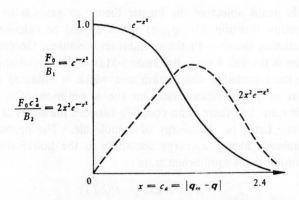

Figure 1.3   Maxwellian distribution function at a given temperature T and a given uniform flow velocity q of a gas

For most of the flow problems of gases, the molecular distribution function F deviates slightly from the Maxwellian distribution function F_0. Hence we write

$$F = F_0(1 + \phi) \tag{1.28}$$

where $|\phi| \ll 1$. Usually the higher order terms of ϕ may be neglected. We may have a linearized equation for ϕ. The coefficient of viscosity and that of the thermal conductivity of a gas are usually determined from the function ϕ. We shall discuss them in chapter VIII.

10. Atomic and Molecular Structures

The energy expression (1.25) may be applied to all kinds of molecules in a gas or a plasma and it represents the kinetic energy of the molecule in translational motion. But most of the molecules have other forms of energies, in addition to the translational energy. In order to know these energies, we have to know the structure of the molecule.

An atom or a monatomic molecule consists of a heavy nucleus of positive charge, around which the electrons revolve. For neutral gases, the net electrical charge is zero. The positive charge of the nucleus is equal and opposite to the total negative charges of electrons around it. A molecule consists of a number of heavy nuclei, the atomic nuclei of the atoms or ions which form the molecule; around these nuclei the electrons revolve. For atoms or monatomic molecules, the kinetic energy due to rotational motion of the molecules is negligibly small; while for a molecule, the kinetic energy due to rotational motion of the molecule is of the same order of magnitude as that of translational motion. In equilibrium conditions, we have the energy of $\frac{1}{2}kT$ per degree of freedom whether it is due to translational motion or rotational motion.

For molecules, the relative motion of the nuclei, i.e., the vibrational mode of internal energy, may be also of the same order of magnitude as that due to translational motion. These vibrational energy becomes important at temperature higher than $2000°K$.

There is a given amount of energy associated with each electron revolving around the nuclei. In order to see clearly such energy associated with an electron, we consider the simplest case: the Bohr hydrogen atom which consists of a small heavy nucleus with an electron moving as a satellite in a circular orbit about the nucleus. The nucleus has a mass approximately equal to that of the hydrogen atom and a positive charge equal in magnitude to the charge of electron. According to the quantum theory of atom,[14] the radius "a_b" of the orbits of the electron is limited to a number of possible stable states which is given by the formula:

$$a_b = a_1 \frac{n^2}{Z} = \frac{n^2 h^2}{4\pi^2 m Z e^2} \qquad (1.29)$$

where Ze is the positive charge of the nucleus and n is known as the first quantum number which may be any positive integer, i.e., 1, 2, $\cdots$, and h is the Planck constant,

$$h = 6.62 \times 10^{-27} \text{ erg} \cdot \text{sec} \qquad (1.30)$$

For every integer n, we have a possible orbit (Fig. 1.4) and the energy level associated with this orbit is E_n which is given by the formula:

$$E_n = -\frac{Ze^2}{2a_b} = -\frac{Z^2 e^4 4\pi^2 m}{2n^2 h^2} \qquad (1.31)$$

where e is the absolute charge of an electron, m is the mass of an electron. The energy scale (Fig. 1.5) is so normalized that its zero point corresponds to n towards infinity. Hence $-E_n$ is the work needed to remove the electron from the state n to rest at infinity and it is the ionization energy from the state n. Normally, the electron is in the ground state which corresponds to $n=1$ and the radius of this ground state is $a_1 = h^2/(4\pi^2 me^2) = 0.528$ Å which is generally known as the Bohr radius. The energy $-E_1$ is the ionization energy from the ground state which is usually expressed in electron volts and referred to as the ionization potential. If an atom receives an amount of energy greater than its ionization energy from the ground state, one of the electron will move away from the influence of the nucleus and the atom becomes an ion. Ordinarily, the electron is in the ground state. If the electron is in

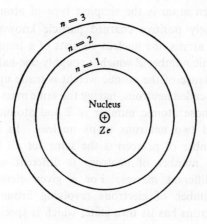

Figure 1.4 Orbits of electron around a nucleus of a hydrogen atom

a state other than the ground state, the atom is said to be in an excited state. For a transition from a state E_n to another state E_m, the energy emitted or absorbed is according to the formula:

$$h\nu = E_n - E_m \qquad (1.32)$$

where ν is the frequency of the photon emitted.

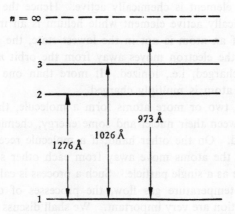

Figure 1.5 Energy levels and the most important lines of the hydrogen spectrum

The hydrogen atom is the simplest type of atom of which the nucleus is a singly positive charged particle known as a proton. For other heavy atoms, the nucleus consists of a number of protons equal to the atomic number Z which is roughly one-half of the atomic weight. The balance of the atomic weight is made up of electrically neutral particles, called neutrons, having the same mass as the protons. Thus helium, whose atomic number is 2 and atomic weight 4 has two protons and two neutrons in its nucleus. In isotopes of an element, the number of protons is the same for all isotopes of the element but the number of neutrons is different so that various isotopes have different masses. For a given element, there are ordinarily Z number of electrons revolving around its nucleus. Each of the electrons has its own state, which is specified by various quantum numbers. No two electrons in an atom can have the same values for all the five quantum numbers. The electrons are normally in various shells according to the total quantum number n. For $n=1$, the shell is called K-shell in which two electrons may exist. The electron in the hydrogen atom and the electrons in the helium atom are usually in the K-shell. For $n=2$, the shell is known as L-shell which will be filled by 8 electrons. For an element, if its electrons fill the shell, is an inert element; while if the outmost shell has less electrons than the number required to fill it, the element is chemically active. Hence the hydrogen atom is a chemically active element while helium is an inert gas. If the electron of an atom is not in the lowest state, the atom is excited. If one of the electron moves away from the orbit $n \to \infty$, the atom is singly charged, i.e., ionized. If more than one electrons move away, the atom is multiply charged.

When two or more atoms form a molecule, there are binding forces between their nuclei and some energy, chemical energy, may be released. On the other hand, if a molecule receives an amount of energy, the atoms move away from each other so that they will not behave as a single particle. Such a process is called dissociation. For high temperature gas flow, the processes of dissociation and recombination are very important. We shall discuss them in chapter V.

11. Specific Heats, Internal Energy, and Enthalpy

If we add an amount of heat dQ to a fluid, its temperature will usually increase by an amount dT. We define a specific heat c_{sp} of the fluid as

$$c_{sp} = \lim_{dT \to 0} \frac{dQ}{dT} = \frac{\partial Q}{\partial T} \qquad (1.33)$$

The specific head c_{sp} depends on the process in which the heat is added; in other words, when we add heat, we have to keep certain thermodynamic characteristics of the fluid unchanged and the value of c_{sp} depends on this characteristic of the fluid. For instance, two most common cases are:

$$c_v = \left(\frac{\partial Q}{\partial T} \right)_v = \text{specific heat at constant volume} \qquad (1.33a)$$

$$c_p = \left(\frac{\partial Q}{\partial T} \right)_p = \text{specific heat at constant pressure} \qquad (1.33b)$$

The specific heats are closely connected with the internal energy of the molecules, which depends greatly on the molecular structure. For a perfect gas, the internal energy per unit mass U_m is a function of temperature T only and we have

$$c_v = \frac{dU_m}{dT} \qquad (1.34)$$

For a monatomic gas at low temperature, the internal energy of the molecules is simply the difference of their total kinetic energy due to molecular motion and the mean kinetic energy of the flow. This type of internal energy is usually referred to as translational internal energy U_{mt} which for per unit mass is

$$U_{mt} = \frac{1}{2} \left(\overline{\xi^2} + \overline{\eta^2} + \overline{\zeta^2} \right) - \frac{1}{2}(u^2 + v^2 + w^2) = \frac{1}{2}\overline{c_a^2} = \frac{3}{2}RT \qquad (1.35)$$

The specific heat at constant volume due to U_{mt} is simply $\frac{3}{2}R$. This part of the internal energy holds true for all kinds of gases or plasmas.

The temperature so defined is referred to as kinetic temperature.

For a polyatomic gas, the internal energy depends on the molecular structure. For simplicity, we consider the case of a diatomic gas (Fig. 1.6). The internal energy consists of six parts as follows:

$$U_m = U_{mt} + U_{mr} + U_{mv} + U_{md} + U_{me} + U_{mi} \qquad (1.36)$$

where

(a) $U_{mt} = \dfrac{3}{2} RT$ for all kinds of gases represents the internal energy due to translational motion of the molecules. It is translational internal energy (Fig. 1.6a).

(b) U_{mr} represents internal energy due to rotational motion of the molecules. It is rotational internal energy (Fig. 1.6b). For a monatomic gas, the moment of inertia of the atom is negligibly small and we may assume $U_{mr} = 0$. For a diatomic gas, the moment of inertia about the axis connecting the two atoms is negligibly small, while those about axes perpendicular to the axis connecting the atoms are not small. We may consider that there are two degrees of freedom due to rotational motion of a diatomic gas. By the principle of equi-partition of energy, i.e., Eq. (1.15), $U_{mr} = RT$ for diatomic gases or a polyatomic gas with linear molecules because there are only two degrees of freedom of rotational motion. $U_{mr} = \dfrac{3}{2} RT$ for polyatomic gas with nonlinear molecules because there are three degrees of freedom of rotational motion in this case.

(c) U_{mv} represents the vibrational motion between atoms in a molecule. It is vibrational internal energy (Fig. 1.6c). For each vibrational mode, there is a fundamental frequency v_j and the corresponding internal energy is

$$U_{mv} = RT \left[\frac{hv_j/kT}{\exp(hv_j/kT) - 1} \right] \qquad (1.37)$$

For complicated molecule there are a number of fundamental frequencies v_j, the total vibrational internal energy U_{mv} is the sum of the internal energy of individual modes given by Eq. (1.37).

(d) U_{md} represents the amount of energy needed to dissociate the atoms of a molecule. It is the dissociation energy

(Fig. 1.6d) (see next section).

(e) U_{me} represents the energy to excite an electron of the molecule from its ground state to a higher state according to Eq. (1.32). It is the electron excitation energy (Fig. 1.6e).

(f) U_{mi} represents the energy to ionize a molecule so that one or more of the electrons move away from the neighborhood of the nucleus. The molecule becomes an ion. It is the ionization energy (Fig. 1.6f).

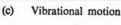

(a) Translational motion —— 3 degrees of freedom along x-, y-, and z-axis

(b) Rotational motion —— 2 degrees of freedom about y- and z-axis

(c) Vibrational motion

(d) Dissociation

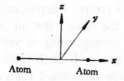
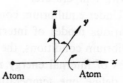

(e) Excited state of an atom

(f) Ionized atom —— an electron moves outside of n = orbit

Figure 1.6 Various modes of internal energy of a diatomic gas

Below $T = 2000\,°K$, the only modes of internal energy of a gas which are important are the translational energy and rotational energy. Under this condition, we have $U_m = U_{mi} + U_{mr}$ and the specific heat of a gas at constant volume is a constant, i.e.,

$$c_v = \frac{3}{2}R \text{ for monatomic gas} \tag{1.38a}$$

$$c_v = \frac{5}{2}R \text{ for diatomic gas} \tag{1.38b}$$

A gas with constant c_v is usually referred to as an *ideal gas*. When the temperature T is above 2000°K, the other modes of internal energy should be considered in the flow problem, and the specific heat at constant volume is a function of temperature. Fig. 1.7 shows a typical variation of specific heat at constant pressure of air with temperature.

Under equilibrium conditions, the temperatures T corresponding to various modes of internal energies are the same. But in non-equilibrium conditions, the temperature T corresponding to various modes of internal energy may not be the same. In such cases, we may define one temperature for each mode of internal energy in the study of the flow problem of a gas. For instance, we have a temperature of vibrational mode T_v which may be different from the kinetic temperature of the gas (see chapter V).

We define an enthalpy H of a gas as

$$H = U_m + \frac{p}{\rho} \tag{1.39}$$

For a perfect gas, we have

$$c_p = \frac{dH}{dT} = c_v + R \tag{1.40}$$

In the fluid dynamic problems, sometimes it is more convenient to use enthalpy instead of internal energy. The ratio of c_p to c_v is usually denoted by the symbol γ, i.e.,

$$\gamma = c_p/c_v \tag{1.41}$$

The ratio γ is used to measure the relative internal complexity of the molecules.

The specific heats of gases are additive. Hence the specific heat of a plasma may be determined from our knowledge of the specific heat of gases because a plasma may be considered as a mixture of several species of gases in which electrons and ions are considered as special kinds of gas.

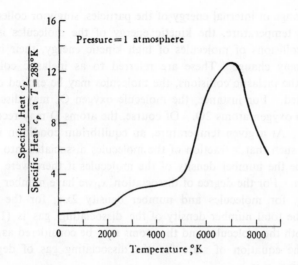

Figure 1.7 Variation of specific heat at constant pressure of air with temperature

The specific heat of a liquid has not been well understood because the kinetic theory of liquid has not been well developed yet. However many liquids, including water and mercury, behave like a solid in that they have atomic heats about six calories/mol per atom, i.e., $c_v = 3R$. Many other liquids may have atomic heats much larger than six cal/mol.

Another interesting point of the heat capacity of liquid in fluid mechanics is that there is a definite relation between the pressure and temperature at the boiling point. In the process of evaporation or condensation, the heat may be absorbed or released. Such a heat is known as latent heat. In the study of the flow of fluids in two phases, this phenomenon should be considered (see chapter IX section 3).

12. Dissociation, Ionization, and Radiation

Let us consider a diatomic gas such as oxygen O_2 or nitrogen N_2. At low temperature, the molecules remain unchanged even though the molecules are constant in collision with one another. Such collisions are referred to as elastic collisions in which there is no change in internal energy of the particles, singly or collectively. At high temperature, the kinetic energy of the molecules is high. In the collisions of molecules of high kinetic energy, their internal energy may change. These are referred to as inelastic collisions. During the inelastic collisions, the molecules may be excited or even dissociated. For instance, the molecule oxygen O_2 may dissociated into two oxygen atoms $2O$. Of course, the atoms O may recombine into O_2. At a given temperature, an equilibrium condition may be reached such that x fraction of the molecules dissociates into atoms. Let n_0 be the number density of the molecules if there were no dissociation. For the degree of dissociation x, we have number density $(1-x)n_0$, for molecules and number density $2xn_0$ for the atoms. Hence the total number density of the dissociating gas is $(1+x)n_0$. Since both the molecules and the atoms may be considered as perfect gases, the equation of state for the dissociating gas of degree of dissociation x is

$$p = \rho R T (1+x) \qquad (1.42)$$

The degree of dissociation x in the equilibrium condition is a function of both the temperature T and the pressure p of the gas. Usually the degree of dissociation x may be expressed in terms of the equilibrium constant K for the dissociation reaction considered, i.e.,

$$\frac{4x^2}{1-x^2} = \frac{K}{p} \qquad (1.43)$$

where K is a function of temperature only.

At very high temperatures, the collisions between particles in a gas may ionize some of the particles. The simplest case is the case of a monatomic gas. Let the degree of ionization of this gas be α. Then the equation of state for the plasma which consists of $(1-\alpha)n_0$ neutral particles, αn_0 ions and αn_0 electrons per unit volume will be

$$p = \rho R T (1+\alpha) \qquad (1.44)$$

The degree of ionization α in equilibrium condition is also a function of temperature and pressure of the gas. For singly ionized plasma of monatomic gas, the degree of ionization α is given by the Saha relation:

$$\frac{\alpha^2 p^2}{1-\alpha^2} = 3.16 \times 10^{-7} \times T^{5/2} \exp[-eV_i/(kT)] \tag{1.45}$$

where p is the pressure in atmospheres, T is the temperature in °K, V_i is the ionization potential of the gas, and eV_i is the ionization energy in ergs.

The degree of ionization of a mixture of gases will be discussed in chapter VI.

In high temperature gases, photons may be emitted or absorbed by the particles of gas and they represent the thermal radiation phenomena. Except at very high temperatures (see chapter VII) the effects of thermal radiation is usually negligible in the flow problems of a fluid. However, when the temperature is very high or the density of the gas is very low, thermal radiation may have predominant influence on the flow problem of gases or plasma. In thermal equilibrium conditions, the spectral energy density of thermal radiation is given by the Planck radiation law:

$$U_\nu = \frac{8\pi h \nu^3}{c^3} \frac{1}{\exp[h\nu/(kT)] - 1} \tag{1.46}$$

where ν is the frequency of the thermal radiation and c is the speed of light. The total energy density of radiation in thermal equilibrium is then:

$$E_R = \int_0^\infty U_\nu \, d\nu = a_R T^4 \tag{1.47}$$

where $a_R = 7.67 \times 10^{-15}$ erg·cm^{-3}·°K^{-4} is known as the Stefan-Boltzmann constant. Since a_R is a very small number, the energy density of radiation E_R is usually negligibly small in comparison with the internal energy of a gas U_m except when the temperature T is very high and the density of the gas is very low (see chapter VII).

13. Viscosity and Rheology

Even though fluid offers no resistance to the deformation of its shape, it does offer resistance to the rate of change of shape. Such a resistance is usually referred to as viscous force. We consider a simple experiment (Fig. 1.8) in which an ordinary fluid such as water or air is placed between two parallel plates at a distance y_0 apart. Let the lower plate be fixed, while the upper plate is moving with a uniform velocity U and parallel to the lower one. A resistance D is experienced, and to a first approximation, the resistance is given by the formula:

$$D = A\mu \frac{U}{y_0} \qquad (1.48)$$

where A is the area of the upper plate and μ is a constant of proportionality which is called the coefficient of viscosity.

Since there is no slip on the plates for ordinary fluids and the fluid is displaced in such a manner that various layers of fluids slide uniformly over one another, the velocity u of a layer of the fluid at a distance y from the lower plate is then

$$u = U \frac{y}{y_0} \qquad (1.49)$$

The type of flow in this experiment is known as laminar flow which occurs at low velocity and small gap y_0, or more precisely at a low Reynolds number which will be discussed in chapter III.

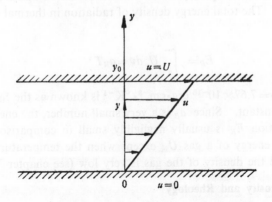

Figure 1.8 Viscous fluid between two parallel plates: **Couette flow**

The result of the above experiment shows that the tangential force per unit area, or the shearing stress τ_v is proportional to the slope of the velocity of the flow field, i.e.,

$$\tau_v = \mu \frac{\partial u}{\partial y} \qquad (1.50)$$

This linear relation is found to be very closely correct for many common fluids. The expression (1.50) can also be derived from the simple kinetic theory of gases as follows:

We consider the case that the mean velocity of the fluid has only the x-component u which is a function of y only. The shearing stress at a point P on a plane perpendicular to the y-axis and in the direction of the x-axis is equal to the sum of the rate of change of x-momentum of all molecules transferred across a unit area of the plane at the point P, i.e.,

$$\tau_v = \Sigma (\text{mass}) c_x c_y \qquad (1.51)$$

Here we may take $c_x = L_{fm} (\partial u / \partial y)$ because we assume that at the last collision the molecule has a mean velocity of the layer u. The free path of the molecule is L_{fm}. The value of c_y may be taken as the peculiar velocity of the molecule. Because the molecules are in random motion, we may assume that one sixth of the mass is transferred from one side of the plane and the same amount from the other side. As a result, the total shearing stress given by the averaging process of Eq. (1.51) is

$$\tau_v = \frac{1}{3} nmc_a L_f \frac{\partial u}{\partial y} \qquad (1.51a)$$

Comparing Eqs. (1.50) and (1.51a), we have

$$\mu = \frac{1}{3} nm\bar{c}_a L_f = \frac{1}{3} \rho \bar{c}_a L_f \qquad (1.52)$$

For a more refined theory, we have similar formula for the coefficient of viscosity of gases with the mean molecular velocity $\bar{c}_a$ and mean free path L_f except that the numerical factor is 0.499 instead of 1/3.

The coefficient of viscosity of a fluid is one of the most important properties in fluid mechanics. Its dimension is mass per unit length per unit time. We may express the coefficient of viscosity in slugs per foot per second or dyne·seconds per square centimeter. The unit of one dyne·second per square centimeter is called a poise. However most of the fluids used in engineering problems such as water or common gases have low viscosities. Hence it is convenient to use centipoise (=0.01 poise) as a unit of coefficient of viscosity. Specifically, the coefficient of viscosity of water at 68.4°F is one centipoise. The value of viscosity in centipoise is an indication of the viscosity of any fluid relative to that of water. For most of the gases, the viscosity is about 1/50 of that of water. For instance, the viscosity of air at 60°F is 0.0180 centipoise.

As the temperature increases, the viscosities of all liquids decrease while those of all gases or vapors increase. Viscosity is due to the cohesion between molecules and also to the interchange of molecules between the layers of different velocities as shown in equation (1.52). For liquids, the force of cohesion predominates and it decreases with increase of temperature, while for gases the interchange of molecules predominates and it increases with temperature. From the simple kinetic theory of gas (1.53), the product nL_f is approximately constant and $\overline{c_a}$ is proportional to the square root of temperature T; hence the coefficient of viscosity is proportional to the square root of T. For more refined theory, the variation of viscosity of gases with temperature is very complicated.[6] For engineering purpose, it is sufficiently accurate to assume that the coefficient of viscosity of gases is proportional to a power of the absolute temperature, i.e.,

$$\frac{\mu}{\mu_0} = \left(\frac{T}{T_0}\right)^n \tag{1.53}$$

where the subscript 0 refers to a certain reference value, n is a factor whose value lies between 1/2 and 1. For air at ordinary temperature, the value of n is usually taken as 0.76. As the temperature of the gas increases, the value of n decreases toward 1/2.

The absolute viscosity of a liquid or a gas is practically independent of pressure for ordinary range in many engineering problems, such as within a pressure of a few atmospheres.   But at very high pressures, the viscosity increases with pressure.

The viscosities for a liquid and its saturated vapor merge at the critical temperature.

The ratio of the coefficient of viscosity to its density of a fluid is known as the coefficient of kinematic viscosity ν_g which is usually denoted by

$$\nu_g = \frac{\mu}{\rho} \tag{1.54}$$

The dimension of kinematic viscosity is velocity times length or square of length per unit time.   In the metric system, the unit of one square centimeter per second is called a stoke.   At temperature 68.4°F, the kinematic viscosity of water is one centistoke.   Because of low densities, the kinematic viscosity of gas is much larger than that of a liquid, even though the viscosity of liquid is usually larger. For air at a pressure of one atmosphere and at a temperature of 68.4°F, the kinematic viscosity is about 18 centistokes.   For gases and vapor, the kinematic viscosity depends on both temperature and pressure.

Fluids with linear relation of viscous stress given by Eq. (1.58) are known as Newtonian fluids.   Ordinary fluids such as water, oil, air, and other gases in which the mean free path is small in comparison with the dimension of the flow field, are Newtonian fluids. Hence the viscous flow of Newtonian fluids has been extensively studied.   However there are many fluids whose viscous forces do not follow the simple linear relation of Eq. (1.50).   Fig. 1.9 shows some relations between shearing stress and the rate of change of velocity for various materials.   For Newtonian fluid, it is a straight line starting from the origin in Fig. 1.9.   For non-Newtonian fluid, it is a curve starting at the origin, the relation between τ_v and $\partial u/\partial y$ is nonlinear.   There are some colloids or plastics which have a small initial friction which is independent of the velocity.   The colloids behave like elastic bodies first and like a Newtonian fluid later.   There are also some non-Newtonian fluids in which the viscosities vary with the rate of deformation.   We shall discuss them in chapter IX.

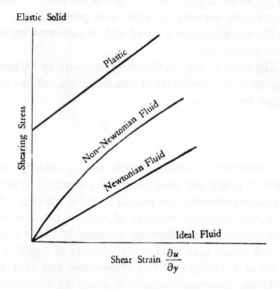

Figure 1.9   Relations between shearing stress and rate of change of velocity for various materials

For rarefied gas, the linear relation (1.50) does not hold.   For very rarefied gases, the shearing stress depends not only on the rate of change of velocity but also on the gradient of temperature.   For rarefied plasma, the shearing stress may depend not only on gas-dynamic variables but also on electromagnetic fields.   However for dense plasma, Eq. (1.58) may be used as a first approximation.   The study of non-Newtonian fluids is known as Rheology which will be discussed in chapter IX.

14. Heat Transfer.   Heat Convection, Heat Conduction, and Thermal Radiation

For the flow of a compressible fluid, we have to study the mechanics of fluid simultaneously with the heat transfer problem. There are three basic modes of heat transfer:

(1)   Heat transfer by convection in fluids in a state of motion,

(2)   Heat transfer by conduction in solids or fluids, and

(3)   Heat transfer by thermal radiation which takes place with no material carrier.

In general, all these three modes of heat transfer occur simultaneously.   Heat convection depends on the velocity field and the specific heat of the fluid.   The flux of heat transferred by convection at a point with velocity q is $\rho c_p T q$.

Heat conduction is a physical property of the fluid.   The mechanism of heat conduction in a fluid is analogous to that producing viscous stresses.   The heat conduction of gases can be analyzed by the kinetic theory of gases.   The quantity of heat Q conducted between two layers of fluid of infinitesimal distance apart may be expressed by the formula:

$$Q = -\kappa \frac{\partial T}{\partial y} \tag{1.55}$$

where y is the normal distance between these two layers and the negative sign is inserted because the heat flows in the direction of decreasing temperature.   The factor κ is known as the coefficient of thermal conductivity or heat conductivity of the fluid which is a physical property of the fluid.   It is possible to derive the expression of κ in terms of other basic physical quantities of gases from the kinetic theory.   By similar arguments as those in the case of analysis of viscous stress (1.52), the heat flux Q by conduction is simply the amount of heat transferred per unit area of the plane perpendicular to y per unit time and we have

$$Q = \frac{1}{3} \rho \bar{c}_a L_f c_p \frac{\partial T}{\partial y} = \kappa \frac{\partial T}{\partial y} \tag{1.56}$$

Comparing Eqs. (1.55) and (1.56), we have

$$\kappa = \frac{1}{3} \rho \bar{c}_a L_f c_p \tag{1.57}$$

where c_p is the specific heat of the gas at constant pressure.   From Eqs. (1.52) and (1.57), we have the nondimensional parameter:

$$P_r = \frac{\mu c_p}{\kappa} = 1 \qquad (1.58)$$

where P_r is known as the Prandtl number of the gas which is a measure of the relative importance of viscous stresses and heat conductivity. From simple kinetic theory, we obtain $P_r = 1$; but actually the Prandtl number of gases is different from unity because of the rotation and vibration of molecules transferring energy. For air, the Prandtl number is about 0.72. The variation of Prandtl number of gases with temperature is small; hence we may take it as a constant in most problems of fluid mechanics.

For gases, the Prandtl number is of the order of unity, whereas for liquid, the Prandtl number may be different considerably from unity. For water, the Prandtl number is about 10, while for mercury, the Prandtl number is only about 0.01.

Thermal radiation is a transfer of heat by waves or photons. It is important at very high temperature. We have to use quantum mechanics to study the detailed interaction of matter and wave. However for macroscopic flow field, thermal radiation effects may be expressed in terms of temperature T and a mean free path of radiation L_R which represents the average distance of absorption of the radiation ray through a medium. We shall discuss the gas dynamics with thermal radiation effects in detail in chapter VII. In order to show some first order effects of radiation phenomena in the flow field, let us consider the case of small mean free path of radiation. Under this condition (see chapter VII) the effects of thermal radiation on the flow field consist of the following three terms:

(1) Radiation pressure. The thermal radiation will exert a radiation pressure p_R in addition to the gas pressure and the radiation pressure is:

$$p_R = \frac{1}{3} a_R T^4 \qquad (1.59)$$

where T is the absolute temperature in °K and a_R is the Stefan-Boltzmann constant equal to 7.67×10^{-15} erg·cm^{-3}·°K^4. Since

a_R is a very small number, at ordinary temperature, the radiation pressure p_R is very small in comparison with the gas pressure. For instance, at $T=10^4$ °K, $p_R=2.5\times10^{-5}$ atmosphere which is usually negligible in comparison to gas pressure. However for plasma, we sometimes consider very high temperatures, say above 10^6 °K. At such high temperature, the influence of radiation pressure p_R may be not negligible. In general case, we may have all six components of radiation stress tensor which will be discussed in chapter VII.

(2) Energy density of radiation. Thermal radiation is a form of energy. There is always a portion of thermal radiation energy in transit around any point in the flow field. Such an energy should be considered as a part of the total energy of the fluid just as the internal energy of the fluid. The radiation energy density per unit volume for the case of small mean free path of radiation is

$$E_R = a_R T^4 \tag{1.60}$$

E_R should be considered in the flow field whenever p_R is not negligible. They are of the same order of magnitude.

(3) Heat flux of thermal radiation. The heat flux of thermal radiation at small mean free path of radiation L_R is given by the formula:

$$q_R = - D_R \nabla E_R \tag{1.61}$$

where

$$D_R = \frac{1}{3}cL_R = \text{Rosseland diffusion coefficient of radiation} \tag{1.62}$$

where $\nabla = i\, \dfrac{\partial}{\partial x} + j\, \dfrac{\partial}{\partial y} + k\, \dfrac{\partial}{\partial z}$ is the gradient operator and i, j, and k are respectively the unit vectors in x-, y-, and z-direction.

Because the velocity of light c is a large quantity, sometimes the heat flux of thermal radiation should be considered even if energy density of radiation is negligible. When the temperature is above 6000°K, the heat flux by thermal radiation should be considered. For low temperatures, the thermal radiation effects are usually negligible. We shall neglect thermal radiation effects in most parts of this book except in chapter VII.

For all real fluids at ordinary conditions, both viscosity and heat conductivity are different from zero but their values are usually small. Hence in many practical flow problems, the effects of viscosity and heat conductivity are negligible except in certain narrow regions known as boundary layers or transition regions. In the major portion of the flow field, we may assume that the fluid is an ideal fluid which is inviscid and nonheat conducting. A large portion of fluid dynamics has been developed under the assumption of ideal fluid, as we shall discuss in later chapters. At very low temperatures near absolute zero degree, the quantum effects begin to be of importance in the properties of fluids. Since near absolute zero helium is the only substance that exists in the liquid state, we are interested in the properties of liquid helium, particularly the liquid helium II which has many peculiar properties among which zero viscosity and zero heat conductivity are two important features. Such a fluid is known as a superfluid which behaves like ideal fluid. We shall discuss some of the properties of a superfluid in chapter IX.

15. Mixture of Fluids. Diffusion and Chemical Reactions

In many flow problems, we have to deal with a mixture of several fluids. Such a mixture may be divided into three groups:

(1) The mixture of gases,
(2) The mixture of gases and liquids, and
(3) The mixture of liquids.

The mixture of gases is the most common case of fluid mechanics because air is a mixture of gases and a plasma may be considered also as a mixture of gases. For a mixture of gases, the number concentration or simply the concentration c_s which is the ratio of the number density of sth species n_s to the total number density n of the mixture, i.e.,

$$c_s = \frac{n_s}{n} \qquad (1.63)$$

is a new variable in the flow problem. However, if c_s remains unchanged in the flow problem, the mixture may be considered as a single gas in the problem of fluid mechanics. Of course, the physical properties of this new gas such as coefficient of viscosity, specific heat, etc., depend on the composition of the mixture.

Let us consider a mixture of N different perfect gases. Under the equilibrium condition, all the species have the same temperature T but with different number density n_s and partial pressure p_s. The total number density n of the mixture at temperature T and total pressure p is

$$n = n_1 + n_2 + \cdots + n_N = \sum_{s=1}^{N} n_s \qquad (1.64)$$

and the pressure of the mixture p is

$$p = \sum_{s=1}^{N} p_s \qquad (1.65)$$

Let m_s be the mass of a particle of the sth species. The mass density of the mixture under temperature T and pressure p is

$$\rho = \sum_{s=1}^{N} \rho_s = \sum_{s=1}^{N} m_s n_s = mn \qquad (1.66)$$

where m is the average mass of a molecule of the mixture which is a function of the concentration c_s and mass m_s. The mass concentration of sth species in the mixture is the ratio of the mass density $\rho_s = m_s n_s$ of the sth species to the total mass density of the mixture and is given as:

$$k_s = \frac{\rho_s}{\rho} = \frac{m_s n_s}{\rho} \qquad (1.67)$$

Since all the species are perfect gas, each one of them satisfies the perfect gas law (1.18), i.e.,

$$p_s = n_s k T_s = n_s k T \qquad (1.68)$$

where we assume $T_s = T$, i.e., all the species have the same temperature T. In most general case, T_s's of the species may differ from each other and also differ from the temperature of the mixture as a whole (see chapter IX).

The summation of all the N equations of (1.68) gives

$$p=nkT=\rho\frac{k}{m}T=\rho RT \tag{1.69}$$

In the most general case, the temperature of the mixture T is defined as $T=(1/n)\ \Sigma n_s T_s$. Hence Eq. (1.69) holds whether the temperatures of all species are the same or not. Even though Eq. (1.69) is identical to Eq. (1.18) for a single gas, the physical meanings of both m and R are different in these two equations. For a single gas in Eq. (1.18), both m and R are constants. For a mixture of gases, both m and R are functions of the composition of the mixture, i.e., of all the c_s's and m_s's. If the composition of the mixture changes in the flow field, both m and R are variables which depend on the local values of c_s.

The composition of the mixture may change due to diffusion phenomena and chemical reactions between various species of the mixture. By diffusion we mean the relative motion of the particles of the different species in the mixture. For simplicity, we consider a mixture of two gases. The mean velocities of the two species may be different. We call these velocities q_1 and q_2 for the species 1 and 2 respectively. The mean velocity of the mixture as a whole is then defined as

$$q=(\rho_1 q_1+\rho_2 q_2)/\rho \tag{1.70}$$

The diffusion velocity for the sth species is then

$$w_s=q_s-q \tag{1.71}$$

When the mass m_s of all the species in the mixture do not differ from one another very much, the diffusion velocity w_s is usually small. For the case of small diffusion velocity, we may express w_s in terms of diffusion coefficient D_{12} and the gradient of the concentration of the species c_s, i.e.,

$$w_s=-D_{12}\nabla c_s \tag{1.72}$$

The value of diffusion coefficient D_{12} depends on the properties of the species and from the simple kinetic theory of section 13, it is easy to show that

$$D_{12} = \frac{1}{3} \, \bar{c}_a L_f \qquad (1.73)$$

For oxygen O_2 and nitrogen N_2, D_{12} is 0.181 cm²/sec at 0°C and 760 mm Hg, while for oxygen O_2 and hydrogen H_2, D_{12} is 0.722 cm²/sec at the same temperature and pressure. From more accurate kinetic theory, we find that w_s depends also on the temperature gradient and the pressure gradient which are known as thermal diffusion and pressure diffusion respectively. However, in ordinary flow problems, the thermal and pressure diffusions are negligibly small. For ionized gas, the diffusion velocity w_s may also depend on the electromagnetic fields. When the diffusion velocity is large, we should not use the expression of diffusion coefficient such as Eq. (1.72) and we should use the multifluid theory which will be discussed in chapter IX.

The molecules of the sth species may be formed or destroyed by chemical reactions. Heat may be released or absorbed during chemical reaction. The study of the mechanism of chemical reaction is a special science known as chemical kinetics, a science of considerable complexity. Here we mention only those points which are useful in the analysis of problems of fluid mechanics. We shall discuss them in chapter V.

The simplest chemical reaction is the first order chemical reaction in which the rate that a gas r is transformed into another gas s is proportional to the number density of the rth species only, i.e., the rate of transformation is

$$K_r = -k_f n_r \qquad (1.74)$$

where k_f is called the specific reaction rate coefficient.

The second order chemical reaction is that the rate that a gas r is transformed into other gases C and D by reaction with another gas s is proportional to the number densities of both rth and sth species, i.e.,

$$K_r = -k_f n_r n_s \qquad (1.75)$$

There is also a third order chemical reaction. In general, the chemical reaction can proceed in both the forward direction as given by Eq.

(1.74) and (1.75) and in reverse direction (i.e., from gas s to gas r) with a rate constant k_b. At thermodynamic equilibrium, there is no net change in composition. The rate constants k_f and k_b are related to each other.

The rate coefficient k_f is a function of temperature. The dependence of k_f on temperature was first found experimentally by Arrhenius and has been confirmed by theoretical studies. The Arrhenius law is

$$k_f = B_0 \exp\left[-E_A/(kT)\right] \tag{1.76}$$

where B_0 is called the frequency factor. In some literature, one writes $B_0 = 1/t_0$, t_0 being the characteristic time of reaction, E_A is the activation energy which is a measure of the energy that a molecule must possess in order to react successfully.

The heat releases per unit mass due to chemical reaction is the change of internal energy of sth gas which should be determined by quantum mechanics. We shall discuss them in chapter V.

The total internal energy of a mixture is the sum of the internal energies of all species in the mixture. Let U_m be the internal energy of the mixture per unit mass and U_{ms} be the internal energy of the sth species per unit mass. The internal energy per unit volume of the mixture is

$$U_m = \sum_{s=1}^{N} U_{ms} \tag{1.77}$$

where from Eq. (1.36), we may write

$$U_{ms} = U_{mso} + Q_s = U_{mso} + U_{msd} + U_{msi} \tag{1.78}$$

where Q_s consists of the dissociation and ionization energies, Eq. (1.77) may be written as

$$U_m = U_{mo} + Q_c = U_{mo} + \sum_{s=1}^{N} m_s n_s Q_s \tag{1.79}$$

where Q_c is the total heat release in the mixture per unit volume due to chemical reaction and U_{mo} represents the other parts of the internal energy of the mixture due to translational, rotational, and vibrational modes. In the analysis of flow problems, it is convenient to consider Q_c as heat produced by external agencies as we shall show in chapter V.

The fluid properties of the mixture such as coefficients of viscosity, heat conductivity, etc., depend on the concentration of the mixture as well as the corresponding properties of individual species. We shall discuss this in chapter IX when we discuss the multifluid theory.

The mixture of two liquids may be expressed in terms of mass concentration k_s because Q_s may be assumed to be independent of pressure and temperature.

The mixture of gases and liquids presents a very complicated phenomena. We shall discuss it in chapter IX under the title of two phase flow. One of the interesting phenomena for the two phase flow is the phase change, such as condensation and evaporation of the vapor in the fluid. For instance, in the actual flow of air, there is normally a certain amount of moisture present which is usually in the vapor state. If the vapor condenses, heat is released to the air equal to the latent heat of vaporization of the vapor L. The Clausius-Clapeyron equation gives the relation of the vaporization process, i.e.,

$$\frac{dp}{dT} = \frac{L}{T(V_{\mathrm{gas}} - V_{\mathrm{liq}})} \tag{1.80}$$

where V_{gas} is the specific volume of the gas and V_{liq}, that of the liquid. For the vaporization of a perfect gas from its liquid, the specific volume in liquid state is negligible; hence we have

$$V_{\mathrm{gas}} = \frac{RT}{p}, \quad V_{\mathrm{liq}} = 0$$

Eq. (1.80) becomes

$$d \log p_s = \frac{1}{R} \frac{L}{T^2} dT \tag{1.81}$$

Eq. (1.81) gives the saturation pressure p_s at which condensation will just begin under equilibrium conditions as a function of temperature if the function of latent heat in terms of temperature $L(T)$ is known. For water, $L = a + bT$ where a and b are constants and the integration of Eq. (1.81) can be carried out immediately.

Another interesting flow problem of two phase flow is the mixture of gas or liquid with solid particles. We shall discuss it in chapter IX also.

16. Electrical Conductivity of a Plasma and Electrical Charge

Let e_s be the electrical charge of a particle of the sth species in a plasma. The excess electrical charge of the plasma as a whole is

$$\rho_e = \sum_{s=1}^{N} \rho_{es} = \sum_{s=1}^{N} e_s n_s \tag{1.82}$$

The electrical current density J in a plasma is

$$J = \sum_{s=1}^{N} \rho_{es} q_s = \sum_{s=1}^{N} \rho_{es} w_s + q \sum_{s=1}^{N} \rho_{es} = I + \rho_e q \tag{1.83}$$

where I is the electrical conduction current and $\rho_e q$ is the electrical convection current.

The electrical conduction current is due to the complicated relative motion of charged particles in a plasma which depends on the electromagnetic fields as well as gasdynamic forces. For a first approximation, we may use the following relation for the electrical conduction current:

$$I = \sigma_e (E + q \times B) = \sigma_e E_u \tag{1.84}$$

where E is the electrical field strength, B is the magnetic induction and σ_e is the electrical conductivity of the plasma. Eq. (1.84) is known as generalized Ohm's law and E_u is the electrical field strength in a moving coordinate. The electrical conductivity σ_e in Eq. (1.84) is a scalar quantity. When the magnetic field strength is large and the density of the ionized gas is small, we should consider the electrical conductivity as a tensor quantity. We shall discuss this point in chapters VI and IX where a general derivation of the equation for electrical current density will be given.

The scalar electrical conductivity σ_e of a fluid is one of the most important physical quantities in magnetofluiddynamics. The electrical conductivity σ_e may be determined by the kinetic theory of gases or by experiments. For slightly ionized gas, we may use the following formula:

$$\sigma_e = \sigma_c = \frac{3.9 \times 10^{-12}\alpha}{Q_e T^{\frac{1}{2}}} \text{ mho/cm} \tag{1.85}$$

where α is the degree of ionization of the plasma and Q_e is the electron-atom collision cross section which is about 10^{-15}cm^2 for air at $5000°K$.

For fully ionized plasma, we may use the following formula:

$$\sigma_e = \sigma_d = \frac{1.56 \times 10^{-4} \times T^{3/2}}{\ln\left(\frac{1.23 \times 10^4 \times T^{3/2}}{n_e^{\frac{1}{2}}}\right)} \text{ mho/cm} \tag{1.86}$$

where T is the temperature in $°K$ and n_e is the number density of electrons per cubic centimeter.

There is no accurate simple formula for the scalar electrical conductivity for the intermediate case of ionization. For a first approximation, we may use the following approximate formula:

$$\frac{1}{\sigma_e} = \frac{1}{\sigma_c} + \frac{1}{\sigma_d} \tag{1.87}$$

The electrical conductivity of ionized gas may be increased by seeding with an additive of low ionization potential (e.g., cesium or potassium). Some typical values of electrical conductivity σ_e of various conductors are given below:

copper	600,000 mho/cm
mercury	10,800 mho/cm
air (seeded with 0.1K) (shock Mach no. 16)	4 mho/cm
air (shock Mach no. 14)	0.5 mho/cm
salt water (saturated at 25°C)	0.25 mho/cm
pure water	0.000,000,2 mho/cm

Actually if there is a magnetic field, the electrical conductivity is different in different direction with respect to the direction of the magnetic field. Hence the electrical conductivity is a tensor quantity. The above scalar electrical conductivity is the value of electrical conductivity without a magnetic field or in the direction of the magnetic field when ion slip is negligible. The tensor electrical conductivity depends on the value of the scalar electrical conductivity and the ratio of electron cyclotron frequency $\omega_c = eB/m_e$ to the collision frequency of electron and ion f. If ω_c/f is negligibly small, the electrical conductivity is the same in all direction and Eq. (1.84) will be a good approximation. We shall discuss this point in more detail in chapters VI and IX.

PROBLEMS

1. The equipartition energy principle of the kinetic theory of gases states that the mean kinetic energy of an element is equal to $\frac{1}{2}kT$ per degree of freedom. Prove this principle.

2. By the principle of equipartition energy, show that the ratio of specific heats of a gas is

$$\gamma = \frac{c_p}{c_v} = \frac{n_1+2}{n_1}$$

   where n_1 is the number of degrees of freedom of a molecule of the gas.

3. Calculate the mean free path of a gas by simple kinetic theory in terms of temperature and density of the gas.

4. What is the Sutherland formula for the coefficient of viscosity of air? Calculate the coefficient of viscosity of air for the temperature range from 0°C to 1000°C by (a) Sutherland formula and (b) by a power law with $n=0.76$ [see Eq. (1.53)].

5. Calculate the Prandtl number P_r from the values of the coefficient of thermal conductivity κ, the coefficient of viscosity, and the specific heat c_p at $T=273°K$ and 1500°K for air. Also calculate the Prandtl number for (a) water and (b) mercury at $T = 273°K$ and $T = 350°K$.

6. From the kinetic theory, derive the Boltzmann law of distribution Eq. (1.24). Then by the definitions of number density, mean velocity, and temperature, derive the Maxwellian distribution (1.26) from Eq. (1.24).

7. From quantum theory, show that the energy levels associated with various orbits of an atom are given by the formula (1.31).

8. Derive the formula for vibrational internal energy from statistical mechanics.

9. Derive the Saha relation (1.45) from statistical mechanics.
10. Derive the Saha relation for a partially ionized gas consisting of electrons, two species of ions, and one kind of neutral particles.
11. Derive the Planck radiation law (1.46).
12. Discuss thermal diffusion and pressure diffusion in a mixture of gases.
13. Derive the Clausius-Clapeyron equation (1.80) from statistical mechanics.
14. Calculate the scalar electrical conductivity from Eqs. (1.85) and (1.86) for the temperature range from 5000°K to 50,000°K.

REFERENCES

[1] Born, M. Atomic Physics. (6th ed.) Hafner Publishing Co., 1957.
[2] Epstein, P. S. Thermodynamics. John Wiley, 1937.
[3] Frost, A. A. and Pearson, R. G. Kinetics and Mechanism. John Wiley, 1953.
[4] Glasstone, S., Laidler, K. J. and Eyring, H. The Theory of Rate Processes. McGraw-Hill, 1941.
[5] Guggenheim, E. A. Thermodynamics. (3rd ed.) North-Holland Publishing Co., 1957.
[6] Hirschfelder, J. O., Curtiss, C. F. and Bird, R. B. Molecular Theory of Gases and Liquids. John Wiley, 1954.
[7] Jakob, M. Heat Transfer. Vol. 1. John Wiley, 1949.
[8] Lamb, H. Hydrodynamics. (6th ed.) Cambridge University Press, 1932.
[9] Marks, L. S. Mechanical Engineer's Handbook. (5th ed.) McGraw-Hill, 1951.
[10] Pai, S.-I. Introduction to the Theory of Compressible Flow. D. Van Nostrand, 1959.
[11] Pai, S.-I. Magnetogasdynamics and Plasma Dynamics. Springer-Verlag, 1962.
[12] Planck, M. The Theory of Radiation. Dover Publications, 1959.
[13] Prandtl, L. Fluid Dynamics. Hafner Publishing Co., 1952.
[14] Schiff, L. I. Quantum Mechanics. McGraw-Hill, 1955.
[15] Streeter, V. L. (ed.) Handbook of Fluid Dynamics. McGraw-Hill, 1961.

STATICS OF FLUIDS

1. Introduction

In the statics of fluids, we consider the case where the fluids are in equilibrium. By equilibrium condition, we mean that the fluid, from macroscopic point of view, has no relative motion within itself, nor with respect to its container, nor with respect to any body submerged in it. Even though the particles of the fluid are still in perpetual motion, the average velocity of the fluid vanishes. According to the principle of mechanics, the resultant forces on any portion of the fluid in equilibrium must vanish, i.e.,

$$\Sigma F_x = 0; \quad \Sigma F_y = 0; \quad \Sigma F_z = 0 \tag{2.1}$$

where F_x, F_y, and F_z are respectively the x-, y-, and z-component of various external force on the system considered. Eq. (2.1) are the fundamental equations for the statics of fluids.

In order to apply Eq. (2.1) to any specific problem, we have to know the forces in that problem. The following are some major forces occurring in the statics of fluids:

(1) Pressure p. Since there is no relative motion of the fluid, the only stress in the fluid is the pressure which is acting normal to the surface considered. In general, the pressure is a function of spatial coordinates, i.e.,

$$p = p(x, y, z) \tag{2.2}$$

The determination of the function of pressure is one of the main problems in the statics of fluids.

(2) Body forces. Body forces are forces which are proportional to the volume or mass of the fluid. The most common body force is the gravitational force F_g, i.e.,

$$F_g = \rho g \tag{2.3}$$

where F_g is the force per unit volume and g is the gravitational acceleration. The gravitational acceleration g is determined by the law of gravitation, i.e.,

$$F = \frac{Gm_1 m_2}{r^2} \tag{2.4}$$

where F is the attracting force between two particles of masses m_1 and m_2 with a distance r between them. G is known as the gravitational constant which is the force with which two spheres of unit mass attract each other through unit distance. The value of the gravitational constant is

$$G = 6.66 \times 10^{-8} \, cm^3 \cdot gr^{-1} \cdot sec^{-2} \text{ in cgs units}$$
$$= 3.44 \times 10^{-8} \, ft^4 \cdot lb^{-1} \cdot sec^{-4} \text{ in engineering units} \tag{2.5}$$

On the earth, the main attracting force is the force between the earth and the object considered. Hence Eq. (2.4) may be written as

$$F = mg \tag{2.6}$$

where m is the mass of the body considered and $g = Gm_{earth}/r^2$. Since the mass of the earth is approximately constant, the gravitational acceleration is in the direction toward the center of the earth and is a function of the altitude from the earth. Common values taken for the earth gravitational acceleration are 32.16 ft/sec² or 9.81 m/sec² or more accurately:

$$g = 32.1721 - 0.08211 \cos 2a_\theta - 0.000,003h \text{ (ft/sec}^2) \tag{2.7}$$

where a_θ is the latitude in degrees and h is the height in feet above sea level. For force per unit volume, Eq. (2.3) replaces Eq. (2.6). On other planets or stars, the attraction force due to the earth is negligible and the attraction on the planet considered becomes the dominant force. Hence we may change the gravitational acceleration by changing the mass in the definition of g.

For plasma, body forces may be produced by the interaction of the charged particles with the electromagnetic fields.

(3) Surface tension of liquid. It is a special force on the free

surface of liquid which will be discussed in section 2.

From Eq. (2.3), we see that the gravitational force depends on the density of the fluid. Thus for the case of a mixture of fluids, the composition of the mixture of fluids plays an important role in the statics. It is the case for the analysis of the atmosphere of earth or of another planet as we shall discuss in section 5.

Another important problem of statics of fluid is the determination of forces acting on the solid surface or body by the fluid. The floating of a ship on water and the floating of a balloon in the air are the practical applications of the determination of forces. Many instruments are designed according to the principles of statics of fluid.

After the equilibrium condition of a fluid is determined, the next interesting question is whether the system is statically stable or not. By static stability, we mean that if we give a small disturbance to a system, it tends to restore to its original position. We shall study various static stability conditions of fluid in the following sections.

The simplest equilibrium condition of a fluid occurs when the fluid and its container are both at rest or in uniform motion. But the equilibrium condition may also occur when the whole system, the fluid and its container, is in a uniform translational acceleration or in a rotational motion. Strictly speaking, these cases belong to the dynamics of fluid, but since simple treatment similar to the cases of no acceleration may be applied to these cases, we shall discuss these cases briefly in section 9.

2. Statics of Liquid

We consider first the equilibrium condition of a perfect fluid or a liquid of which the density is a constant. If there is no body force in the fluid, pressure will be the only force in the fluid. We consider a circular column of fluid of cross-sectional area A (Fig. 2.1). If we assume that the area A is so small, and then the pressure p on the area may be considered as a constant, which has the value at the point of the fluid considered. Since the fluid is in equilibrium, the external forces on the column in the direction of its axis must be in equilibrium, i.e.,

$$p_1 A = p_2 A, \text{ or } p_1 = p_2 \tag{2.8}$$

Since the orientation of the column in Fig. 2.1 is arbitrary, we conclude that without body force the pressure in a perfect fluid in equilibrium is the same everywhere and in all directions.

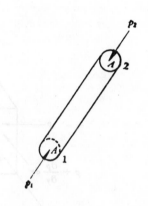

Next we consider the gravitational force in addition to the pressure force of the fluid in equilibrium. The gravitational force acts in the direction of the gravitational acceleration g. For simplicity, we denote the direction of g as the z-axis. Hence there will be the gravitational force of the z-direction only but not in the x- or y-direction. The equilibrium of force in the z-direction on a parallelepiped in a fluid (Fig. 2.2) gives

Figure 2.1 Equilibrium condition of a perfect fluid without body force

$$-\left(p + \frac{\partial p}{\partial z} dz \right) dx\,dy + p\,dx\,dy - \rho g\,dx\,dy\,dz = 0$$

or

$$\frac{\partial p}{\partial z} = -\rho g \tag{2.9}$$

Similarly the equilibrium conditions in the x- and y-direction give

$$\frac{\partial p}{\partial x} = \frac{\partial p}{\partial y} = 0 \tag{2.10}$$

In general, Eqs. (2.9) and (2.10) may be written in the vector form:

$$\nabla p = -\rho g \tag{2.11}$$

The gravitational acceleration g may be directed in any arbitrary direction with respect to the (x, y, z) coordinate system.

It is usually convenient to define a potential ϕ_g for the gravitational field such that

$$\nabla\phi_g = g \qquad (2.12)$$

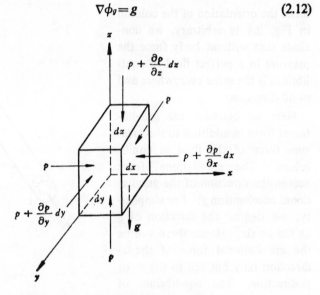

Figure 2.2   A parallelepiped of a fluid

For the system of Eq. (2.9), we have

$$\frac{\partial\phi_g}{\partial z} = g \qquad (2.13)$$

Then the integration of Eq. (2.9) with the help of Eq. (2.13) gives

$$p + \rho\phi_g = \text{constant} \qquad (2.14)$$

Eq. (2.14) shows that for an incompressible fluid, the pressure is constant on the equipotential surface.

From Eq. (2.6), we see that the gravitational acceleration g is not a constant and that it is inversely proportional to the square of the distance from the center of the earth.   We may write

$$g = g_0\left(\frac{R_0}{R_0+h}\right)^2 \qquad (2.15)$$

where g is the gravitational acceleration at an altitude h from the surface of the earth, g_0 is the value of g on the surface of earth whose radius is R_0. If we take $h=z$, the integration of Eq. (2.13) with the help of relation (2.15) gives

$$\phi_g = \frac{zg_0}{1 + \dfrac{z}{R_0}} \tag{2.16}$$

If the altitude is much smaller than the radius of the earth, i.e., $z \ll R_0$, we have

$$\phi_g = g_0 z \tag{2.17}$$

where z is measured from the surface of the earth and positive upward. Eq. (2.14) with the relation (2.17) becomes

$$p = p_0 - \rho g_0 z \tag{2.18}$$

where p_0 is the pressure at $z=0$. Eq. (2.18) as well as Eq. (2.11) are the fundamental equations of hydrostatic pressure. These are many practical applications based on these pressure equations (see Problems).

From Eq. (2.18), we may calculate the buoyancy force on a submerged body. If we divide a submerged body into many infinitesimal small columns of cross-sectional area dA and of height $(z_2 - z_1)$ as shown in Fig. 2.3, the resultant buoyancy force on each of the columns is the net upward pressure force, i.e.,

$$dF = (p_2 - p_1)\,dA = \rho g_0 (z_2 - z_1)\,dA \tag{2.19}$$

The total buoyancy force is the sum of all of these dF over the whole body, i.e.,

$$F = \int dF = \int \rho g_0 (z_2 - z_1)\,dA = \rho g_0 V_0 \tag{2.20}$$

where V_0 is the volume of the submerged body, i.e.,

$$V_0 = \int (z_2 - z_1)\,dA \tag{2.21}$$

Eq. (2.20) is known as the Archimedes principle which states

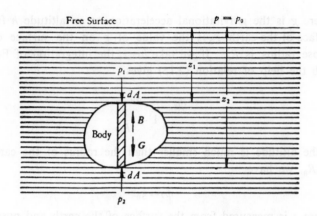

Figure 2.3   Archimedes principle (a submerged body)

that the buoyancy force on a submerged body is equal to the weight
of the fluid displaced.   The Archimedes principle is applied also
to a floating body (Fig. 2.4).   Usually the weight of the air displaced by the body is negligibly small in comparison to that of the
liquid.   For a floating body, the buoyancy force is practically equal
to the weight of the liquid displaced.   The center of buoyancy B
is the center of gravity of the displaced fluid which may be different
from the center of gravity G of the solid body itself.   The stability
of the submerged bodies and floating bodies depends on the relative
positions of the center of buoyancy B and the center of gravity G.
The weight of the body acts at the center of gravity of the body G
and the buoyancy force acts at the center of buoyancy B of the body.
For equilibrium, the line BG must be in the direction of the gravitational acceleration.   The center of buoyancy of the body should be
in such a location with respect to the center of gravity of the body
that the body is statically stable.

We consider the case of a submerged body first (Fig. 2.5).   The
condition for static stability of the body is that the center of gravity
should be directly below the center of buoyancy.   After a small
disturbance of an angle θ, the restoring moment will be

$$M_0 = Wz_0 \sin \theta = Wz_0 \theta \qquad (2.22)$$

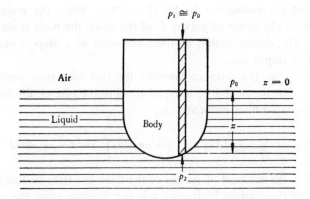

$p_1 \cong p_0$

Air

p_0 $z = 0$

Liquid

Body

z

p_2

Figure 2.4 Archimedes principle (a floating body)

where W is the weight of the body and z_0 is the distance between G and B. If B is below G, the moment of Eq. (2.22) will move the body further away from its equilibrium position.

For a floating body, we have to consider the metacenter of the body. In the equilibrium condition, the center of buoyancy is at the point B while in a small disturbed condition, the center of buoyancy is at the point B' because the shape of the volume of the displaced fluid changes. The lines of buoyancy forces through B and B' intersect at the point M which is called the metacenter and MG is

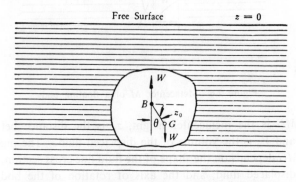

Free Surface $z = 0$

W

B

z_0

θ G

W

Figure 2.5 Restoring moment of a submerged body

known as the metacenter height (Fig. 2.6). When the metacenter M is above the center of gravity G of the body, the body is statically stable. The determination of the metacenter of a ship is very important in ship design.

Let e_B be the distance between the two buoyancy centers B and B'. The moment We_B must be equal to the sum of the moment of all displaced fluid elements, i.e.,

$$We_B = \int \rho g_0 \, dA \cdot y \, d\theta \cdot y = \rho g_0 \, d\theta \int y^2 \, dA = \rho g_0 \, d\theta \cdot I \quad (2.23)$$

where I is the moment of inertia of the fluid line (free surface) cross section of the floating body and y is the distance from the axis of rotation of the body in the disturbance. Since $W = V_0 \rho g_0$ and $e_B = BM \cdot d\theta$, Equation (2.23) gives

$$MG = \frac{1}{V_0} - z_0 \quad (2.24)$$

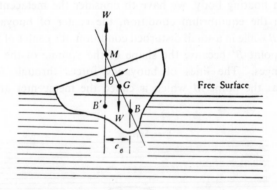

Figure 2.6   Metacenter M of a floating body

where MG is the metacenter height, z_0 is the distance between G and B, and V_0 is the volume of the displaced fluid. The metacenter height MG depends on the shape of the fluid line cross section, volume of the displaced fluid, and the axis of rotation of the body. For instance, if the body is a prism of rectangular cross section with length L, width b, and depth d, the metacenter height for disturbance about the longitudinal axis with y in the direction of b is different from that about the transverse axis with y in the direction of L. In this case, the moment of inertia of the fluid line is $I_L = Lb^3/12$ for the longitudinal axis and is $I_b = L^3b/12$ for the transverse axis. I_b is larger than I_L if L is larger than b.

The metacenter height may be determined experimentally by moving a weight W_0 along the y-direction by a distance y_0. If the difference of inclination of the body is $d\theta$, the metacenter height is then

$$MG = \frac{W_0 \, y_0}{W \, d\theta} \tag{2.25}$$

The molecules of a substance usually attract one another. These attraction forces are different for different substances. Hence on the surface of separation between two media (solid and liquid, liquid and liquid, or liquid and gas) there is a surface force because of the unsymmetrical attraction force. Ordinarily these surface forces are called surface tensions. For instance, the free surface of a liquid, i.e., the contact surface of a liquid with a gas, has a surface tension so that the surface of the liquid tends to be a minimal surface. A drop of liquid under no external force will be spherical. Surface tension will also occur on the surface of separation of two nonmiscible liquids as well as on the surface between a solid and a liquid. For miscible liquids such as water and alcohol, a compression force may occur instead of a tension. The surface tension becomes very significant when a liquid is in a tiny tube known as capillary tube. Hence we often call the surface tension as capillary force.

Surface tension appears when the surface is curved so that a pressure difference may arise due to the surface tension. We consider a curved surface with a pressure difference $(p_1 - p_2)$ as shown in Fig. 2.7. Let the surface tension be T_s and the radii of the curvature of the surface be r_1 and r_2. If we assume that the area of the surface is infinitesimally small, the equilibrium of forces normal to the surface gives

$$p_1 - p_2 = T_s \left(\frac{1}{r_1} + \frac{1}{r_2} \right) \tag{2.26}$$

For the case of two immiscible liquids separated by a curved surface, we have $p_1 = p_0 - \rho_1 gz$ and $p_2 = p_0 - \rho_2 gz$. Then Eq. (2.26) becomes

$$\frac{1}{r_1} + \frac{1}{r_2} = gz \frac{\rho_2 - \rho_1}{T_s} \tag{2.27}$$

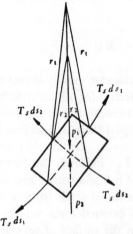

Figure 2.7 Equilibrium condition between surface tension and pressure difference

When the boundaries of three fluids meet, they must meet at certain definite angles such that the three surface tensions are in equilibrium (Fig. 2.8). However if one of the surface tension is larger than the sum of the other two, an equilibrium condition will never be possible. For instance, it is not possible to reach an equilibrium condition for the three fluids: air, mineral oil, and water.

If one of the fluids of Fig. 2.8 is replaced by a solid wall, we need to consider only the equilibrium condition along the surface of the solid wall because there is no freedom of movement normal to the solid wall. Let fluid 3 be replaced by a solid wall (Fig. 2.9). The equilibrium condition gives

$$T_{12} \cos \alpha + T_{23} = T_{13} \qquad (2.28)$$

or

$$\cos \alpha = \frac{T_{13} - T_{23}}{T_{12}} \qquad (2.28a)$$

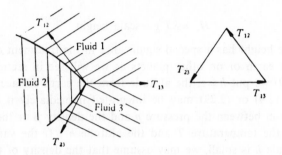

Figure 2.8   Equilibrium of surface tensions among three fluids

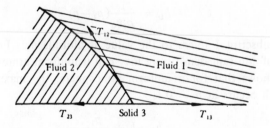

Figure 2.9   Equilibrium of surface tensions between two fluids and a solid wall

If T_{23} is greater than T_{13}, the angle α will be greater than 90°. For instance, when air, mercury, and glass meet, the angle of α is greater than 90°. If $T_{13} - T_{23}$ is greater than T_{12}, no equilibrium condition is possible. In this case, the surface of the solid will be covered by the liquid. For instance, a petrol covers the surface of a glass.

3.   Ideal Gases

Eq. (2.9) holds true for both the incompressible fluid and the compressible fluid. The only difference is that for the compressible fluid the density is a function of the pressure and the temperature of the fluid. For an ideal gas with the equation of state (1.18), Eq.

(2.9) may be written as

$$\frac{dp}{p} = -\frac{dh}{RT/g} = -\frac{dh}{H_c} \tag{2.29}$$

where

$$H_c = RT/g = \text{scale height} \tag{2.30}$$

The scale height has a special significance in the treatment of atmosphere on earth or on other planets as we will see in section 8. In Eq. (2.29) we put $h = z$, the altitude from the earth surface.

Eq. (2.9) or (2.29) may be integrated immediately if we know the relation between the pressure p and the density ρ or the relation between the temperature T and the altitude h. If the variation of the altitude h is small, we may assume that the density of the gas is a constant, Eq. (2.29) gives approximately

$$h - h_1 = \frac{1}{\rho_1 g_0}(p_1 - p) \tag{2.31}$$

where subscript 1 refers to the value at altitude h_1. For a column of gas of constant density, we may call it uniform atmosphere. Eq. (2.31) shows that for a uniform atmosphere, the maximum height h_m which occurs at $p = 0$ is

$$h_m = h_1 + \frac{p_1}{\rho_1 g_0} \tag{2.32}$$

Another interesting case is the isothermal case where $T = T_1 = $ constant. For isothermal atmosphere, Eq. (2.29) gives

$$h - h_1 = (h_m - h_1) \ln (p_1/p)$$

or

$$p = p_1 \exp\left(-\frac{h - h_1}{h_m - h_1}\right) \tag{2.33}$$

In general, we may use the polytropic relation between pressure p and density ρ, i.e.,

$$p\rho^{-n_0} = \text{constant} = K = p_1 \rho_1^{-n_0} \tag{2.34}$$

where n_0 is a constant. For the adiabatic case, we have $n_0 = \gamma = $

c_p/c_v. Substituting Eq. (2.34) into Eq. (2.29) and integrating, we have

$$h - h_1 = \frac{n_0}{n_0 - 1} (h_m - h_1) \left[1 - \left(\frac{p}{p_1} \right)^{(n_0-1)/n_0} \right] \qquad (2.35)$$

Eq. (2.35) is sometimes referred to as the relation for polytropic atmosphere.

4. Mixture of Gases

The actual atmosphere consists of a mixture of several gases. The density and pressure of this mixture are given by Eqs. (1.66) and (1.69) respectively. If the diffusion of various species of the mixture is negligible, the mixture may be considered as a single gas in the analysis of vertical distribution of temperature and pressure in the atmosphere. Eq. (2.29) may be written as

$$\frac{dp}{p} = \frac{dn}{n} + \frac{dT}{T} = -\frac{dh}{H_c} \qquad (2.29a)$$

where n is the number density of the mixture.

For a sufficiently thin layer of atmosphere, we may assume that both gravitational acceleration g and the temperature gradient

$$\lambda = -\frac{dT}{dh} = \text{lapse rate in the atmosphere} \qquad (2.36)$$

are constant, we have then

$$\frac{dH_c}{dh} = \frac{k}{mg} \frac{dT}{dh} = -\frac{k}{mg} \lambda = \text{constant} = b \qquad (2.37)$$

It should be noticed that in general the value of the lapse rate λ varies from place to place and time to time in the atmosphere. The lapse rate λ corresponds to the adiabatic process is

$$\lambda_a = 0.0054\,°F/ft \qquad (2.36a)$$

The polytropic index n_0 of Eq. (2.34) can be determined from the lapse rate, because Eqs. (2.29a) and (2.37) give

$$\frac{dp}{p} = \frac{1}{b} \frac{dT}{T} \qquad (2.38)$$

Integration of Eq. (2.38) gives

$$T = (\text{constant}) \; p^b \tag{2.39}$$

But the polytropic law (2.34) gives

$$T = (\text{constant}) \; p^{(n_0 - 1)/n_0} \tag{2.40}$$

From Eqs. (2.39) and (2.40), we have

$$n_0 = \frac{1}{1 - b} = \frac{1}{1 - \dfrac{\lambda k}{mg}} \tag{2.41}$$

If the actual atmosphere has a lapse rate smaller than λ_a, the atmosphere is stable because the air which moves adiabatically from one altitude to another will be forced back to its original altitude. On the other hand, if the actual atmosphere has a lapse rate larger than λ_a, the atmosphere is unstable because the air which moves adiabatically from one altitude to another will move further away from its original altitude. The adiabatic atmosphere is in neutral equilibrium.

In terms of the polytropic index n_0, we have:

for $n_0 < \gamma$, the atmosphere is stable;

for $n_0 > \gamma$, the atmosphere is unstable and

for $n_0 = \gamma$, the atmosphere is neutral stable.

With the approximation of Eq. (2.37), we have the conventional exponential decrease of pressure and number density as follows:

$$\frac{p}{p_1} = \left(\frac{H_c}{H_{c_1}} \right)^{-1/b} \tag{2.42a}$$

$$\frac{n}{n_1} = \left(\frac{H_c}{H_{c_1}} \right)^{-(1+b)/b} \tag{2.42b}$$

If there is steady diffusion among the species of the mixture, the concentration of each species changes with altitude as in the case of very high altitude (see section 8). In general we should not use the simple expression similar to equations (2.42) for the change of pressure and number density n with altitude. We shall discuss this case in the next section. However, for a steady state of diffusion, it is convenient to use the following equation for a first approximation:

$$\frac{dp_s}{p_s} = \frac{dn_s}{n_s} + \frac{dT}{T} = -\frac{dh}{H_{cs}} \qquad (2.43)$$

where subscript s refers to the value of sth species in the mixture and

$$H_{cs} = \frac{kT}{m_s g} = \text{scale height for } s\text{th species} \qquad (2.44)$$

Integration of Eq. (2.43) gives

$$\frac{n_s}{n_{s_1}} = \left(\frac{H_{cs}}{H_{cs_1}}\right)^{-(1+b_s)/b_s} \qquad (2.45)$$

where $b_s = -[k\lambda/(m_s g)]$. Eq. (2.45) shows that each species has its own concentration or pressure gradient associated with its own mass according to its own scale height gradient b_s. In an atmosphere in diffusive equilibrium, the variation of number density (n_s) is known directly in terms of the corresponding quantities from another species (n_r) because

$$b_s m_s = b_r m_r \qquad (2.46)$$

5. Diffusion and Dissociation in Atmosphere

At low altitude, i.e., the homosphere (see section 8), the concentration of various species in the atmosphere are approximately constant, the vertical distribution of temperature can be determined accurately by measuring the vertical pressure distribution by Eq. (2.42a). However it is not true when the diffusion and dissociation phenomena are important in the atmosphere such as in the heterosphere. We have to use new formula by taking diffusion and dissociation into consideration. In order to illustrate the effects of diffusion and dissociation, we consider the atmosphere consisting of two species: One is a molecular species whose values are indicated by the subscript M and the other is the atomic species whose values are indicated by the subscript A. Hence the number densities of these two species are n_M and n_A while their diffusion velocities are w_M and w_A. Since we consider the atmosphere in static equilibrium, the resultant mass velocity of the mixture of molecules and atoms must be zero, i.e.,

$$n_M w_M + n_A w_A = nV = 0 \qquad (2.47)$$

and

$$n = n_M + n_A \qquad (2.48)$$

and V is the mass velocity of the mixture as a whole, which is zero here. For static equilibrium, the temperature of these two species must be the same and equal to the temperature of the mixture T. Now we have seven variables: T, p_A, p_M, n_A, n_M, w_A, and w_M. The fundamental equations are:

(1) Equations of state for each species:

$$p_A = k n_A T; \quad p_M = k n_M T \qquad (2.49)$$

(2) Equations of continuity for each species

$$\frac{dw_A n_A m_A}{dz} = M_A; \quad \frac{dw_M n_M m_M}{dz} = M_M \qquad (2.50)$$

where M_A and M_M are respectively the mass rate of formation of atoms and molecules respectively and m_A and m_M are the mass of atom and molecule respectively.

(3) The diffusion velocity equation of each species:

$$w_s = -\frac{D_s}{k_s}\frac{dk_s}{dz} + \frac{D_p}{k_s p}\frac{dp}{dz} - \frac{D_T}{k_s T}\frac{dT}{dz} \qquad (2.51)$$

where subscript s may be A or M, $k_s = m_s n_s / \rho =$ mass concentration of species s, $\rho = n_A m_A + n_M m_M$, D_s is the diffusion coefficient of sth species, D_p is the coefficient of pressure diffusion and D_T is the thermal diffusion coefficient.

(4) The energy equation of the mixture

$$\kappa \frac{dT}{dz} = \sum_s m_s n_s w_s H_s + Q_0 \qquad (2.52)$$

where H_s is the enthalpy of sth species, Q_0 is the heat flux due to external agency in the atmosphere. The first term in the right hand side of Eq. (2.52) is the heat transfer due to diffusion and the term on the left hand side of Eq. (2.52) is the heat conduction term.

A complete solution of Eqs. (2.49) to (2.52) is very complicated because we have to know the detailed conditions in the atmosphere.

Sometimes, the static solution may not exist so that we have to introduce the time variation terms in the equations and the problem is no longer a static one. For instance in reference [5], Nicolet discussed some cases in the atmosphere and showed that the vertical velocity of the minor constituent in the atmosphere may increase almost exponentially with altitude.

Another interesting factor in the atmosphere is the amount of water vapor in the atmosphere. As long as the water is in the gaseous state, its effect on the atmosphere is negligible. However at any given temperature, there is a definite limit to the content of water vapor in the atmosphere. We say that the air is saturated when it contains this limiting amount of water vapor. As soon as this limit is passed, some of the water vapor will be condensed and formed into tiny water drops which forms a cloud. If the drops are large enough, we have rain. When the water vapor condenses, it releases heat. It would change the temperature in the atmosphere and affect the stability of the atmosphere. The detailed study of the stability of the atmosphere with the effects of condensation and evaporation of water vapor is one of the most interesting problems in meteorology.

Even though the complete study of the diffusion effects given by Eqs. (2.49) to (2.52) is very complicated (we shall discuss it more in chapter V), the equilibrium condition of heterogeneous fluid due to different composition of fluid is easy to analyze. The density of the mixture is a function of the concentration of its constituents. Let us consider a salt solution with variable concentration. Let c_s be the concentration of the salt in the fluid. The density of the solution is a function of the concentration. If the concentration of the solution is a known function of altitude z, we have

$$c_s = c_s(z), \quad \rho = \rho(c_s) = \rho(z) \tag{2.53}$$

Substituting Eq. (2.53) into Eq. (2.9), we have

$$p = p_0 - \int_0^z g\rho(z)\,dz \tag{2.54}$$

The stability condition for the case of variable concentration of an incompressible fluid is that

$$\frac{d\rho}{dz} < 0 \qquad (2.55)$$

Another case of a heterogeneous fluid is the equilibrium of two nonmiscible fluids of different density, one above the other. Under only the gravitational force, in equilibrium condition, the separation surface of the two fluids must be a horizontal plane ($z=$constant), so that on this plane the density is constant. Without loss of generality we let this separation surface by $z=0$. We have then

$$z=+0, \quad \rho=\rho_1=\text{density of the upper fluid}$$
$$z=-0, \quad \rho=\rho_2=\text{density of the lower fluid}$$

The stability condition of equilibrium for this case is $\rho_2 > \rho_1$, i.e., the heavier fluid should be under the lighter fluid. Let us image that at one point of the separation surface, the surface is elevated by a small height h. The pressure on the side of the fluid 1 is $p_1 = p_0 - g\rho_1 h$ while that on the side of the fluid 2 is $p_2 = p_0 - g\rho_2 h$. The difference of pressure across the separation surface at $z=h$ is

$$p_2 - p_1 = hg(\rho_1 - \rho_2) \qquad (2.56)$$

Only when $\rho_2 > \rho_1$, the resultant pressure $p_1 - p_2$ will force the distorted surface to its original position $z=0$.

6. Plasma and Electrically Conducting Fluid

Under the influence of gravitational force only, the statics of a plasma or an electrically conducting fluid can be treated in the exact same manner as that for neutral fluid discussed above. However, for electrically conducting fluids, besides the gravitational force, sometimes we have to consider the electromagnetic forces. Hence we have to consider the resultant effects of the gravitational and electromagnetic effects. In order to discuss the effect of the electromagnetic forces on the statics of a plasma, we shall consider the simple case of isothermal plasma. The pressure due to gravitational force, known as gravitational pressure p_g, may be eliminated from our equations of forces by the following considerations:

From Eq. (2.31), we assume that the variation of height is small and that the density of the plasma is approximately constant and we have

$$p_g = p_0 - \rho g z \tag{2.57}$$

The total pressure of the plasma may be written as

$$p_t = p_g + p = (p_0 - \rho g z) + p \tag{2.58}$$

where p is the pressure of the plasma due to causes other than gravitational force. If the variation of altitude is small, we may consider $\rho g z = $ constant and then Eq. (2.58) becomes

$$p_t = \text{constant} + p \tag{2.59}$$

Since the constant in Eq. (2.59) will have no influence in the mechanics of fluid, we may put it to be zero for convenience. Thus

$$p_t = p \tag{2.60}$$

This concept of static pressure p will be used whenever the gravitational force or its variation is negligible in the mechanics of fluid.

Under the concept of Eq. (2.60), the equation of statics of an electrically conducting fluid or a plasma is

$$\nabla p = J \times B + \rho_e \nabla \phi_e \tag{2.61}$$

where

$$J = \nabla \times H = \text{electrical current density} \tag{2.62a}$$

$$E = \nabla \phi_e = \text{electrical field strength} \tag{2.62b}$$

$$B = \mu_e H = \text{magnetic induction} \tag{2.62c}$$

$$\rho_e = e(n_+ - n_-) = \text{excess electric charge} \tag{2.62d}$$

H is the magnetic field strength, μ_e is the magnetic permeability, ϕ_e is the electric potential, e is the absolute electric charge, n_+ is the number density of ions which is assumed to be singly charged, and n_- is the number density of electrons. We shall discuss these equations more in chapter VI. In general, we have to solve Eq. (2.61) with equations of energy and state. But for isothermal case, we may consider Eq. (2.61) only.

First we consider the case of no electrical current, i.e., $J=0$. Eq. (2.61) becomes

$$\nabla(p+\rho_e\phi_e)=0 \qquad (2.63)$$

where we assume $\rho_e=$ constant. Eq. (2.63) is identical to Eq. (2.14). Hence we may draw a similar conclusion by replacing ϕ_g by ϕ_e and ρ by ρ_e. The distance z is the distance along the normal of an equi-potential surface. In fact we may draw a general conclusion that a stable configuration of a heterogeneous fluid is possible only the body force has a potential. The surface of equi-potential is also the surface of equal pressure or that of equal density.

Another interesting case is that the force due to the variation of magnetic field is the dominant force which is the case for many astrophysical problems. Then we have

$$\nabla p=\mu_e(\nabla\times H)\times H=\mu_e(H\cdot\nabla)H-\frac{1}{2}\mu_e\nabla(H\cdot H) \qquad (2.64)$$

Eq. (2.64) gives the variation of pressure according to the variation of the magnetic field strength H.

The lines in space with the same direction of H are usually referred to as the lines of force of the magnetic field. If the strength of the magnetic field is constant along its lines of force, we have

$$(H\cdot\nabla)H=0 \qquad (2.65)$$

Then Eq. (2.64) gives

$$p+\frac{1}{2}\mu_eH^2=p+p_H=\text{constant} \qquad (2.66)$$

The term $p_H=\frac{1}{2}\mu_e H^2$ is known as the magnetic pressure. Eq. (2.66) shows that under the present condition, the sum of the fluid pressure p and the magnetic pressure p_H is a constant. The simplest condition which satisfies Eq. (2.65) is that the lines of force are straight and parallel.

7. Rarefied Gas

In the last five sections, we assumed that the gas is not rarefied and we used the macroscopic properties such as pressure and temperature to describe the situation of the fluid. However, when the gas is highly rarefied such as the case of outer space, it is not sufficient to use the macroscopic properties to describe the conditions of the gas and we have to consider the discrete properties of gas particles. The macroscopic equation such as Eq. (2.11), etc., can not be used to describe the statics. We have to use the Boltzmann equation (see chapter VIII) to describe the mechanics of gas. We shall discuss this problem in detail in chapter VIII. However one of the simplest assumptions which has often been used to describe the properties of a gas in equilibrium is the Maxwellian distribution of Eq. (1.26). For the static problem, we have $q=0$ but the number density n and temperature T are given values, then

$$F_0 = n \left(\frac{m}{2\pi kT} \right)^{3/2} \exp \left(-\frac{mq_m^2}{2kT} \right) \qquad (2.67)$$

For a mixture of gases, we may use this Maxwellian distribution function for each species with their own mass m_s, number density n_s, and temperature T_s. Such an analysis has been used to discuss the composition of atmosphere at an extremely high altitude.

8. Atmosphere on Earth

Our previous treatment may be used to analyze the physical constitution and chemical composition in the atmosphere of earth or other planet. It would be of interest to give a brief description of earth's atmosphere and how we may apply the above analysis to the earth atmosphere.

The earth's atmosphere consists of various types of gases, ions, and electrons. The composition varies with altitude as well as time and place. However it is accurate enough to divide the earth's atmosphere into various regions in which different laws of variation of pressure and temperature with altitude hold. There are two major regions: one is known as homosphere which covers the altitude approximately below 85 kilometers and in which the composition of the earth atmosphere remains the same, and the other is known as heterosphere which covers the altitude above 85 kilometers and in which the composition of the earth atmosphere changes with altitude. In each of these two main regions, we may also divide them into several subregions according to the variation of temperature with altitude as follows:

(A) Homosphere

Homosphere may be divided into three subregions: troposphere, stratosphere, and mesosphere (see Fig. 2.10).

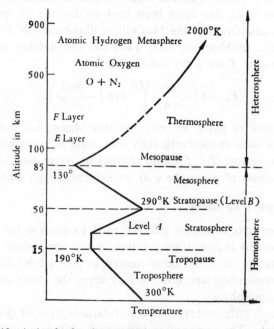

Figure 2.10   A sketch of various regions of the earth's atmosphere (not in scale). (The temperatures are just some representative values)

(1)   Troposphere. At low altitude, from sea level to an altitude of about 10 kilometers, the composition of the air is approximately constant and the temperature decreases with altitude. It is known as the troposphere. The lapse rate λ in the troposphere is almost a constant even though its value varies from place to place and time

to time. Dry atmosphere at sea level consists of 78.03% of molecular nitrogen N_2, 20.99% of molecular oxygen O_2, 0.93% of argon A, 0.030% of carbon dioxide CO_2, and 0.01% of molecular hydrogen H_2 by volume and a slight amount of neon Ne, helium He, krypton Kr, and xenon Xe. For many practical purposes, the physical properties of air are computed by assuming that the air is simply composed of 78.12% of N_2, 20.95% of O_2, and 0.93% of A at sea level by volume. For low altitude, a standard atmosphere has been adopted in order to facilitate the comparison of test date in different atmosphere conditions. The standard atmosphere for the troposphere is based on the following assumptions:

    (a) The air is a perfect gas with a gas constant $R = 53.33$ ft/°F.

    (b) The pressure of air at sea level is $p_0 = 29.921$ in Hg.

    (c) The temperature of air at sea level is $T_0 = 59$°F.

    (d) In the troposphere, the lapse rate is a constant of a value $-0.003,566$°F/ft.

    (e) The troposphere ends when the temperature reaches $T_s = -67$°F.

At the end of the troposphere, there is a region known as the tropopause which is a short region between troposphere and stratosphere. The height of the tropopause depends greatly on the latitude. It may be below 10 km in the polar region and above 15 km in the equatorial belt.

    (2) Stratosphere. The stratosphere covers the region from the tropopause up to an altitude of about 50 km. In this region, the temperature does not decrease with increase of altitude. First the temperature remains constant and then increases with altitude. In the old definition of standard atmosphere, we assume that the temperature remains constant in the whole stratosphere. But from recent data, we know that the temperature in the stratosphere increases with altitude to a peak of the order of 270°K. Sometimes we may divide the stratosphere into two sub-regions. In the lower region, the temperature is constant up to a level A and in the upper region the temperature increases with altitude up to the stratopause which is also known as level B.

(3) Mesosphere. It is the region above the stratopause up to an altitude of about 85 km, in which the temperature decreases again with increase of altitude. The minimum temperature is of the order of 150°K. The end of mesosphere is known as the mesopause. In the mesosphere, the photochemical action is very important.

(B) Heterosphere

In the heterosphere the composition of the air varies greatly with altitude and the dissociation and ionization processes are important. The science which studies the upper atmosphere of the heterosphere is known as Aeronomy. The heterosphere may be divided into five subregions: thermosphere, ionosphere, metasphere, protosphere, and exosphere.

(1) Thermosphere and ionosphere. Above the mesopause, the temperature of the air again increases with altitude up to a maximum value on the order of 1500°K to 2000°K. This region is known as the thermosphere in which dissociation and ionization processes are important and the composition of air is no longer constant. At the lower end of the thermosphere at an altitude of about 100 km, the dissociation of oxygen takes place. Hence we cannot assume that the composition of the air remains constant. At an altitude of around 500 km, most of the molecular oxygen have dissociated and we may assume that the major constituents of the air are molecular nitrogen and atomic oxygen. At higher altitudes, the dissociation of molecular oxygen and the recombination of atomic oxygen are not the only chemical reactions. There are many other chemical reactions which take place. For instance, the atomic oxygen may attach to molecular oxygen to form ozone. The molecular nitrogen may dissociate into atomic nitrogen. The atomic oxygen and atomic nitrogen may form nitric oxide. The water vapor in the air may dissociate into atomic hydrogen and hydroxyl (OH). The carbon dioxide may dissociate into atomic oxygen and carbon monoxide. Many other chemical reactions take place.

From mesopause and up, ionization of air takes place. Even below mesopause (80 km), there are some free electrons in the atmosphere in what is known as the *D*-layer with free electrons with a number density of 10^2 to 10^4 electrons per cubic centimeter. As the altitude increases, the electron density increases to a maximum of the order of 10^5/cc at what is known as E-layer and which is at about 120 km. After a slight drop of electron density, the electron density increases again with altitude to a greater maximum of the order of 10^6/cc at the F-layer and which is at about 300 km. As the altitude further increases, the electron density decreases. The region in which there are a considerable number of free electrons is known as the ionosphere. In the ionosphere, the diffusion as well as photochemical and photoelectric actions are very important. The thermosphere extends a few hundred kilometers above the F-layer of ionosphere until the temperature of the atmosphere reaches a maximum of 1500 to 2000°K.

(2) Exosphere —— metasphere and protosphere. The temperature of the atmosphere above the thermosphere remains almost constant for a considerable altitude. This isothermal region was called the exosphere because it was thought that in this region the laws of gas kinetics no longer applied. In this region, the particles will suffer few collisions and when they move upward, they may escape from the earth gravitational field. Hence the name exosphere is used. In this region, we should not consider the air as a continuum and should consider the discrete character of the particles. Hence we have to use the rarefied gasdynamics to study the motion of the particles, particularly using the free molecule flow analysis as we shall discuss in chapter VIII.

In the lower portion of the exosphere, the air is still mainly un-ionized. We should use the free molecule flow of neutral particles with gravitational force as the main body force to study the dynamic process of the atmosphere. This region is known as metasphere. In the upper portion of the exosphere, the gas particles of the atmosphere are almost fully ionized and the protons are more abundant than the neutral hydrogen. This region is known as the protosphere. In the protosphere, we should consider both the gravitational force and the electromagnetic force to study the dynamic process of the atmosphere. In the exosphere, we should find out the molecular distribution function and then determine the statistical average of various properties of the atmosphere.

9. Equilibrium of Moving Systems

If we consider the fluid and its container as a system, our previous results (sections 2 to 6) hold true for the cases that the system is at rest or in a uniform motion. However, equilibrium condition may exist in other cases of moving system too.

Equilibrium condition may occur for a system moving with a constant linear acceleration a_c. By d'Alembert's principle, the system may be considered as in equilibrium if we add the inertial force per unit volume $F_i = \rho a_c$ to the system. Our analysis of the statics of the moving system with constant translational acceleration is exactly the same as that of the gravitational force by simply adding the body force F_i on the fluid.

Another interesting case is that the whole system is rotating about an axis with a constant angular velocity ω. For simplicity, we consider the fluid as an incompressible fluid and the axis as the vertical axis z. The acceleration due to rotation is the centrifugal force acceleration, i.e.,

$$a_c = i_r \rho \omega^2 r \qquad (2.68)$$

where i_r is the unit vector in the radial direction and $r^2 = x^2 + y^2$.

The equilibrium condition gives

$$\nabla p = -\rho g i_z + i_r \rho \omega^2 r \qquad (2.69)$$

where i_z is the unit vector in the vertical (z) direction.

The z- and r-components of force of Eq. (2.69) are respectively

$$\frac{\partial p}{\partial z} = -\rho g \qquad (2.70a)$$

$$\frac{\partial p}{\partial r} = \rho \omega^2 r \qquad (2.70b)$$

Integration of Eqs. (2.70) gives

$$p = p_0 - \rho g z + \frac{1}{2} \rho \omega^2 r^2 \qquad (2.71)$$

where p_0 is the pressure at $z=0$ and $r=0$.

PROBLEMS

1. The two-liquid manometer shown in Fig. 2.11 is sensitive to small pressure differences. Derive a formula for the difference of p_1-p_2 in terms of the heights h_1, h_2 and h_3 and the densities ρ_1 and ρ_2. If $h_3=h_1+h_2$ and the liquid 1 is water and the liquid 2 is an oil of specific gravity of 0.83, what would be the pressure difference p_1-p_2 when h_1 is one inch? The specific gravity of a substance is the ratio of its density to that of water.

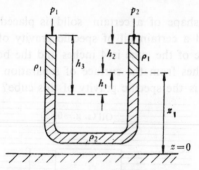

Figure 2.11   Two-liquid manometer

2. For the hydraulic press of Fig. 2.12, the area of A_2 is one square inch and the force F_2 is 100 pounds. What will be the area of piston 1 if the force F_1 is 3 tons?

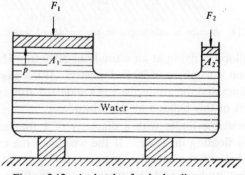

Figure 2.12   A sketch of a hydraulic press

3. Find the total forces and their locations on the side wall *AB* and *A'B'* of the tank of Fig. 2.13.

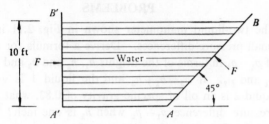

Figure 2.13   Pressure force on side walls of a water tank

4. A cubic shape of a certain solid is placed in the middle of water and a certain oil of specific gravity of 0.83 (Fig. 2.14). If the side of the cube is 4 inches and the bottom of the cube is 2.5 inches from the surface of separation of the water and oil, what is the specific gravity of this cube?

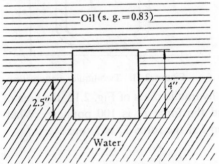

Figure 2.14   A cube is submerged in the middle of water and oil layers

5. A balloon is flying at an altitude of 5000 ft. The shape of the balloon is a sphere of 20 ft diameter. If the gas in the balloon is helium and the atmosphere is a standard atmosphere, estimate the lift of this balloon at that altitude.

6. A circular cylinder with a diameter of 1 foot and a length of 3 ft is floating in water. If the weight of the cylinder is such that half of the cylinder is under the water (Fig. 2.15), find the longitudinal and the transverse metacenter heights of the

cylinder if the center of gravity of the cylinder is on the surface of the water.

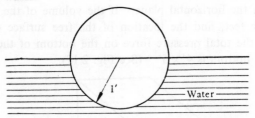

Figure 2.15 Metacenter height of a circular cylinder

7. A prism of a cross section as shown in Fig. 2.16 and of a length of 10 ft is floating on water. Find the longitudinal and the transverse metacenter heights if the center of gravity is 3 inches above the surface of the water.

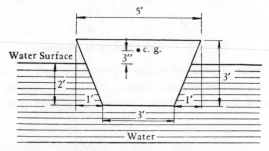

Figure 2.16 Metacenter height of a prism of 10 feet long

8. The sea-level condition of air is exactly that of a standard atmosphere. Find the height of the atmosphere if we assume that:
    (1) The density of the air is constant;
    (2) the density of the air varies according to an adiabatic process.

·9. The sea-level condition of air is exactly that of a standard atmosphere. Find the pressure of the air at 5000 ft if:
    (1) The atmosphere is a standard atmosphere;
    (2) the atmosphere varies according to an adiabatic process;
    (3) the atmosphere varies according to an isothermal process.

10. A rectangular tank of water is in an airplane climbing at a constant acceleration of 10 ft/sec² at an angle of 30 degrees from the horizontal plane. If the volume of the water is 120 cubic feet, find the location of the free surface of the water and the total pressure force on the bottom of the tank. The tank bottom is 4′ × 5′. (See Fig. 2.17)

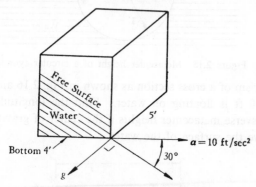

Figure 2.17   Water tank in an airplane

11. A cylindrical tank of water is rotating about its vertical axis with a constant angular velocity of $\omega = 10$ turns/sec. Find the shape of the free surface of water in the tank (Fig. 2.18).

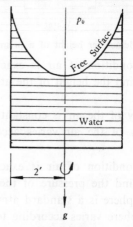

Figure 2.18   A rotating cylindrical tank of water

12. A bucket of water of one pound is on the edge of a wheel of 2 ft diameter rotating at a constant angular velocity of 50 rpm. Show that the resultant force on the water always passes through a given point P on the vertical axis OP. Find the location of this point P (Fig. 2.19).

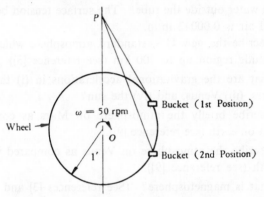

Figure 2.19 A bucket on a rotating wheel

13. A column of plasma is in equilibrium with a tangential magnetic field as shown in Fig. 2.20. If the plasma pressure inside the column is one atmosphere pressure, find the strength of the magnetic field in the vacuum space immediately outside the plasma column.

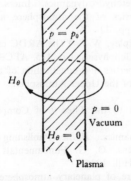

Figure 2.20 Self-pinch plasma

14. The tank of Fig. 2.13 is filled with salt water of density $\rho = \rho_0(1-Az)$ up to the same height as shown in the figure. Calculate the force on the side walls AB and $A'B'$ and their location.

15. If a capillary tube of diameter of 0.01 inch is inserted in water, find the height of water in the tube above the free surface of the water outside the tube. The surface tension between water and air is 0.00042 lb/in.

16. Describe the new U.S. standard atmosphere which covers the altitude region up to 700 km (see reference [3]).

17. What are the gravitational accelerations in (i) the moon, (ii) Mars, (iii) Venus, and (iv) the sun?

18. Describe briefly the atmosphere on Mars as compared with that on earth (see reference [9]).

19. Describe the atmosphere on Venus as compared with that on earth (see reference [9]).

20. What is magnetosphere? (See references [3] and [9])

21. Discuss some possible chemical reactions in earth atmosphere at a height of about 100 km (see reference [5]).

REFERENCES

[1] Chapman, S. "The thermosphere — The earth's outermost atmosphere" Chapter I in Physics of the Upper Atmosphere. Academic Press, 1960, pp. 1—16.

[2] Cowling, T. G. Magnetohydrodynamics. Interscience Publishers, 1957.

[3] Johnson, F. S. "Physics of the atmosphere and space", *Astronautics*, Vol. 7, No. 11, 1962, pp. 72—75.

[4] Minzner, R. A. and Ripley, W. S. "The ARDC model atmosphere 1959", Air Force Surveys in Geophysics, No. 115, AFCRC, 1959.

[5] Nicolet, M. "The properties and constitution of the upper atmosphere", Chapter II in Physics of the Upper Atmosphere. Academic Press, 1960, pp. 17—71.

[6] Pai, S.-I. Introduction to the Theory of Compressible Flow. D. Van Nostrand, 1959.

[7] Prandtl, L. Fluid Dynamics. Hafner Publishing Co., 1952.

[8] Prandtl, L. and Tietjens, O. G. Fundamentals of Hydro- and Aeromechanics. McGraw-Hill, 1934.

[9] Rasool, S. I. "Structure of planetary atmosphere", *AAIA Jour.*, Vol. 1, No. 1, 1963, pp. 6—19.

[10] Roberts, H. E. "The earth's atmosphere", *Aeronautical Engineering Review*, Vol. 8, No. 10, 1949, pp. 18–31.

[11] Spitzer, L., Jr. Physics of Fully Ionized Gases. Interscience Publishers, 1956.

FUNDAMENTALS OF FLUID DYNAMICS

1. Microscopic and Macroscopic Points of View

Even though a fluid is composed of a large number of particles, ordinarily the fluids are so dense that we may consider them as continuous media in our treatment of fluid mechanics. However, if the fluid is very rarefied, one would expect that the coarse molecular structure would affect the flow phenomena and in the extremely rarefied gas, the gas should behave like individual particles. There are two different approaches in the theoretical investigations of the flow of a fluid: the microscopic point of view and the macroscopic point of view. The microscopic treatment is the more accurate way of analysis in which the motion of individual particles and their interactions are analyzed. Such an analysis is known as the kinetic theory of fluids. Because of many physical and mathematical difficulties, it is not possible at present to treat the flow problem of fluids exactly by molecular theory. Many simplified assumptions about the molecular forces and the collision phenomena have to be made in the formulation of the theory[14] and the resultant equations can only be solved approximately. The most successful molecular theory of fluids is the kinetic theory of gases. Recently considerable efforts have been made to develop the kinetic theory of plasma[1]. The basic equation of the kinetic theory of gases is known as the Boltzmann equation which is a nonlinear partial differentio-integral equation. At present time it is not possible to solve the Boltzmann equation even for rather simple practical flow problems. We would not expect to use the molecular theory of fluids to analyze flow problems in the near future. But, the Boltzmann equation serves two important aspects in the study of gasdynamics. In the first place, the fundamental equations of gasdynamics may be derived from the Boltzmann equation as a first approximation. Thus we may have some guides about the validity of the fundamental equations for a macroscopic description of the fluid flow from the analysis of Boltzmann equation. In the second place, the Boltzmann equation may give valuable information on the transport coefficients such as the coefficient of viscosity, heat conductivity, etc. In the macroscopic analysis, these transport coefficients are simply introduced as known functions of physical quantities of gasdynamics such as temperature and pressure, etc. We shall discuss the microsonic treatment of gas flow in chapter VIII.

For many practical problems, the description of the motion of the fluid by molecular theory such as in terms of the Maxwellian-Distribution function (chapter I section 9) is too detailed to be useful. We do not care about the motion of individual particles in the fluid and are interested in the resultant effects due to the motion of a large number of particles. In other words, we are interested in the macroscopic quantities only, such as pressure, density, temperature, flow mean velocities, electrical current density, etc. (see chapter I). It is possible to postulate the fundamental equations for the mechanics of fluids based on the conservation laws of mass, momentum, energy, and charge in terms of these macroscopic quantities only without considering the fluids as composed of individual particles. These fundamental equations should, of course, be consistent with those derived from the microscopic description. A major portion of our knowledge of fluid mechanics is based on these macroscopic equations. We shall use these equations in this book except as otherwise specified.

The fundamental equations of fluids from the macroscopic point of view are known as the Navier-Stokes equations which are useful in the analysis of ordinarily dense fluid. For rarefied gas, these equations should be modified and sometimes, they should not be used at all. Hence in the study of fluid mechanics, we should have a criterion to determine the degree of rarefication of the fluid. For liquid, the Navier-Stokes equations can always be used to analyze the flow problems. For a gas or a plasma, the degree of rarefication may be expressed by the nondimensional parameter known as the Knudsen number K_f which is defined as

$$K_f = \frac{L_f}{L} \qquad (3.1)$$

where L_f is the mean free path of the gas or plasma and L is the characteristic length of the flow field. The calculation of the mean free path will be discussed in chapter VIII. When the Knudsen number is not negligibly small, the flow should be considered as a rarefied gas flow. When the Knudsen number is negligibly small, the fluid may be considered as a continuum medium. The choice of the characteristic length L depends on the problems considered. Hence we may choose the typical dimension of the body as L as we study the forces on this body in a gas flow. We may use the boundary layer thickness on a body as L when we are interested in the skin friction and heat transfer through the boundary layer. When we investigate the transition region in a shock wave, the thickness of the shock wave may be used as L. Because of the various choice of the characteristic length L, whether a gas flow should be considered as rarefied or not depends on the particular problem considered. Once the value of L is chosen, the Knudsen number K_f tells us the degree of rarefication.

From the kinetic theory of gases, the mean free path of a neutral gas is defined approximately by the following formula:

$$\nu_g = \frac{1}{2} L_f \bar{c}_a \qquad (3.2)$$

where ν_g is the coefficient of kinematic viscosity and $\bar{c}_a$ is the mean molecular velocity which is related to the sound speed "a" of the gas by the formula:

$$a = \bar{c}_a (\gamma \pi / 8)^{\frac{1}{2}} \qquad (3.3)$$

where γ is the ratio of the specific heat at constant pressure c_p to that at constant volume c_v and sometimes known as isentropic exponent. From Eqs. (3.1) to (3.3), we have

$$K_f = \frac{L_f}{L} = 1.255\sqrt{\gamma} \, M / R_e \qquad (3.4)$$

where $M = U/a$ is the Mach number of the flow field, U is the typical velocity of the flow field and $R_e = UL/\nu_g$ is the Reynolds number of the flow field. Eq. (3.4) shows that the Knudsen number becomes important when the Mach number M is large and the Reynolds number R_e is small. Since the modern trend for the flow problems is towards high speed, i.e., high M and high altitude, i.e., low R_e, rarefied gasdynamics becomes important, particularly in space flight.

For a plasma, besides the mean free path L_f, there are two other characteristic lengths which also have influence on the rarefied effects of the corresponding flow problems. These are the Debye length and the Larmor radius.

The Debye length L_D is obtained from the condition that the electrical potential energy is equal to the thermal energy and it is given by the formula:

$$L_D = [kT\epsilon/(ne^2)]^{\frac{1}{2}} \tag{3.5}$$

where k is the Boltzmann constant, n is the number density of the plasma, T is the absolute temperature of the plasma, e is the absolute electric charge, and ϵ is the inductive capacity. The nondimensional parameter

$$K_D = L_D/L = \text{Electrical Knudsen number} \tag{3.6}$$

is a measure of the distance over which the excess electric charge may be different appreciably from zero.

The Larmor radius L_L may be defined as follows:

$$L_L = \frac{a}{\omega_c} = \frac{am}{eB} \tag{3.7}$$

where

$$\omega_c = \frac{eB}{m} = \text{Larmor frequency} = \text{cyclotron frequency} \tag{3.8}$$

B is the magnitude of the magnetic induction and m is the mass of the charged particle. In Eq. (3.7), we use the sound speed a to represent the particle velocity of the plasma. The Larmor radius is a measure of the radius of the helical path of the charged particle of mass m and charge e in a magnetic induction B. The nondimensional parameter

$$K_L = \frac{L_L}{L} = \text{magnetic Knudsen number} \qquad (3.9)$$

shows the rarefied effect of a plasma under the influence of a magnetic field. If K_L is large, we should consider the discrete effect of the charged particles under the influence of a magnetic field. There are practical cases where the influence of a magnetic Knudsen number is more important than the Knudsen number such as the case when the ion Larmor radius is smaller than the mean free path of a plasma.[1] We shall discuss such cases in chapters VI and IX.

Except as otherwise specified, we assume that all the three Knudsen numbers K_f, K_D, and K_L are negligibly small so that the fluid may be considered as a continuous medium.

2. Flow Regimes

It is advisable to divide the flow field into various regimes according to the value of the Knudsen number. The two extreme regimes are well defined, i.e., if the Knudsen number is negligibly small, the flow is in the continuum regime in which the classical fluid dynamics is one of the main topics; and if the Knudsen number is much larger than unity, the flow is in the free molecule regime in which the gas particles should be considered to be free in motion without the influence of other particles. It should be noticed that even in the continuum regime, the structure of the molecules may have some influence on the flow field, particularly when chemical reaction and ionization are important.

The intermediate regime of a rarefied gas flow is rather complicated. One would expect that the flow would gradually change from the continuum flow regime into the free molecule regime as the value of the Knudsen number increased. The division of this intermediate regime into several regimes is usually arbitrary. There are several ways to make this division. One of this type of division proposed by Tsien[2] is that the intermediate regime may be divided into two regimes, i.e., the slip flow regime and the transition regime according to the values of Mach number M and Reynolds number R_e. For low Reynolds number case, the value L may take the typical dimension of the body in the flow or the typical dimension of a channel in the case of internal flow. Hence for the low Reynolds number case, the expression (3.4) may be used to show the relation between Knudsen number K_f and the Mach number M and Reynolds number R_e. However for the high Reynolds number flow, it is well known that the viscous effect is limited in the boundary layer, and the boundary layer thickness δ should be used as a typical length. Hence the Knudsen number should be defined as

$$K_f = \frac{L_f}{\delta} = \frac{L}{\delta} \frac{L_f}{L} \cong \frac{M}{(R_e)^{\frac{1}{2}}} \qquad (3.10)$$

where $\delta/L \cong 1/(R_e)^{\frac{1}{2}}$. Hence for the large Reynolds number case, the relation (3.10) should be used.

According to Tsien's classification, we have

Regimes	Range of M and R_e
Free molecule flow	$M/R_e \geq 3$
Transition region	$M/R_e \leq 3$ & $M/(R_e)^{\frac{1}{2}} \geq 0.1$
Slip flow	$0.1 \geq M/(R_e)^{\frac{1}{2}} \geq 0.01$
Continuum flow	$M/(R_e)^{\frac{1}{2}} \leq 0.01$

In the slip flow regime, the fluid still behaves as a continuum medium, but there is a slip of velocity on the solid boundary and a jump in temperature on the solid boundary due to the rarefied effects. In the transition regime, the flow depends greatly on the discrete character of the fluid. Our present knowledge of the flow in the transition regime is very limited. Finally when the Knudsen number is much larger than unity, the collision between the molecules is negligible and we have the free molecule regime. Except in chapter VIII, we consider only the fluid mechanics in the continuum regime which is the most important domain and which is the case extensively studied. In chapter VIII, we shall discuss other methods of division of flow regime and the treatment of flow problems in various flow regimes.

3. Kinematics of Fluids

Even though the fluid is considered as a continuum medium,

it is still convenient to use the concept of fluid particle or fluid element in fluid mechanics. The fluid particle or fluid element is a group of molecules of a fluid in a small volume which may be considered as a point from a practical point of view (see chapter I, section 6). Such a group of molecules may be observed in the actual flow of fluid. For instance, when we put a small amount of dye in a flow of liquid, we may see clearly the motion of colored liquid spots which may be regarded as fluid elements. Various techniques of visualization of flow pattern verify the existence of fluid particles or fluid elements. One of the objects of fluid mechanics is to predict the motion of fluid elements under various conditions. The first question is: how do we describe the motion of such fluid elements? There are two methods of describing the fluid motion: the Lagrangian method and the Eulerian method, although both methods are in reality due to Euler.[6]

(1) Lagrangian method. In this method, we are concerned with the history of the individual fluid elements. We want to know their velocities, accelerations, and other properties at various places and different times. If at any given time $t=t_0$, a fluid element has Cartesian coordinates (x_0, y_0, z_0), then at time $t=t$, it will move to the position (x, y, z). It is evident that the coordinates (x, y, z) of the fluid element are functions of the initial coordinates (x_0, y_0, z_0) and the time t. We may write

$$\left.\begin{array}{l} x=F_1(x_0, \ y_0, \ z_0, \ t) \\ y=F_2(x_0, \ y_0, \ z_0, \ t) \\ z=F_3(x_0, \ y_0, \ z_0, \ t) \end{array}\right\} \qquad (3.11)$$

Without loss of generality, the three initial coordinates (x_0, y_0, z_0) representing the name of the fluid element can also be considered as Cartesian coordinates. Eq. (3.11) gives the position of all fluid particles at different times. The first partial derivatives of Eq. (3.11) with respect to time give the velocity components of the fluid elements and the second partial derivatives with respect to time of Eq. (3.11) give the corresponding accelerations.

(2) Eulerian method. In this method, we are concerned with

what is happening at a given time t at a point (x, y, z) in the space filled with the fluid. The velocity components of the fluid at any time and at any point in space may be written as

$$u=f_1(x, y, z, t) \\ v=f_2(x, y, z, t) \\ w=f_3(x, y, z, t) \quad\quad (3.12)$$

Here the coordinates (x, y, z) represent a fixed point in space and not the location of a given fluid element.

The acceleration of the fluid element used in the Eulerian method may be derived in the following manner:

Consider a fluid element at the location (x, y, z) and at time t having an x-component of velocity

$$u=f(x, y, z, t) \quad\quad (3.13)$$

At time $t+\delta t$, i.e., a short interval δt later, this element will have moved to the position $(x+u\delta t, y+v\delta t, z+w\delta t)$ and will have an x-component $u+\delta u$. Hence we have

$$u+\delta u=f(x+u\delta t, y+v\delta t, z+w\delta t, t+\delta t)$$

$$=f(x, y, z, t)+\left(u \frac{\partial f}{\partial x} +v \frac{\partial f}{\partial y} +w \frac{\partial f}{\partial z}+\frac{\partial f}{\partial t}\right)\delta t$$

$$+\text{higher order terms in } \delta t \quad\quad (3.14)$$

The x-component of acceleration of the fluid at point (x,y,z) is the limit:

$$\lim_{\delta t \to 0} \frac{\delta u}{\delta t}=\frac{Du}{Dt}=\left(\frac{\partial}{\partial t} +u \frac{\partial}{\partial x} +v \frac{\partial}{\partial y} +w \frac{\partial}{\partial z}\right)u \quad\quad (3.15)$$

The symbol

$$\frac{D}{Dt} = \frac{\partial}{\partial t} +u \frac{\partial}{\partial x} +v \frac{\partial}{\partial y} +w \frac{\partial}{\partial z} = \frac{\partial}{\partial t} + (q \cdot \nabla) \quad\quad (3.16)$$

is known as the total differential with respect to time or the material differential with respect to time. It gives the rate of change of a physical quantity following the path of the fluid element. In the

Eulerian method, we should use this differential with respect to time when we describe the time rate of change of any quantity of the fluid.   In Eq. (3.16), q is the velocity vector of the fluid which has the components u, v, and w, i.e.,

$$q = iu + jv + kw \qquad (3.17)$$

and the operator del ∇ is

$$\nabla = i \frac{\partial}{\partial x} + j \frac{\partial}{\partial y} + k \frac{\partial}{\partial z} \qquad (3.18)$$

where i, j, and k are respectively the unit vectors in the x-, y-, and z-direction.

There is a definite relation between the Lagrangian and the Eulerian methods.   For each fluid element, we have

$$\left. \begin{aligned} \frac{dx}{dt} &= u = f_1(x,\ y,\ z,\ t) \\[2mm] \frac{dy}{dt} &= v = f_2(x,\ y,\ z,\ t) \\[2mm] \frac{dz}{dt} &= w = f_3(x,\ y,\ z,\ t) \end{aligned} \right\} \qquad (3.19)$$

If we solve Eqs. (3.19) with the initial conditions $t=t_0$, $x=x_0$, $y=y_0$ and $z=z_0$, we will have Eqs. (3.11).   Hence in principle, the Lagrangian method of description can always be derived from the Eulerian method of description by the help of Eqs. (3.19).   In most engineering problems, we are interested in the pressure, velocity, and other physical quantities of the fluid at certain given points in space, such as on the surface of a body in the fluid.   Since the Eulerian method gives us these results directly, ordinarily we use the Eulerian method. In this book we always use the Eulerian method except for special cases where the Lagrangian method is more convenient.

In the Eulerian method, the basic variables are the velocity components of the fluid.   We would like to know the deformation of the velocity field of a fluid flow.   Any deformation of a continuum may be accomplished by two successive processes which are independent from each other if the second order quantities are neglected. The first is a simple extension or compression, that is, the normal strain, and the second is a shearing strain which measures the change of skewness of the element. The strain tensor of a fluid has the following nine components:

$$
S = \begin{vmatrix}
\epsilon_x , & \dfrac{1}{2}\,\gamma_{xy} , & \dfrac{1}{2}\,\gamma_{xz} \\[2mm]
\dfrac{1}{2}\,\gamma_{yx} , & \epsilon_y , & \dfrac{1}{2}\,\gamma_{yz} \\[2mm]
\dfrac{1}{2}\,\gamma_{zx} , & \dfrac{1}{2}\,\gamma_{zy} , & \epsilon_z
\end{vmatrix}
\tag{3.20}
$$

where

$$
\epsilon_x = \frac{\partial u}{\partial x} , \quad \epsilon_y = \frac{\partial v}{\partial y} , \quad \epsilon_z = \frac{\partial w}{\partial z} , \quad \gamma_{xy} = \frac{\partial u}{\partial y}+\frac{\partial v}{\partial x} = \gamma_{yx},
$$

$$
\gamma_{yz} = \frac{\partial v}{\partial z} + \frac{\partial w}{\partial y} = \gamma_{zy} , \quad \gamma_{zx} = \frac{\partial w}{\partial x} + \frac{\partial u}{\partial z} = \gamma_{xz}
$$

and ϵ_x is x-wise normal strain, etc., and γ_{xy} is the x-y shearing strain, etc.

The vorticity ω of the fluid is the curl of the velocity vector, i.e.,

$$
\omega = \nabla \times q = i\omega_x + j\omega_y + k\omega_z
\tag{3.21}
$$

where

$$
\omega_x = \frac{\partial w}{\partial y} - \frac{\partial v}{\partial z} , \quad \omega_y = \frac{\partial u}{\partial z} - \frac{\partial w}{\partial x} , \quad \omega_z = \frac{\partial v}{\partial x} - \frac{\partial u}{\partial y}
$$

and Gibbs vector notations are used.

The vorticity of the flow field is closely related to the circulation of the flow. The circulation Γ is defined as the line integral of the velocity q along a closed curve:

$$
\Gamma = \oint_C q \cdot ds = \oint_C (u\,dx + v\,dy + w\,dz) = \oint_A \omega \cdot dA
\tag{3.22}
$$

where ds is the differential arc length of a closed curve C in the fluid and A is the area enclosed by the curve C. The Stokes theorem

(3.22) is used to transform the line integral into the surface integral (see problem 5).

A large class of flow problems is known as irrotational flow in which the fluid elements do not rotate. In other words, the vorticity of the flow field is zero except at some singular points. Furthermore, we shall show that under certain conditions, the circulation does not change with time. Hence under these conditions, the circulation within a closed curve will remain zero if the flow is started from rest. This is the reason why the irrotational flow is very important because it occurs quite often.

For irrotational flow, we may introduce a velocity potential ϕ such that

$$q = \nabla \phi \qquad (3.23)$$

where ϕ is a scalar function of x,y,z, and t. Since

$$\omega = \nabla \times q = \nabla \times \nabla \phi = 0 \qquad (3.24)$$

the condition of irrotationality is satisfied for a potential flow. For an irrotational flow, a velocity potential ϕ always exists. Hence we sometimes call the irrotational flow the potential flow. It is more convenient to consider a single scalar function ϕ than the vector quantity q.

Since the velocity vector q is the gradient of the velocity potential, the streamline is perpendicular to the equi-potential line. The streamlines are lines parallel to the velocity vector in the space at any given time. Let the coordinates of a streamline be

$$dr = i\,dx + j\,dy + k\,dz \qquad (3.25)$$

This streamline is parallel to the velocity vector at time t_0, i.e., $q(r, t_0)$ where $r = ix + jy + kz$. The condition of parallelism of dr and q gives

$$dr \times q = 0 \qquad (3.26)$$

or

$$dx \ : \ dy \ : \ dz = u \ : \ v \ : \ w \qquad (3.27)$$

For an unsteady flow, the streamlines change their shape with time while for a steady flow the streamlines have definite shape in space which show the directions of the fluid flow. The concept of streamlines may be used for any vector field. For instance, if we consider q as the magnetic field, the corresponding streamlines are the lines of force of the magnetic field.

If we draw a small closed curve in the fluid and draw from each point on this closed curve a streamline, we have a streamtube. For an unsteady flow, the shape of this streamtube may change with time but for a steady flow, the shape of the streamtube is fixed in space; thus the streamtube behaves like a real tube because no fluid will cross the wall of the streamtube and the fluid flows along the tube. The concept of the streamtube is very useful in the analysis of fluid mechanics because along a streamtube we have only one component of velocity which is in the direction of the streamtube. Some simplifications may be obtained by considering the streamtubes in the fluid flow.

There are two other kinds of lines which are closely related to streamlines. They are called the path lines of the fluid elements and streak lines. The path line of a fluid particle or element is given by Eq. (3.11). In a steady flow, it is identical to the streamline passing through the point (x_0, y_0, z_0). The streak line is the line joining all fluid elements which pass through a given point in space. In a steady flow, the streak line also coincides with the streamline. In an unsteady flow, these three kinds of lines (streamlines, path lines, and streak lines) are in general different. Streak lines are very useful in experiments of flow visualization.

4. Conservation of Mass. Equation of Continuity and Diffusion Equations

One of the fundamental relations of fluid mechanics is the equation of continuity which is based on the conservation of mass of the fluid. Let u, v, w, and ρ be the x-, y-, and z-component of velocity and the density of the fluid respectively at a point $P\ (x,y,z)$ in space and at a given time t. Let us draw a parallelepiped with sides dx, dy, and dz from P as shown in Fig. 3.1. The mass flowing out of the parallelepiped in the x-direction is

$$\left(\rho u + \frac{\partial \rho u}{\partial x} \, dx \right) dy \, dz - \rho u \, dy \, dz = \frac{\partial \rho u}{\partial x} \, dx \, dy \, dz \qquad (3.28)$$

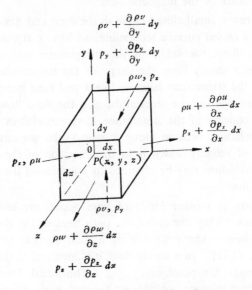

Figure 3.1   Parallelepiped in a fluid

We have similar expressions for the mass flow in the y- and the z-directions. The total amount of mass flowing out of the parallelepiped is then

$$\left(\frac{\partial \rho u}{\partial x} + \frac{\partial \rho v}{\partial y} + \frac{\partial \rho w}{\partial z} \right) dx \, dy \, dz \qquad (3.29a)$$

By the conservation of mass, the expression of (3.29a) must be equal to the time rate of change of mass in the parallelepiped which consists of two parts: one is the change of density of the fluid in the parallelepiped and the other is the increase of mass due to a source σ_0 per unit volume, i.e.,

$$-\frac{\partial \rho}{\partial t}\,dx\,dy\,dz+\sigma_0\,dx\,dy\,dz \qquad (3.29b)$$

By the conservation of mass, the expression of (3.29a) must be equal to the expression of (3.29b). Thus we have the equation of continuity:

$$\frac{\partial \rho}{\partial t}+\frac{\partial \rho u}{\partial x}+\frac{\partial \rho v}{\partial y}+\frac{\partial \rho w}{\partial z}=\sigma_0 \qquad (3.30)$$

In vector form, Eq. (3.30) becomes

$$\frac{\partial \rho}{\partial t}\;+\;\nabla\cdot(\rho q)\;=\;\sigma_0 \qquad (3.31)$$

For most problems in fluid mechanics, there is no source term σ_0, i.e., $\sigma_0=0$. We shall assume that $\sigma_0=0$ except where otherwise specified.

Eq. (3.31) with $\sigma_0=0$ may be written in the following form:

$$\frac{D\rho}{Dt}=\frac{\partial \rho}{\partial t}+(q\cdot\nabla)\rho=-\rho\,(\nabla\cdot q) \qquad (3.32)$$

Eq. (3.32) is the equation of continuity of fluid mechanics which will be used in most parts of this book.

For incompressible fluid, the density is a constant and Eq. (3.32) becomes

$$\nabla\cdot q\;=\;\frac{\partial u}{\partial x}+\frac{\partial v}{\partial y}+\frac{\partial w}{\partial z}=0 \qquad (3.33)$$

The scalar product of the operator ∇ with the vector q is known as the divergence of q which represents the flux of q flowing out of a unit volume in space. It is interesting to notice the essential difference between the flow of an incompressible fluid and that of a compressible fluid from Eqs. (3.32) and (3.33). In the incompressible flow case, Eq. (3.33) shows that we may find the velocity field q from the kinematic consideration only. But in the compressible flow case, we have to consider simultaneously the kinematics and dynamics of the fluid in order to determine the velocity field, because the variable density ρ couples the kinematics with the dynamics through the equations of motion and other fundamental equations, as we shall discuss in detail later.

In many problems of fluid flow, the fluid is actually a mixture of several species, e.g., air is a mixture of oxygen, nitrogen, and other gases. For the flow of a mixture of gases, new variables, in the form of concentration of any one species in the mixture and the diffusion velocities of the species enter our problem. For a mixture of N-species, we may treat our flow problems by the multi-fluid theory (chapter IX, section 4) in which we consider the partial variables of each species, such as velocity vector q_s, temperature T_s, pressure p_s, and density ρ_s for sth species where $s=1,2,\cdots, N$. We may derive the equations of continuity for each species in the same manner as that for the mixture as a whole, Eq. (3.30), i.e.,

$$\frac{\partial \rho_s}{\partial t} + \frac{\partial \rho_s u_s}{\partial x} + \frac{\partial \rho_s v_s}{\partial y} + \frac{\partial \rho_s w_s}{\partial z} = \sigma_s \qquad (3.34)$$

If we use Eq. (3.34) in our analysis, we have to know the partial velocity u_s^i, i.e., $u_s^1 = u_s$, $u_s^2 = v_s$, and $u_s^3 = w_s$. The partial velocity u_s^i may be expressed in terms of the velocity of the mixture as a whole u^i and the diffusion velocity w_s^i, cf. Eq. (1.71):

$$w_s^i = u_s^i - u^i \qquad (3.35)$$

When the diffusion velocity is small as it happens in most cases (chapters V and IX), we replace the exact differential equation for diffusion velocity by some simplified relations in terms of diffusion coefficients. These diffusion coefficients may be obtained from the kinetic theory of gases and their relations with number concentration c_r, pressure p and temperature T and the diffusion velocity w_s^i are given by the following formula:

$$\rho_r w_r^i = \sum_{q \neq r} \rho D_{qr} \frac{\partial c_q}{\partial x^i} + D_p \frac{\partial \ln p}{\partial x^i} - D_T \frac{\partial \ln T}{\partial x^i} \qquad (3.36)$$

where the number concentration of species r is the ratio of the number density of rth species n_r to the number density of the mixture as a whole n, i.e.,

$$c_r = \frac{n_r}{n} \tag{3.37}$$

and the mass concentration k_r of the rth species is

$$k_r = \frac{\rho_r}{\rho} = \frac{m_r n_r}{\rho} \tag{3.38}$$

D_{qr} is the diffusion coefficient between the rth and qth species in the mixture. For a binary mixture, the first term on the right hand side of Eq. (3.36) becomes

$$(k_r w_r^i)_c = -D_{qr} \frac{\partial c_r}{\partial x^i} \tag{3.39}$$

The binary diffusion coefficient D_{qr} is one of the most important diffusion coefficients in fluid dynamics. D_p is known as the coefficient of pressure diffusion and D_T is the coefficient of thermal diffusion. The diffusion velocity w_r^i may depend on other physical phenomena. For instance, the diffusion velocity of charged particles in a plasma depends mainly on the electromagnetic field and we should add the corresponding terms in the right hand side of Eq. (3.36). Since the importance of various terms in the diffusion velocity equation is different in different case, we may retain only the most important term in each case. For the flow where chemical reaction is important, we may use Eq. (3.36) or even Eq. (3.39). For a plasma, the diffusion velocity of charged particles is closely associated with the electrical conductivity, other formula such as generalized Ohm's law should be used. We shall discuss it in chapter VI.

Substituting Eqs. (3.35) and (3.36) into Eq. (3.34) and, using the equation of continuity (3.32), we have the diffusion equation of fluid dynamics which will be further studied in chapter V as follows:

$$\rho \frac{Dk_r}{Dt} = -\sum_{q \neq r} \frac{\partial}{\partial x^j} \left(\rho D_{qr} \frac{\partial c_q}{\partial x^j} \right) - \frac{\partial}{\partial x^j} \left(D_p \frac{\partial \ln p}{\partial x^j} \right)$$

$$+ \frac{\partial}{\partial x^j} \left(D_T \frac{\partial \ln T}{\partial x^j} \right) + \sigma_r \tag{3.40}$$

The source term σ_r depends on the chemical reactions and processes of ionization which will be discussed later.

5.   Stream Function

Two of the most common types of flow in the study of fluid dynamics are the steady flows of two dimensional case and those of axisymmetric cases. In these two cases, the equation of continuity (3.32) may be written in the following form:

$$\frac{\partial \rho u}{\partial x} + \frac{\partial \rho v}{\partial y} + \delta \frac{\rho v}{y} = 0 \qquad (3.41)$$

where $\delta = 0$ for the two dimensional case and $\delta = 1$ for the axisymmetric case. Here x and y denote respectively the distance along and perpendicular to the flow axis and u and v are respectively the x- and y-components of velocity.

From Eq. (3.41), we may define a streamfunction ψ such that

$$\frac{\partial \psi}{\partial y} = \rho u y^\delta , \quad -\frac{\partial \psi}{\partial x} = -\rho v y^\delta \qquad (3.42)$$

It is evident that Eq. (3.41) is automatically satisfied by the streamfunction ψ. In many problems of fluid dynamics as we shall see later, it is more convenient to consider the streamfunction ψ as the unknown quantity instead of the two velocity components u and v.

For the surface $\psi = $ constant, we have

$$d\psi = \frac{\partial \psi}{\partial x}(dx)_\psi + \frac{\partial \psi}{\partial y}(dy)_\psi = 0$$

or

$$\left(\frac{dy}{dx}\right)_\psi = -\frac{\partial \psi / \partial x}{\partial \psi / \partial y} = \frac{v}{u} \qquad (3.43)$$

where subscript ψ means the value along the curve $\psi = $ constant. Eq. (3.43) shows that the slope of the curve $\psi = $ constant is the same as the direction of the velocity vector q. Consequently $\psi = $ constant lines are the streamlines and this is the reason that ψ is called streamfunction.

For the unsteady three dimensional flow, we may use two streamfunctions ψ and ϕ to define the flow field as follows:

$$\rho u = \frac{\partial \psi}{\partial y}, \quad \rho v = -\frac{\partial \psi}{\partial x} - \frac{\partial \phi}{\partial z} - \frac{\partial}{\partial t}\left(\int \rho dy \right),$$

$$\rho w = \frac{\partial \phi}{\partial y} \tag{3.44}$$

Substituting Eq. (3.44) into Eq. (3.32), the equation of continuity is automatically satisfied. When we use the streamfunctions as the unknowns in our analysis of fluid dynamic problems, we need not consider the equation of continuity which is automatically satisfied.

6. Equations of Motion

The fundamental equation of dynamics is the Newton's second law of motion:

$$\text{Force} = \text{mass} \times \text{acceleration} \tag{3.45}$$

If we apply Eq. (3.45) to a fluid element of unit volume, we have

$$\rho \frac{Dq}{Dt} = F \tag{3.46}$$

Eq. (3.46) is the equation of motion of fluid dynamics from the Eulerian point of view which is one of the fundamental equations of fluid dynamics. The force F is the force per·unit volume. We have to use the total acceleration Dq/Dt given by Eq. (3.15) or (3.16) in the equation of motion (3.46).

The force F may be divided into two classes: one is due to the inhomogeneous state of the stresses in the fluid, i.e., the surface forces, and the other is the body force.

Since the fluid does offer resistance to the deformation of a velocity field, there is nonuniform distribution of stresses in the fluid in motion. The stresses are the internal forces per unit area of the fluid. Consider again the small parallelepiped of Fig. 3.1; let p_x be the force per unit area on the surface $0yz$ which is perpendicular to the axis x. Then the force per unit area on the surface parallel to $0yz$ but at a distance dx from $0yz$ is, after the higher order terms are neglected,

$$p_x + \frac{\partial p_x}{\partial x}dx$$

Similarly p_y and p_z are the force per unit area on the surfaces $0xz$ and $0xy$ respectively. On the corresponding surfaces at dy and dz from the original surface will be the forces

$$p_y + \frac{\partial p_y}{\partial y}dy \quad \text{and} \quad p_z + \frac{\partial p_z}{\partial z}\,dz$$

respectively. Each of the forces p_x, p_y, and p_z may be resolved into three components along the x-, y-, and z-axis. We may write:

$$\left.\begin{array}{l} p_x = i\sigma_x + j\tau_{xy} + k\tau_{xz} \\ p_y = i\tau_{yx} + j\sigma_y + k\tau_{yz} \\ p_z = i\tau_{zx} + j\tau_{zy} + k\sigma_z \end{array}\right\} \tag{3.47}$$

where the symbol σ represents the normal stress, i.e., the stress perpendicular to the surface considered; hence σ_x is the normal stress on the surface perpendicular to the x-axis. The symbol τ represents the shearing stress, i.e., the stress in the surface considered. The first subscript to τ refers to the direction of the axis perpendicular to the surface considered and the second subscript refers to the direction of the force in the surface. Thus τ_{xy} denotes the component of shearing stress in the surface perpendicular to the axis x and in the direction of the y-axis.

If we reduce the size of the parallelepiped indefinitely, it will tend to the point P as a limit from the macroscopic point of view, and then Eq. (3.47) gives the stresses at the point $P(x, y, z)$ as a limit. The stress at a point in the fluid is a tensor quantity of second order which has nine components given in Eq. (3.47). In tensor notation, we may write the ijth component of the stress tensor as τ^{ij} which is in general a function of the spatial coordinates and time. It is easy to calculate the resultant force due to the nonhomogeneous distribution in the stresses. Consider the force on the parallelepiped of Fig. 3.1 in the x-direction due to the stress σ_x, τ_{yx}, and τ_{zx}. We have

$$\left(\sigma_x + \frac{\partial \sigma_x}{\partial x}dx\right) dy\, dz - \sigma_x\, dy\, dz + \left(\tau_{yx} + \frac{\partial \tau_{yx}}{\partial y}dy\right) dx\, dz$$

$$- \tau_{yx}\, dx\, dz + \left(\tau_{zx} + \frac{\partial \tau_{zx}}{\partial z}dz\right)dx\, dy - \tau_{zx}\, dx\, dy$$

$$= \left(\frac{\partial \tau_x}{\partial x} + \frac{\partial \tau_{yx}}{\partial y} + \frac{\partial \tau_{zx}}{\partial z}\right) dx\, dy\, dz$$

or the x-wise force per unit volume due to the variation of stresses is

$$X_v = \frac{\partial \tau_x}{\partial x} + \frac{\partial \tau_{yx}}{\partial y} + \frac{\partial \tau_{zx}}{\partial z} \tag{3.48a}$$

Similarly the y- and z-components of the force per unit volume due to the nonhomogeneous state of stresses are respectively

$$Y_v = \frac{\partial \tau_{xy}}{\partial x} + \frac{\partial \sigma_y}{\partial y} + \frac{\partial \tau_{zy}}{\partial z} \tag{3.48b}$$

$$Z_v = \frac{\partial \tau_{xz}}{\partial x} + \frac{\partial \tau_{yz}}{\partial y} + \frac{\partial \sigma_z}{\partial z} \tag{3.48c}$$

Hence the force due to the nonhomogeneous state of stresses is

$$F_v = iX_v + jY_v + kZ_v \tag{3.49}$$

with the ith component F_v^i as follows:

$$F_v^i = \frac{\partial \tau^{ij}}{\partial x^j} \tag{3.49a}$$

where the summation convention is used.

For an ideal fluid, the shearing stresses are zero. By definition, the pressure is taken as the negative value of the normal stress. Then in an ideal fluid at any point in space we have

$$\left.\begin{array}{l} \sigma_x = \sigma_y = \sigma_z = -p \\ \tau_{xy} = \tau_{yx} = \tau_{yz} = \tau_{zy} = \tau_{xz} = \tau_{zx} = 0 \end{array}\right\} \tag{3.50}$$

In many fluid dynamic problems, Eq. (3.50) is a good approximation for the actual fluid, particularly for those thin fluids such as water and gaseous fluid and in the region where there is no large variation of velocities. For instance, in the calculation of the pressure over a body in a thin fluid or the wave motion in a thin fluid, Eq. (3.50) is usually used.

For actual fluids, the shearing stresses are not zero. Thus we have all the nine components of the stress tensor τ^{ij}. It can be shown that, of the six shearing stress components, those which have the same suffices but in reversed order are equal. It follows from the equilibrium of moment on an element in the fluid. If we take the parallelepiped of Fig. 3.1 sufficiently small, the body forces may be neglected in comparison with the surface force. Then the equilibrium of moment about the z-axis gives

$$\tau_{xy} \, dy \, dz \, dx = \tau_{yx} \, dx \, dz \, dy$$

Hence

$$\tau_{xy} = \tau_{yx} \tag{3.51a}$$

Similarly we have

$$\tau_{yz} = \tau_{zy} \tag{3.51b}$$

$$\tau_{zx} = \tau_{xz} \tag{3.51c}$$

Thus we need to consider only six stress components, σ_x, σ_y, σ_z, τ_{xy}, τ_{yz}, and τ_{xz}.

In general, we should consider the stresses as unknowns which are governed by certain differential equations (see chapter VIII). However, for most of the fluid dynamic problems, we may use some simple relations between the stresses and the strains (3.20), which have been discussed in chapter I section 13. For many thin fluids, such as water and gaseous fluids, where the mean free path L_f is small, the following Navier-Stokes relations have been used:

$$\left. \begin{array}{ll} \sigma_x = -p + \lambda(\nabla \cdot q) + 2\mu\epsilon_x, & \tau_{xy} = \mu\gamma_{xy} \\ \sigma_y = -p + \lambda(\nabla \cdot q) + 2\mu\epsilon_y, & \tau_{yz} = \mu\gamma_{yz} \\ \sigma_z = -p + \lambda(\nabla \cdot q) + 2\mu\epsilon_z, & \tau_{zx} = \mu\gamma_{zx} \end{array} \right\} \tag{3.52}$$

or in vector and tensor notation as

$$\tau^{ij}=\mu\left(\frac{\partial u^i}{\partial x^j}+\frac{\partial u^j}{\partial x^i}\right)+\left[-p+\lambda\left(\frac{\partial u^k}{\partial x^k}\right)\right]\delta^{ij} \qquad (3.52a)$$

where $\delta^{ij}=0$ if $i\neq j$ and $\delta^{ij}=1$ if $i=j$. The notations in Eq. (3.52) are: p is the hydrostatic pressure discussed in chapter I section 6, μ is the ordinary coefficient of viscosity discussed in chapter I section 13, and λ is the second coefficient of viscosity. The only restriction on the existance of two independent coefficients of viscosity in order that the general viscous dissipation function

$$\Phi=2\mu\left[\left(\frac{\partial u}{\partial x}\right)^2+\left(\frac{\partial v}{\partial y}\right)^2+\left(\frac{\partial w}{\partial z}\right)^2+\frac{1}{2}\gamma_{xy}^2+\frac{1}{2}\gamma_{yz}^2\right.$$

$$\left.+\frac{1}{2}\gamma_{zx}^2\right]+\lambda\left(\frac{\partial u}{\partial x}+\frac{\partial v}{\partial y}+\frac{\partial w}{\partial z}\right)^2=(\tau^{ij}+p\delta^{ij})\frac{\partial u^i}{\partial x^j} \qquad (3.53)$$

must never be negative are

$$\mu\geqq0, \quad 3\lambda+2\mu\geqq0 \qquad (3.54)$$

because it is against the second law of thermodynamics if the dissipation function Φ is negative.

The relation $3\lambda+2\mu=0$ is true for a monoatomic gas, which may be derived from the simple kinetic theory of gas. For ordinary fluid dynamics problems, the relation $3\lambda+2\mu=0$ is good. We use this relation in this book unless otherwise specified. With this relation, we have only one viscosity coefficient μ in our stress-strain relation which has been discussed in chapter I section 13.

The second coefficient of viscosity plays an important role in the explanation of the results of sound dispersion measurement.

The most common body force is the gravitational force of Eq. (2.11) which is

$$F_g=\rho g=\rho\nabla\phi_g \qquad (3.55)$$

where ϕ_g is the potential of the gravitational field.

Another body force is the electromagnetic force F_e on an electrically conducting fluid which will be discussed in chapter VI.

When the thermal radiation is important, we have the stresses due to thermal radiation and the corresponding force F_R due to these radiation stresses. In general, the force F may be written as follows:

$$F = F_v + F_g + F_e + F_R \qquad (3.56)$$

The equation of motion (3.46) with the expression of force F (3.56) should be solved with other fundamental equations. However, some important relations may be derived from the equation of motion under simple conditions.

Let us consider the case of an inviscid fluid in which the stress tensor is given by Eq. (3.50) and the only body force is the gravitational force F_g. The equation of motion in vector form is:

$$\rho \frac{Dq}{Dt} = \rho g - \nabla p = \rho \nabla \phi_g - \nabla p \qquad (3.57)$$

In Cartesian coordinates, the equations of motion are

$$\frac{Du}{Dt} = \frac{\partial \phi_g}{\partial x} - \frac{1}{\rho} \frac{\partial p}{\partial x} \qquad (3.57a)$$

$$\frac{Dv}{Dt} = \frac{\partial \phi_g}{\partial y} - \frac{1}{\rho} \frac{\partial p}{\partial y} \qquad (3.57b)$$

$$\frac{Dw}{Dt} = \frac{\partial \phi_g}{\partial z} - \frac{1}{\rho} \frac{\partial p}{\partial z} \qquad (3.57c)$$

Eq. (3.57) may also be written in the following form:

$$\frac{\partial q}{\partial t} + \nabla \left(\frac{1}{2} q^2 \right) - q \times \omega = \nabla \phi_g - \frac{1}{\rho} \nabla p \qquad (3.58)$$

where ω is the vorticity given by Eq. (3.21).

For irrotational flow with $\omega = 0$, Eq. (3.58) can be integrated. By the help of the mathematical identity:

$$\nabla \phi \cdot dr = \frac{\partial \phi}{\partial x} dx + \frac{\partial \phi}{\partial y} dy + \frac{\partial \phi}{\partial z} dz = d\phi \qquad (3.59)$$

if we multiply dr scalarly to Eq. (3.58) for irrotational flow and integrate over the whole space, we have

$$\frac{\partial \phi}{\partial t} + \frac{1}{2} q^2 + \phi_g + \int \frac{dp}{\rho} = f(t) \qquad (3.60)$$

where the arbitrary function of time $f(t)$ could be absorbed into the velocity potential ϕ without altering the relation of the velocity and the velocity potential (3.23). Thus Eq. (3.60) becomes

$$\frac{\partial \phi}{\partial t} + \frac{1}{2} q^2 + \phi_g + \int \frac{dp}{\rho} = \text{constant} = B_0 \qquad (3.61)$$

Eq. (3.61) is known as Bernoulli's equation. The constant B_0 is called Bernoulli's constant. We shall show later that isentropic flow is irrotational. For isentropic flow, we have

$$p = (\text{constant}) \cdot \rho^{\gamma} \qquad (3.62)$$

For incompressible fluid and steady flow, Eq. (3.61) becomes

$$\frac{1}{2} \frac{q^2}{g} + z + \frac{p}{\rho g} = \frac{B_0}{g} = \frac{1}{2} \frac{q_1^2}{g} + z_1 + \frac{p_1}{\rho g} \qquad (3.63)$$

Bernoulli's equation in the form of Eq. (3.63) has been extensively used in hydraulics. The subscript 1 refers to the value at a certain reference point. The physical significance of the three terms on the left hand side of Eq. (3.63) is as follows:

(1) the term $q^2/(2g)$ is known as the velocity head of the fluid which is the height from a fluid element at rest must fall under the influence of gravity g in order to attain a velocity q;

(2) the value z is simply the height of the point considered;

(3) the term $p/(\rho g)$ is known as the pressure head which is the height reached by a column of fluid under the action of pressure p against gravity.

From the above three terms, we may call the constant B_0/g as the total head of the flow because it is the sum of the three heads. We may interpret the constant B_0/g (usually the variation of g is small and it may be considered as a constant) as the height of the column of fluid over some reference section where both the pressure and the velocity of the fluid are zero.

For an incompressible fluid, we may divide the pressure p into two parts: one is the gravitational pressure p_g and the other is the static pressure p_s, i.e., $p = p_g + p_s$. The gravitational pressure p_g is the pressure of the fluid caused by the weight of the fluid; hence

$$p_g = \text{constant} - \rho g z \qquad (3.64)$$

Substituting Eq. (3.64) into Eq. (3.63), we have

$$\frac{1}{2}\rho q^2 + p_s = \text{constant} = p_t = \rho B_0 - \text{constant} \qquad (3.65)$$

where p_t is known as stagnation pressure or total pressure. Eq. (3.65) has been extensively used in hydraulics where we simply use the pressure p for the static pressure p_s without concerning the gravitational force.

For a compressible fluid, we cannot divide the pressure p into two unrelated parts as in the case of an incompressible fluid. For an isentropic flow of a gas, Eq. (3.61) for a steady flow is

$$\frac{1}{2}\frac{q^2}{g} + z + \frac{\gamma p}{(\gamma-1)\rho g} = \frac{B_0}{g} = \frac{1}{2}\frac{q_1^2}{g} + z_1 + \frac{\gamma p_1}{(\gamma-1)\rho_1 g} \qquad (3.66)$$

When the difference of height $(z - z_1)$ in the flow field is negligibly small in comparison with other terms, Eq. (3.66) becomes

$$\frac{1}{2}q^2 + \frac{\gamma p}{(\gamma-1)\rho} = \frac{1}{2}q_1^2 + \frac{\gamma p_1}{(\gamma-1)\rho_1} = \text{constant} = B'_0 \qquad (3.67)$$

Eq. (3.67) has been extensively used in ordinary gasdynamics where the gravitational force may not be considered. If the difference $z - z_1$ is not negligibly small such as in the study of dynamic meteorology, we have to consider the gravitational force in fluid dynamics of a compressible fluid.

For rotational flow where the velocity potential ϕ does not exist, we cannot obtain a simple relation similar to the Bernoulli's equation (3.61) which holds for the whole flow field. However, for steady flow case, some simple relations may be obtained which hold along a streamline. Let s be an elementary arc along a streamline and the velocity vector q has the same direction as s. For a steady flow, the equation of motion in the direction of s is as follows:

$$q\frac{\partial q}{\partial s} = \frac{\partial \phi_g}{\partial s} - \frac{1}{\rho}\frac{\partial p}{\partial s} \qquad (3.68)$$

Integrating Eq. (3.68) along a streamline s, we have

$$\frac{1}{2}\,q^2 - \phi_g + \int \frac{dp}{\rho} = C \qquad (3.69)$$

where C is a constant for a given streamline.

For a mixture of several species, we may derive the equations of motion for each species in terms of the velocity vector for each species which will be discussed in chapter IX.

7. Kelvin's Theorem

We have mentioned several times the irrotational flow of a fluid. The question is: under what conditions may an irrotational flow of a fluid exist? The question may be answered by the famous Kelvin's theorem which concerns the time rate of change of circulation in an inviscid fluid. The total rate of change of circulation Γ in an inviscid fluid in which all the body forces are negligible and only the pressure gradient ∇p is considered in the equation of motion is:

$$\frac{D\Gamma}{Dt} = \oint \frac{Dq}{Dt} \cdot ds + \oint q \cdot \frac{Dds}{Dt} = \oint \frac{Dq}{Dt} \cdot ds + \oint q \cdot dq$$

$$= \oint \frac{Dq}{Dt} \cdot ds + \frac{1}{2} \oint dq^2 = \oint \frac{Dq}{Dt} \cdot ds = \oint \left(-\frac{1}{\rho}\,\nabla p \right) \cdot ds \quad (3.70)$$

where the Euler's equation (3.57) without gravitational force has been used. If we apply the Stokes' theorem to transform the right-hand side of Eq. (3.70) into a surface integral, we have

$$\frac{D\Gamma}{Dt} = -\oiint \left[\nabla\left(\frac{1}{\rho}\right) \times \nabla p \right] \cdot dA \qquad (3.71)$$

For an incompressible fluid, $\rho =$ constant and $\nabla(1/\rho) = 0$. For a barotropic fluid where the pressure p is a function of density ρ only, i.e., $p = p(\rho)$, we have

$$\nabla \times \left(\frac{1}{\rho}\,\nabla p \right) = \nabla \times \left(\frac{1}{\rho}\frac{dp}{d\rho}\,\nabla\rho \right) = \nabla \times \nabla f(\rho) = 0 \quad (3.72)$$

Hence for an incompressible fluid or for a barotropic fluid, Eq. (3.70) becomes

$$\frac{D\Gamma}{Dt} = 0 \qquad\qquad (3.73)$$

Eq. (3.73) is Kelvin's theorem which states that for an incompressible and inviscid fluid as well as for an inviscid barotropic fluid, the circulation along a closed fluid curve remains constant in time. One special case for barotropic fluid is the isentropic flow. Hence for an isentropic flow of an inviscid fluid, the circulation along a closed fluid curve remains constant in time.

The circulation is closely connected with vorticity by equation (3.22). If the whole flow field is irrotational at any given time, the circulation in the whole flow field at this time will be zero by Eq. (3.22). If Kelvin's theorem applies, the circulation will be zero at all times. Furthermore, when the flow starts from rest, initially the vorticity is everywhere zero and the flow is irrotational. There is a very large class of physically important isentropic flows of an inviscid fluid which are irrotational. This is the reason that irrotational flow has been extensively investigated.

Kelvin's theorem may not hold true for many other fluid dynamic problems such as viscous fluid nor for some cases where the electro-magnetic forces are important. The viscosity of a fluid will produce vorticity in the fluid. However, for fluid of small viscosity, the viscous effect is limited in a narrow region near the boundary of a solid body or other transitional region. As a result, for a large region far away from the body or transitional region, the fluid may be considered as an inviscid fluid and Kelvin's theorem holds true.

It is possible that the flow is irrotational in a portion of the flow field and rotational in another portion of the flow field. For fluid of small viscosity, the diffusion of vorticity from one portion of the flow field to another is small. Thus Kelvin's theorem may apply to a large portion of the actual flow field for a sufficient long time in many practical cases.

8. Momentum Theorems for Steady Flow

If we solve the fundamental equations of fluid mechanics, i.e., equations of motion, the equation of continuity, etc., from given boundary conditions, we know the detailed picture of the whole flow field. Sometimes this detailed picture of the flow field is too complicated to be useful. For many engineering flow problems, it is sufficient to know the general relations of the resultant forces and moments of the whole field without knowing the detailed picture of the flow field. The momentum theorems are very useful for these cases especially for the steady flow problems, because we may draw many general conclusions about the flow simply from the conditions along the boundary of the flow field.

The momentum theorems are simply the application of Newton's laws of motion to a finite region of fluid. In other words, the time rate of change of momentum is equal to the resultant force acting on the system and the time rate of change of angular momentum is equal to the moments acting on the system. We are going to derive the momentum theorems for the steady flow of a fluid as follows:

The equation of motion of a fluid under arbitrary surface and body forces F_i may be written in the following form:

$$L_i = \frac{\partial \rho u_i u_k}{\partial x_k} - \frac{\partial T_{ik}}{\partial x_k} - F_i = 0 \tag{3.74}$$

where $i, k = 1, 2$, or 3 representing the three directions of the Cartesian coordinates. T_{ik} is the ikth component of the stress tensor, u_i is the ith component of velocity and F_i is the ith component of the body force. Eq. (3.74) is simply the ith component of the equation of motion. The summation convention is used in Eq. (3.74), i.e.,

$$\frac{\partial \rho u_i u_k}{\partial x_k} = \frac{\partial \rho u_i u_1}{\partial x_1} + \frac{\partial \rho u_i u_2}{\partial x_2} + \frac{\partial \rho u_i u_3}{\partial x_3} \tag{3.75}$$

Now we consider a region of the flow field enclosed by a surface S (Fig. 3.2). Inside the surface S, there may be some solid bodies. The volume V of the fluid within the surface S is the volume enclosed by the surface S minus the volume of the solid bodies. Eq. (3.74) holds true for every element of the fluid in the volume V. Now we integrate Eq. (3.74) over the volume V and have

$$\oint_V \left(\frac{\partial \rho u_i u_k}{\partial x_k} - \frac{\partial T_{ik}}{\partial x_k} - F_i \right) dV = 0 \tag{3.76}$$

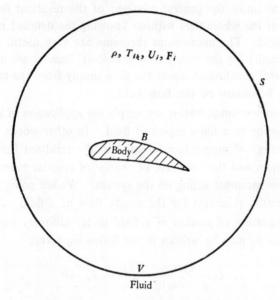

Figure 3.2   Control surface in a flow field

Applying Gauss's theorem to the first two terms of Eq. (3.74), we have

$$\oint_A \rho u_i u_k n_k dA = \oint_A T_{ik} n_k dA + \oint_V F_i dV \tag{3.77}$$

where the area A is the surface of the volume V which consists of the surface S and the surface of the bodies B inside the volume enclosed by S, n_k is the kth component of the outward normal of the surface A of the volume V. It should be noticed that a part of the surface S may be a solid surface. On the surface of a solid body, the normal component of the velocity is zero, i.e., $u_k n_k = 0$. Eq. (3.77) is the general momentum theorem of fluid dynamics.

Let us consider a special case of inviscid fluid without body force, i.e., $F_i = 0$, $T_{ik} = -p\delta_{ik}$, where $\delta_{ik} = 1$ if $i = k$ and $\delta_{ik} = 0$ if

$i \neq k$. Eq. (3.77) becomes

$$P_i = \oint_B p\delta_{ik}\, n_k\, dA = - \oint_S (\rho u_i u_k + p\delta_{ik})\, n_k\, dA \quad (3.78)$$

where P_i is the ith component of the force of the fluid acting on the body B. Now we apply the momentum theorem (3.78) to the problem of a sudden enlargement of a pipe. If the flow of a fluid in a cylindrical pipe with velocity q_1 passes into a wide cylindrical pipe (Fig. 3.3), the jet flow from the first pipe will mix with the surrounding fluid of the second pipe. After mixing, the final flow will be almost uniform again at section 2 with an average velocity q_2. The increase of pressure due to the mixing may be calculated by means of the momentum theorem (3.78) without going into the details of the process. Considering the control surface consisting of a cross section 1 of the original pipe and another section 2 of the widened pipe and the surface of the pipe, the momentum theorem gives

$$p_2 - p_1 = \rho q_2\,(q_1 - q_2) \quad (3.79)$$

where the fluid is assumed to be incompressible. The pressure force on the pipe wall does not contribute to the momentum in the direction of the axis of the pipe. If the mixing takes place in a gradually widening pipe, by Bernoulli's theorem (3.65), the pressure at section 2 should be p'_2 such that

$$p'_2 - p_1 = \frac{1}{2}\rho\,(q_1^2 - q_2^2) \quad (3.80)$$

We thus see that sudden widening of the pipe causes a loss of pressure

$$p'_2 - p_2 = \frac{1}{2}\rho\,(q_1 - q_2)^2 \quad (3.81)$$

The detailed motion of the flow during the mixing period is very difficult to analyze. It should be noticed that in the calculation of Eq. (3.79), we assume that the velocity and the pressure are constant at each section. As a result, the surface integral is simply the area times the momentum flux per unit area or the pressure at that section.

For the change of angular momentum, we consider the moment

about the origin of our coordinate system and integrate the following expression over the volume V of the fluid:

$$x_j L_k - x_k L_j = 0 \qquad (3.82)$$

Hence

$$\oint_V (x_j L_k - x_k L_j) dV = \oint_V \left\{ \frac{\partial}{\partial x_m} \left[(\rho u_k u_m - T_{km}) \, x_j \right] - x_j F_k \right.$$

$$\left. - \frac{\partial}{\partial x_m} \left[x_K (\rho u_j u_m - T_{jm}) \right] + x_k F_j \right\} dV = 0 \qquad (3.83)$$

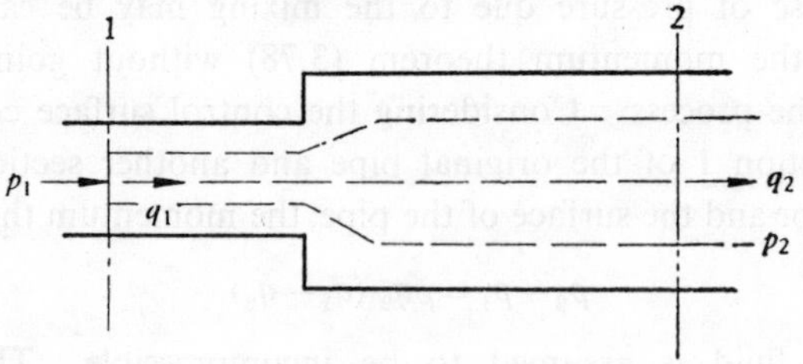

Figure 3.3 A sudden widening of a pipe

Applying Gauss's theorem to Eq. (3.83), we have

$$\oint_A [x_j(\rho u_k u_m - T_{km}) - x_k(\rho u_j u_m - T_{jm})] n_m \, dA = \oint_V (x_j F_k - x_k F_j) \, dV$$

$$(3.84)$$

Eq. (3.84) is the general angular momentum theorem of fluid dynamics.

Let us consider a special case of inviscid fluid without body force, i.e., $F_i = 0$ and $T_{ik} = -p\,\delta_{ik}$. Eq. (3.84) becomes

$$M_i = \oint_B (x_j p \delta_{km} - x_k p \delta_{jm}) \, n_m dA$$

$$= - \oint_S [\, x_j(\rho u_k u_m + p\delta_{km}) - x_k(\rho u_j u_m + p\delta_{jm}) \,] \, n_m dA \qquad (3.85)$$

where M_i is the moment about the ith axis of the force of the fluid acting on the body B. Eq. (3.85) gives the well known Euler's turbine theorem:

Consider a water turbine in Fig. 3.4. The mass of water flowing through a turbine is ρQ per second. Let q_1 be the resultant velocity of the water at the entrance section, which makes an angle θ_1 with the direction of motion of the wheel at a radius r_1. The corresponding values at the exit are q_2, θ_2 and r_2. The difference of the pressure at the entrance and at the exit is negligibly small. Hence the moment about the axis of the wheel is equal to the rate of change of angular momentum of the fluid, i.e.,

$$M_0 = \rho Q(q_1 r_1 \cos \theta_1 - q_2 r_2 \cos \theta_2) \qquad (3.86)$$

The most efficient working condition is the case $\cos \theta_2 = 0$ and the corresponding moment is $Q\, q_1 r_1 \cos \theta_1$. The amount of work supplied to the turbine is then $\rho Q\, q_1 r_1 \cos \theta_1 \cdot \Omega$, where Ω is the angular velocity of the wheel of the turbine.

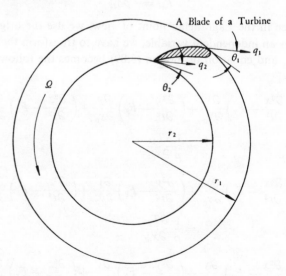

Figure 3.4   Flow through a blade of a water turbine

9. Equations of Motion from Lagrangian Point of View

In Lagrangian method, the independent variables are the initial coordinates $r_0(x_0, y_0, z_0)$ and time t.   Let the coordinates of a fluid particle at time t be $r(x, y, z)$.   The acceleration of this fluid particle is

$$\frac{\partial^2 r}{\partial t^2} = i \frac{\partial^2 x}{\partial t^2} + j \frac{\partial^2 y}{\partial t^2} + k \frac{\partial^2 z}{\partial t^2} \tag{3.87}$$

Newton's law of motion applying to a fluid particle of a unit volume gives

$$\frac{\partial^2 r_i}{\partial t^2} = F_i - \frac{\partial T_{ik}}{\partial x_k} \tag{3.88}$$

where $x=x_1$, $y=x_2$, and $z=x_3$, i, $k=1$, 2, or 3. F_i is the ith component of the body force on the fluid element and T_{ik} is the ikth component of the stress tensor and for inviscid fluid, we have simply

$$T_{ik} = - p\delta_{ik} \tag{3.89}$$

Since in the Lagrangian point of view, we use the original position r_0 as an independent variable, we have to transform the variables x_i to x_{0i} and equation of motion (3.89) becomes the following equations:

$$\left(\frac{\partial^2 x}{\partial t^2} - F_x\right)\frac{\partial x}{\partial x_0} + \left(\frac{\partial^2 y}{\partial t^2} - F_y\right)\frac{\partial y}{\partial x_0} + \left(\frac{\partial^2 z}{\partial t^2} - F_z\right)\frac{\partial z}{\partial x_0}$$
$$= \frac{1}{\rho}\frac{\partial T_{1k}}{\partial x_{0k}} \tag{3.90a}$$

$$\left(\frac{\partial^2 x}{\partial t^2} - F_x\right)\frac{\partial x}{\partial y_0} + \left(\frac{\partial^2 y}{\partial t^2} - F_y\right)\frac{\partial y}{\partial y_0} + \left(\frac{\partial^2 z}{\partial t^2} - F_z\right)\frac{\partial z}{\partial y_0}$$
$$= \frac{1}{\rho}\frac{\partial T_{2k}}{\partial x_{0k}} \tag{3.90b}$$

$$\left(\frac{\partial^2 x}{\partial t^2} - F_x\right)\frac{\partial x}{\partial z_0} + \left(\frac{\partial^2 y}{\partial t^2} - F_y\right)\frac{\partial y}{\partial z_0} + \left(\frac{\partial^2 z}{\partial t^2} - F_z\right)\frac{\partial z}{\partial z_0}$$

$$=\frac{1}{\rho}\frac{\partial T_{3k}}{\partial x_{0k}} \tag{3.90c}$$

where $x_{01}=x_0$, $x_{02}=y_0$, and $x_{03}=z_0$.

Ordinarily, the equation of motion in Eulerian point of view is preferred because it gives directly the results for many engineering problems such as velocity and pressure, etc., at a given point in space. However, in some problems, the Lagrangian method may be preferable such as diffusion problems or propagation of strong shock in which the streamlines or path lines play an important role.[12] We shall discuss them in later chapters.

10. Equation of Energy

Another important equation in the fluid dynamics of a compressible fluid is the equation of energy. For an incompressible fluid of constant transport coefficients, the energy equation is decoupled from the equations of motion. Hence we may calculate the velocity field without considering the equation of energy. Hence in classical hydrodynamics, one might not discuss the energy equation. In a compressible fluid, the energy equation is coupled with other fundamental equations of fluid dynamics, and we have to solve the energy equation simultaneously with other fundamental equations of fluid dynamics.

The law of conservation of energy requires that the difference in the rate of supply of energy to a volume V fixed in space with a surface S and the rate at which energy goes out through the surface S must be equal to the net rate of increase of energy in this volume. Thus the law of conservation of energy gives the following equation of energy:

$$\int_V \frac{\partial Q}{\partial t}\,dV + \int_S u_i(\tau_{ij}n_j)\,dS - \int_S E_t\rho u_j n_j\,dS + \int_S \kappa\frac{\partial T}{\partial x_i}n_i dS$$

$$=\frac{\partial}{\partial t}\int_V \rho E_t\,dV \tag{3.91}$$

where $\partial Q/\partial t$ is the rate of heat produced per unit volume in this volume V by external agency, E_t is the total energy of the system per unit mass

$$E_t = \frac{1}{2} u_i u_i + K + U_m + E_R/\rho \qquad (3.92)$$

where $\frac{1}{2} u_i u_i$ is the kinetic energy per unit mass, K is the potential energy per unit mass. U_m is the internal energy per unit mass and E_R is the radiation energy per unit volume (see chapter VII). The τ_{ij} is the ijth component of the stress tensor which consists of the viscous stresses and the radiation stresses, n_i is the ith component of the outer normal of the surface S, and κ is the coefficient of thermal conductivity. The first two terms on the left hand side of Eq. (3.91) are respectively the rate at which the energy is produced by external agencies and the rate at which heat is produced by the viscous and radiation stress tensors in contact with outside; the next two terms represent respectively the rate of energy loss by heat convection and heat conduction. By Gauss's theorem:

$$\int_S A \cdot n_j dS = \int_V \frac{\partial A}{\partial x_j} \, dV \qquad (3.93)$$

we transform the surface integrals into volume integrals and Eq. (3.91) becomes

$$\int_V \left\{ \frac{\partial Q}{\partial t} + \frac{\partial u_i \tau_{ij}}{\partial x_j} - \frac{\partial E_t \rho u_j}{\partial x_j} + \frac{\partial}{\partial x_j} \left(\kappa \frac{\partial T}{\partial x_j} \right) - \frac{\partial \rho E_t}{\partial t} \right\} dV = 0 \quad (3.94)$$

Since the volume is arbitrarily chosen, we conclude that the integrand of Eq. (3.94) itself must be zero. The energy equation is then

$$\frac{\partial Q}{\partial t} + \frac{\partial u_i \tau_{ij}}{\partial x_j} - \frac{\partial E_t \rho u_j}{\partial x_j} - \frac{\partial \rho E_t}{\partial t} + \frac{\partial}{\partial x_j} \left(\kappa \frac{\partial T}{\partial x_j} \right) = 0 \quad (3.95)$$

The following relations help us to simplify the energy equation:

$$\frac{\partial \rho E_t u_j}{\partial x_j} + \frac{\partial \rho E_t}{\partial t} = E_t \left(\frac{\partial \rho}{\partial t} + \frac{\partial \rho u_j}{\partial x_j} \right) + \rho \left(\frac{\partial E_t}{\partial t} + u_j \frac{\partial E_t}{\partial x_j} \right)$$

$$= \rho \frac{DE_t}{Dt} = \rho \left(u_j \frac{Du_j}{Dt} + u_j \frac{\partial K}{\partial x_j} + \frac{DU_m}{Dt} + \frac{D(E_R/\rho)}{Dt} \right) \quad (3.96)$$

where the equation of continuity (3.32) and the definition of total derivative (3.16) have been used. We also use the fact that $\partial K/\partial t = 0$. From the equation of motion (3.46) and (3.56), i.e.,

$$\rho \frac{Du_i}{Dt} = \frac{\partial \tau_{ij}}{\partial x_j} + \rho \frac{\partial \phi_g}{\partial x_i} + F_e^i \quad (3.97)$$

where $K = -\phi_g$, we may obtain the equation for the kinetic energy of the fluid by the scalar product of Eq. (3.97) with the velocity vector u_i, i.e.,

$$\rho u_i \frac{Du_i}{Dt} = u_i \frac{\partial \tau_{ij}}{\partial x_j} + \rho u_i \frac{\partial \phi_g}{\partial x_j} + u_i F_e^i \quad (3.98)$$

The left hand side of Eq. (3.98) is the rate of change of the kinetic energy $\frac{1}{2} u_i u_i$ of the fluid. Hence Eq. (3.98) shows how the kinetic energy varies in the flow field. Sometimes Eq. (3.98) is simply called the energy equation. But it should be noticed that this kinetic energy equation is simply another form of the equation of motion and it can not be used as an independent relation to replace the true energy equation (3.95). We may use it to simplify Eq. (3.95).

The third relation is

$$\frac{\partial u_i \tau_{ij}}{\partial x_j} = u_i \frac{\partial \tau_{ij}}{\partial x_j} + \tau_{ij} \frac{\partial u_i}{\partial x_j}$$

$$= \rho u_i \left(\frac{Du_i}{Dt} + \frac{\partial K}{\partial x_i} - \frac{F_e^i}{\rho} \right) + \tau'_{ij} \frac{\partial u_i}{\partial x_j} - p_t \frac{\partial u_i}{\partial x_i} \quad (3.99)$$

where $\tau_{ij} = \tau'_{ij} - p_t \delta_{ij}$ and τ'_{ij} consists of the viscous stresses and the radiation stresses and the total pressure $p_t = p + p_R$, the gas pressure p and the radiation pressure p_R. Except in chapter VII, we shall neglect the radiation stresses and the radiation pressure p_R. Then the dissipation function Φ (3.53) is simply:

$$\Phi = \tau'_{ij} \frac{\partial u_i}{\partial x_j} \quad (3.100)$$

Finally from the equation of continuity (3.32), we have

$$p \frac{\partial u_i}{\partial x_i} = - p \frac{1}{\rho} \frac{D\rho}{Dt} = p\rho \frac{D(1/\rho)}{Dt} \qquad (3.101)$$

Now substituting Eq. (3.98) to (3.101) into Eq. (3.95), we have the energy equation as follows:

$$\rho \left[\frac{DU_m}{Dt} + p \frac{D(1/\rho)}{Dt} \right] = \frac{\partial}{\partial x_i} \left(\kappa \frac{\partial T}{\partial x_i} \right) + \Phi - u_i F_e^i + \frac{\partial Q}{\partial t} \qquad (3.102)$$

The last two terms on the right hand side of Eq. (3.102) may be combined as we shall discuss them in chapter VI.

The two terms of the left hand side of Eq. (3.102) may be simplified for special fluid. For a perfect gas, we have by Eq. (1.39)

$$\rho \left[\frac{DU_m}{Dt} + p \frac{D(1/\rho)}{Dt} \right] = \rho \frac{DH}{Dt} - \frac{Dp}{Dt} \qquad (3.103)$$

then the equation of energy becomes

$$\rho \frac{DH}{Dt} = \frac{Dp}{Dt} + \frac{\partial}{\partial x_i} \left(\kappa \frac{\partial T}{\partial x_i} \right) + \Phi + \frac{\partial Q}{\partial t} - u_i F_e^i \qquad (3.104)$$

For an incompressible fluid with constant viscosity and heat conductivity coefficients, the energy equation becomes

$$\rho \frac{DU_m}{Dt} = \kappa \nabla^2 T + \Phi - u_i F_e^i + \frac{\partial Q}{\partial t} \qquad (3.105)$$

In ordinary fluid dynamics, we assume that the temperatures of all species of a mixture are the same and hence we need one energy equation for the mixture as a whole. In certain cases, the temperatures of all species may not be the same. For instance, in a plasma, the electron temperature may differ greatly from that of heavy particles. Then we have used the temperature T_s for each species s and we have to use one energy equation for each species which will be discussed in detail in the multifluid theory of fluid dynamics in chapter IX section 4.

11. Fundamental Equations of Fluid Dynamics

The basic fluid dynamic problems are the determinations of the velocity field u_i and the states of the fluid: its pressure p, density ρ, and temperature T at all times t and in all space x_i. There are six unknowns u_i, p, ρ, and T with four independent variables t and x_i in the general case. Hence we need to find six basic equations for these six unknowns. These basic equations are:

(1) Equation of state which connects the pressure p, density ρ and temperature T of the gas such as Eq. (1.18) or (1.20). Ordinarily we use Eq. (1.18), i.e.,

$$p = \rho R T \tag{3.106a}$$

(2) Equation of continuity which expresses the conservation of mass of the medium such as Eq. (3.32), i.e.,

$$\frac{\partial \rho}{\partial t} + q \cdot \nabla \rho + \rho (\nabla \cdot q) = 0 \tag{3.106b}$$

(3) Equations of motion which are generally three in number and which express the conservation of momentum such as Eq. (3.46), i.e.,

$$\rho \frac{Dq}{Dt} = - \nabla p + \nabla \cdot \tau' + F_b \tag{3.106c}$$

where τ'_{ij} is the viscous stress tensor in which we neglect the radiation stresses and in which the relation $3\lambda + 2\mu = 0$ is used and F_b is the body force which includes F_g and F_e etc.

(4) Equation of energy which expresses the conservation of energy such as Eq. (3.95), i.e.,

$$\frac{\partial Q}{\partial t} + \frac{\partial u_i \tau_{ij}}{\partial x_j} + \frac{\partial}{\partial x_j} \left(\kappa \frac{\partial T}{\partial x_j} \right) = \rho \frac{DE_t}{Dt} \tag{3.106d}$$

In order to solve the above six basic equations (3.106), there is much additional informations required. For instance, we have to know the variation of stress tensor τ_{ij} with the above six unknowns. For instance we may use the Navier-Stokes relations (3.52) for the stress-strain relations. If the Navier-Stokes equations do not hold, we have to find an equation of the stress tensor and solve it simultaneously with Eqs. (3.106). Even when the Navier-Stokes relations hold, we have to give relation of the coefficient of viscosity with respect to the state variables of the fluid such as temperature and density as we have discussed in chapter I section 13.

Similarly for the heat conduction terms, we may use the Fourier relation (1.55) and some definite relation between the coefficient of heat conductivity and the state of the fluid. If the Fourier law does not hold, we have to find an equation for the heat conduction flux and solve this equation simultaneously with the basic equations of fluid dynamics (3.106).

There are many other relations which are needed to solve the interaction between ordinary gasdynamics and other branches of physics. For a plasma, the electromagnetic force F_e depends on the electromagnetic fields and we have to solve the equations of electromagnetic fields with the basic equations of fluid dynamics (3.106) which will be discussed in chapter VI.

When the chemical reactions occur in the flow field, we have to use an equation or other simple formula for each chemical reaction. Furthermore, due to chemical reaction, the composition of the fluid may change and then we have to consider the diffusion phenomena by means of diffusion equations (3.40). We shall discuss the treatment of these equations in chapter V.

When the temperature is very high and the density is low, thermal radiation becomes an important mode of heat transfer. We have to consider the thermal radiation effects. Since thermal radiation effects may be expressed in terms of specific intensity of thermal radiation, we have to solve an equation which governs the specific intensity of thermal radiation, the equation of radiative transfer, simultaneously with the basic equations of fluid dynamics (3.106). We shall discuss the treatment of these equations in chapter VII.

In the rarefied gas, the basic equations of fluid dynamics (3.106) may not be good approximations to the actual flow conditions. We shall discuss the basic equations of fluid dynamics derived from the kinetic theory of gases point of view in chapter VIII. For a first approximation, it can be shown that the basic equations (3.106) may be derived from the Boltzmann equation of kinetic theory of gases which will be shown in chapter VIII. At the same time, we shall also discuss the cases where the classical basic equations (3.106) do not hold, particularly for the transition flow region and the free molecular flow region. We shall also show in chapter VIII how to obtain the relations between the transport coefficients, coefficient of viscosity, etc., and the state variables.

There are many other cases where the basic equations of fluid dynamics (3.106) are not sufficient or should be modified such as the cases of two phase flow, superfluid flow, multifluid theory, and relativistic fluid mechanics and biomechanics. We shall discuss those cases in chapter IX.

12. Boundary Conditions and Initial Conditions

For every particular problem of fluid dynamics, we have certain initial and boundary conditions which define this particular problem. Our problem is to find solutions of the basic equations of fluid dynamics (3.106) together with proper auxiliary equations such as electromagnetic field equations in plasma dynamics or other equations which satisfy these initial and boundary conditions.

For simplicity, let us consider first the case where the classical basic equations (3.106) are sufficient and where we consider the six unknowns u_i, p, ρ, T only. By initial conditions, we mean the velocity distribution and the state of the fluid at the whole flow field at an initial time $t=0$. Customarily in fluid dynamics, we do not give the spatial distribution of these initial condition but we only require that the initial values be consistent with the boundary conditions for $t=0$ and the basic equations. Hence we need to examine the boundary conditions only. In the case of a dynamic system with a finite number of degrees of freedom, the motion of a particle is determined by its initial position and velocity. For a continuous medium which has an infinite number of degrees of freedom, the motion of the medium is determined not only by the initial conditions but also by the boundary conditions, such as, for example, conditions of the velocity on the boundary of a domain considered at all times.

The boundary conditions of the gasdynamic field depend on the mean free path of the gas particles. When the mean free path is negligibly small, the gas may be considered as a continuum. For

a continuum, the no-slip boundary condition is a good approximation. Under this condition, when we have a surface separating a
body and a fluid or two fluids, the velocity components, the stresses,
the heat flux, and the temperature are all continuous. In most of
the problems discussed in this book, we consider the cases that the
gas, liquid, or plasma may be considered as a continuum and these
no-slip boundary conditions will be used.

When the mean free path of a gas is not small, even though the
gas may still be considered as a continuum, the basic equations
(3.106) hold true but there will be slip on the boundary. The velocity
and the temperature of the gas at the wall may be different from
those of the wall. Hence there will be a velocity slip and temperature jump across the boundary. This is known as slip flow in rarefied
gasdynamics which will be discussed in chapter VIII.

When the mean free path of the gas is larger than the dimension
of the flow field, the gas cannot be considered as a continuum. We
then have the free molecular flow. The boundary condition in the
free molecule flow depends on the smoothness of the surface. When
the free molecules strike the wall, they may reflect diffusely or specularly. In general, part of the molecules will reflect diffusely and part
of the molecules will reflect specularly. We shall discuss it more in
chapter VIII.

Even though the no-slip boundary conditions are the correct
boundary conditions for the gasdynamic field when the gas is considered as a continuum and when the mean free path of the gas is
small, such no-slip conditions may be relaxed under certain conditions, particularly in the flow field far away from the boundary or
transition region at very high Reynolds numbers. Under this condition, where the viscous effect is small, we may assume that the gas
or the fluid is inviscid. In inviscid fluid, surface of discontinuity is
allowable. The no-slip conditions should be relaxed. Two types of
surface of discontinuity may occur in an inviscid flow field: the
shock wave and the vortex surface. Across a shock wave, there is
a discontinuity in the normal velocity component but a continuity in
the tangential velocity component. On the other hand, across
a vortex sheet, there is a discontinuity in tangential velocity

component but a continuity in the normal velocity component. For a solid body in an inviscid flow field, we assume that a vortex sheet surrounds this body, which represents the boundary layer region in the limiting case, and then the normal velocity component relative to the body is zero but the tangential velocity component may not be zero. Thus we relax the no-slip condition on the tangential velocity on a solid body in an inviscid fluid.

When we study complicated flow problems where in addition to the basic equations of fluid dynamics (3.106), we have to consider other physical phenomena, we have to give corresponding boundary conditions on these additional physical quantities. For instance, in magnetofluid dynamics, we have to give the boundary conditions on the electromagnetic fields; in radiation gasdynamics, we have to give the boundary conditions of the specific intensity of radiation; in the flow with chemical reactions, we have to give the boundary conditions of concentration of various species in the fluid. We shall discuss these boundary conditions in later chapters.

13. Various Types of Flows

Even though the basic equations of fluid dynamics (3.106) are only six in number, they are nonlinear differential equations. There is no general way to solve these equations for reasonable complicated boundary conditions. For practical applications, we have to simplify these equations by proper approximations. In the following, we list a few of these approximations:

(1) Incompressible fluid. For liquids, it is always a good approximation in fluid dynamics by assuming that the density of the liquid is a constant. For gases and plasmas, when the Mach number, the ratio of the flow velocity to the local sound speed (see chapter IV), is small, we may also assume that the density of the gas or plasma is a constant and independent of the velocity. For an incompressible fluid of ordinary fluid dynamics, we need to consider five unknowns, the velocity components u_i, pressure p, and temperature T. The equation of state is replaced by $\rho=$constant. For an incompressible fluid, we may further assume that the transport coefficients such as coefficient of viscosity and coefficient of heat

conductivity, etc., are constant.   Under this condition, the equation
of continuity and the equations of motion are decoupled from the
energy equation.   We may then solve the velocity components u_i
and the pressure from the equation of continuity and the equations
of motion without considering the temperature distribution, i.e.,

$$\nabla \cdot q = 0 \tag{3.107a}$$

$$\rho \frac{Dq}{Dt} = - \nabla p + \mu \nabla^2 q \tag{3.107b}$$

For simplicity, in this section, we consider only the classical fluid
dynamics without any body forces and without any interaction with
other physical phenomena.   The interaction with other physical
phenomena will be discussed in later chapters.   To illustrate the
simplification due to various approximations, it is sufficient to consider
only the classical fluid dynamics.

It is interesting to note that the solution of an incompressible
flow is much simpler than the corresponding compressible flow case.
When the viscous forces are negligibly small, Eq. (3.107b) can be
integrated and we have the Bernoulli's equation (3.63).   We have a
relation between the pressure and velocity u_i.   As a result, for inviscid and incompressible flow, we need to solve the equation of
continuity for the velocity distribution, which is a linear differential
equation and for which the general solution is well known.   For a
compressible and inviscid flow, the interaction between the velocity
and state variables of the fluid cannot be neglected and we usually
have a set of nonlinear partial differential equations for which there
is no general method of solution.

For an incompressible and viscous flow, we have to solve Eqs.
(3.107) simultaneously.   In general, we have to solve a set of nonlinear partial differential equations which are known as Navier-Stokes
equations of an incompressible fluid, i.e., Eqs. (3.107).   Special
treatises should be referred to for the solution of this set of equations.[10]   We shall further discuss the properties of these equations
later.

After we obtain the distributions of velocity and pressure of the flow field, we may solve the energy equation (3.95) or (3.105) for the temperature distribution with these known distributions of velocity and pressure. One should notice the difference of energy equation for a truly incompressible fluid from that of a gas of which the Mach number is small. For a truly incompressible fluid, the density ρ is a constant and the energy equation is Eq. (3.105), i.e.,

$$\rho c_v \frac{DT}{Dt} = \kappa \nabla^2 T + \Phi \qquad (3.108)$$

where we assume that both the specific heat at constant volume c_v and the coefficient of heat conductivity κ are constant and that there is no body force nor heat addition term. Eq. (3.108) is a linear partial differential equation for temperature T because the dissipation function Φ is a known function after we obtain the velocity distribution.

For a gas of small Mach number, the corresponding energy equation is Eq. (3.104), i.e.,

$$\rho c_p \frac{DT}{Dt} = \kappa \nabla^2 T + \Phi \qquad (3.109)$$

The difference between Eqs. (3.108) and (3.109) is that we have to use c_p in Eq. (3.109) for the case of a gas of small Mach number and to use c_v for a truly incompressible fluid.

(2) Inviscid fluid. For many fluids of practical importance, such as water, air, and many gases, the coefficient of viscosity is rather small. Hence the viscous forces are usually smaller than the inertial forces or the pressure gradient except where the variation of velocity is large. Hence in a major portion of the flow field, we may neglect the viscous stresses and the basic equations of fluid dynamics are greatly simplified. The basic equations for an inviscid and compressible fluid are then:

$$p = \rho RT \qquad (3.110a)$$

$$\frac{1}{\rho} \frac{D\rho}{Dt} = -\nabla \cdot q \qquad (3.110b)$$

$$\rho \frac{Dq}{Dt} = - \nabla p \qquad (3.110c)$$

$$\rho \frac{DH}{Dt} = \frac{Dp}{Dt} \qquad (3.110d)$$

In general, we should solve all the equations of (3.110) for q, p, ρ, and T. As we have mentioned, for the case of an incompressible fluid, Eq. (3.110b) gives the velocity distribution. After the velocity distribution is known, Eq. (3.110c) gives the pressure distribution. Finally Eq. (3.110d) gives the temperature distribution. The same procedure may be used for the compressible fluid of small Mach number.

For a compressible fluid of which the Mach number is not negligibly small, we have to solve Eqs. (3.110) simultaneously. However, there are some other approximations which may be used. For an incompressible fluid or a barotropic fluid with the help of Kelvin's theorem, the flow may be considered as a potential flow or irrotational flow. We may introduce the velocity potential ϕ of Eq. (3.23). For an incompressible fluid, the equation of continuity of an irrotational flow gives

$$\nabla \cdot q = \nabla^2 \phi = \frac{\partial^2 \phi}{\partial x^2} + \frac{\partial^2 \phi}{\partial y^2} + \frac{\partial^2 \phi}{\partial z^2} = 0 \qquad (3.111)$$

Hence the velocity potential of an incompressible fluid satisfies the Laplace equation. The analysis of the well known harmonic functions may be used to solve Eq. (3.111) which has been extensively discussed.[6]

For a compressible fluid, we may have potential flow too. For instance, in many isentropic flow, potential flow exists. Eq. (3.110d) in terms of entropy S is simply as follows:

$$\frac{DS}{Dt} = 0 \qquad (3.112)$$

One of the solutions of Eq. (3.112) is that $S=$constant, i.e., the entropy is constant throughout the flow field, i.e., isentropic flow. For isentropic flow, the pressure is a function of density only, i.e.,

Eq. (2.34), Kelvin's theorem holds and we have irrotational flow. For irrotational flow of a compressible fluid, Bernoulli's theorem (3.61) holds. Differentiating Eq. (3.61) with respect to t, we have

$$\frac{\partial^2 \phi}{\partial t^2} + q \cdot \frac{\partial q}{\partial t} + a^2 \frac{1}{\rho} \frac{\partial \rho}{\partial t} = 0 \qquad (3.113)$$

where

$$a = (dp/d\rho)^{1/2} = \text{local sound speed} \qquad (3.114)$$

The equation of continuity (3.32) may be written as

$$\frac{1}{\rho} \frac{\partial \rho}{\partial t} + \nabla^2 \phi + \frac{1}{\rho} q \cdot \nabla \rho = 0$$

By the help of equation of motion, we have

$$q \cdot \frac{1}{\rho} \nabla \rho = \frac{1}{a^2} q \cdot \frac{1}{\rho} \nabla p = \frac{1}{a^2} q \cdot \left\{ -\frac{\partial q}{\partial t} - (q \cdot \nabla) q \right\}$$

Then Eq. (3.113) becomes

$$\frac{1}{a^2} \frac{\partial^2 \phi}{\partial t^2} + \frac{2}{a^2} q \cdot \frac{\partial q}{\partial t} = \nabla^2 \phi - \frac{1}{a^2} q \cdot [(q \cdot \nabla) q] \qquad (3.115)$$

or

$$\left(1 - \frac{u^2}{a^2}\right) \frac{\partial^2 \phi}{\partial x^2} + \left(1 - \frac{v^2}{a^2}\right) \frac{\partial^2 \phi}{\partial y^2} + \left(1 - \frac{w^2}{a^2}\right) \frac{\partial^2 \phi}{\partial z^2} - 2 \frac{uv}{a^2} \frac{\partial^2 \phi}{\partial x \partial y}$$

$$- 2 \frac{vw}{a^2} \frac{\partial^2 \phi}{\partial y \partial z} - 2 \frac{wu}{a^2} \frac{\partial^2 \phi}{\partial z \partial x}$$

$$= \frac{1}{a^2} \left\{ \frac{\partial^2 \phi}{\partial t^2} + 2u \frac{\partial^2 \phi}{\partial x \partial t} + 2v \frac{\partial^2 \phi}{\partial y \partial t} + 2w \frac{\partial^2 \phi}{\partial z \partial t} \right\} \qquad (3.115a)$$

Eq. (3.115) is the differential equation for the velocity potential ϕ of a compressible fluid. For an incompressible fluid, $a = \infty$, Eq. (3.115) reduces to Eq. (3.111). Eq. (3.115) is a nonlinear differential equation of which there is no general method of solution as in the case of Laplace equation (3.111). However, the solution of Eq. (3.115) has been extensively discussed in standard textbook of gasdynamics.[12]

According to Kelvin's theorem, the total circulation and the vorticity within a closed fluid curve will not be produced in the flow of a barotropic fluid starting from rest. But vorticity may be produced in the flow of an inviscid and compressible fluid where the flow is not isentropic such as the case behind a curved shock or where heat is added, i.e., diabatic flow. There is a definite relation between the vorticity and entropy change. Let us consider the case of steady flow of an inviscid fluid with heat addition. The energy equation is as follows:

$$q \times \omega = T\nabla S - \nabla H_S \qquad (3.116)$$

where

$$H_S = H + \frac{1}{2} q^2 = \text{stagnation enthalpy} \qquad (3.117)$$

Eq. (3.116) shows that the vorticity ω may be produced by the change of entropy S or by the change of stagnation enthalpy H_S which is produced by heat addition. For a two dimensional or axisymmetric flow with a streamfunction ψ, Eq. (3.116) becomes

$$\omega = \frac{p}{R} y^\delta \frac{\partial S}{\partial \psi} - \rho y^\delta \frac{\partial H_S}{\partial \psi} \qquad (3.118)$$

For adiabatic flow, the stagnation enthalpy is constant in the whole field. Eq. (3.118) then becomes

$$\omega = \frac{p}{R} y^\delta \frac{\partial S}{\partial \psi} \qquad (3.119)$$

From Eq. (3.112), we know that the entropy is constant along a streamline for adiabatic steady flow except when the streamline is intersected by a shock wave. A finite jump in entropy occurs across a shock wave. The value of vorticity ω remains unchanged along a streamline as long as the streamline is not intersected by a shock wave. When a uniform supersonic flow passes over a body, shock wave may occur in the flow field. The uniform flow is irrotational. Hence the flow remains irrotational up to the first shock. After the curved shock, $\partial S/\partial \psi$ is not zero. Hence the flow field becomes rotational after the curved shock and then the vorticity remains proportional to the pressure along each streamline.

(3) Viscous flow. When the viscous forces are not negligible, the flow is usually rotational. It is interesting to know what are the relations between the viscous flow field and the corresponding potential flow field of an inviscid fluid. For simplicity, let us consider a viscous incompressible fluid. The velocity components of the fluid u_i and the pressure p are governed by the Navier-Stokes equations:

$$\frac{\partial u_k}{\partial x_k} = 0 \qquad (3.120a)$$

$$\frac{\partial u_i}{\partial t} + u_k \frac{\partial u_i}{\partial x_k} = -\frac{1}{\rho} \frac{\partial p}{\partial x_i} + \nu_g \frac{\partial^2 u_i}{\partial x_k^2} \qquad (3.120b)$$

where $\nu_g = \mu/\rho$ is the coefficient of kinematic viscosity.

The only difference between Eqs. (3.120) and those of an inviscid fluid is the viscous force term, i.e., the last term in Eq. (3.120b). However, for a potential flow, the viscous term in Eq. (3.120b) is identical to zero, i.e.,

$$\frac{\partial^2 u_i}{\partial x_k^2} = \frac{\partial^2}{\partial x_k^2} \left(\frac{\partial \phi}{\partial x_i} \right) = \frac{\partial}{\partial x_i} \left(\frac{\partial^2 \phi}{\partial x_k^2} \right) = 0 \qquad (3.121)$$

Eq. (3.121) means that the solution of an irrotational flow of an inviscid and incompressible fluid satisfies the Navier-Stokes equations (3.120) of a viscous fluid. The viscous forces which are individually present in the actual fluid are collectively in equilibrium. Because of this fact, we have that the potential flow solution is a good approximation of the actual flow at large distance from a solid wall or other transition region.

Even though the solution of a potential flow satisfies the differential equations of motion of a viscous fluid, it cannot satisfy all the boundary conditions of a viscous fluid. Because in the potential flow problem, we simply let the coefficient of viscosity be zero, the order of the differential equation (3.120) is lowered by one. The arbitrary constants in the solution will also be reduced by one. As a result, we have to relax one of the boundary conditions of the original problem. In the actual fluid, usually both the tangential and the normal velocity components on the surface of a solid wall are zero relative to the wall owing to the no-slip conditions. But in the potential flow problem, the motion is completely determined by the condition of zero normal velocity component on the solid surface, and there is no possibility of making the tangential velocity component vanish simultaneously with the normal velocity component. Hence the solution of the Navier-Stokes equations represents a flow field in general and is different from the corresponding potential flow, especially near a solid boundary or other transition region. Hence we may use the assumption of an inviscid fluid for the flow field far away from a solid body but some other approximations for the flow field near the surface of a solid boundary.

(4) Boundary layer flow. When the Reynolds number $R_e = UL/\nu_g$, where U and L are respectively the characteristic velocity and length of a flow field, and ν_g, the typical coefficient of kinematic viscosity, is high, the viscous effects are confined to a very thin boundary layer whose thickness tends to zero as the coefficient of kinematic viscosity ν_g goes to zero. Navier-Stokes equations (3.120) can be simplified by such a consideration that the viscous effects in the boundary layer may be analyzed. For the sake of clarity, let us consider two dimensional flow over a flat plate. Let x be the direction along the plate and also that of the main flow (Fig. 3.5). Let y be the direction perpendicular to the plate. As far as the region

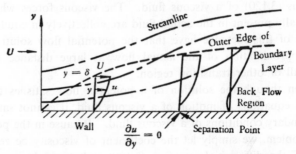

Figure 3.5 Velocity distributions and flow pattern in a boundary layer with adverse pressure gradient

of the boundary layer over the plate is concerned, the y-dimension is much smaller than the x-distance. We may define a boundary layer thickness δ such that when $y=\delta$, the x-wise velocity component is almost equal to the free-stream velocity $U(x)$ which may be a given function of x. The thickness of the boundary layer δ is proportional to the square root of the coefficient of the kinematic viscosity ν_g of the fluid. We are interested in the region $0 \leq y \leq \delta$. For a fluid of vanishingly small viscosity such as water or air, the y-dimension of the boundary layer is usually very small for engineering problems. To see what is inside the boundary layer, we must amplify the y-dimension of the boundary layer as ν_g decreases. In the theory of the boundary layer flow,[10] we know that if we put

$$\lim_{\nu_g \to 0} \left(\frac{y}{\sqrt{\nu_g}} \right) = \text{finite distance} = \eta \qquad (3.122)$$

into Navier-Stokes equations, we have the boundary layer equations by neglecting the higher order terms. These boundary layer equations give a good approximation for the flow in the boundary layer. Many flow phenomena which cannot be described by Euler's equations of an inviscid fluid can be studied by Prandtl's boundary layer equations. The order of the boundary layer equations is higher than that of the inviscid fluid equation; hence, the no-slip boundary conditions on the solid walls can be satisfied by the solution of the boundary layer equations. In the boundary layer equations, the inertial force is of the same order of magnitude as the viscous force. The boundary layer equations will give us the skin friction of a body in a flow field which has been checked very well from experimental results. By Reynolds analogy (see chapter IV), the boundary layer equations will give good results about the heat transfer from the body.

Another interesting phenomenon is the separation of flow from the surface of a body in the case of adverse pressure gradient. This problem cannot be predicted in the theory of an inviscid fluid. In the boundary layer theory, the slope of the velocity on the wall will decrease downstream in the adverse pressure gradient case. The separation point is the point where $\partial u / \partial y = 0$ on the wall. Backward flow will develop downstream from the separation point. The boundary layer theory may predict this separation point with good accuracy. When separation occurs, the drag of the body will be affected greatly. There was a famous experiment on the drag coefficient of a sphere which is defined as

$$C_D = \frac{\text{drag}}{\left(\frac{1}{2} \rho U^2\right)\left(\frac{1}{4} \pi d^2\right)} \tag{3.123}$$

where U is the velocity of a uniform stream in which the sphere of a diameter d is situated. The drag of the sphere is essentially the pressure drag which is due to the difference of pressure in the front side of the sphere from that on the back side. Prandtl found that C_D of a sphere is 0.5 in his wind tunnel while Eiffel found that it is 0.1 in his experiment. At first there were some arguments about the accuracy of their experimental results. Finally Prandtl explained this phenomenon by considering the separation in the boundary layer flow from the sphere. In Prandtl's experiments, the flow in the boundary layer of the sphere is laminar (see section 5) and it separates at an earlier point on the sphere (Fig. 3.6a). The wake region is large and the drag is also large. In Eiffel's experiments, the flow in the boundary layer of the sphere is turbulent and it separates at a later point on the sphere (Fig. 3.6b). The wake region is small and the drag is also small. If we plot the drag coefficient C_D of the sphere against the Reynolds number $R_e = Ud/\nu_g$ as in Fig. 3.6c, there is a critical Reynolds number R_{ec}, above which $C_D = 0.1$ and below which $C_D = 0.5$. The critical Reynolds number depends on the free stream turbulence. The higher the free stream turbulence is, the lower the value of the critical Reynolds number will be. The free stream turbulence in the Eiffel's wind tunnel is higher than that in Prandtl's wind tunnel. Hence the critical Reynolds number of the sphere in Eiffel's wind tunnel, say R_{ec_1}, is lower than that in Prandtl's wind tunnel, say R_{ec_3}. As a result, even their tested Reynolds numbers of the sphere are the same, say R_{ec_2}, they obtained different values for the drag of the sphere (Fig. 3.6c). Now the drag of a sphere may be used as an indicator of the free stream turbulence in a low speed wind tunnel.

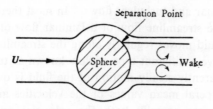

(a)   Laminar boundary layer on a sphere

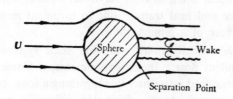

(b)   Turbulent boundary layer on a sphere

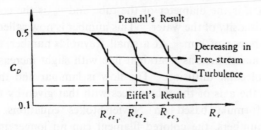

(c)   Drag coefficient of a sphere

Figure 3.6   Flow patterns over a sphere in a uniform stream U and its drag coefficient

The concept of boundary layer was first introduced by Prandtl in 1904, who opened a new era for fluid dynamics of a viscous fluid. Even though the original concept of the Prandtl boundary layer flow was for an incompressible fluid, the boundary layer phenomena occur in a compressible fluid too. We have both the boundary layer of velocity and that of temperature. When the fluid is electrically conducting, we may have the boundary layer of magnetic field too, which will be discussed in chapter VI.

(5) Laminar and turbulent flows. In most theoretical analysis, we consider the streamline flow or a laminar flow of a viscous fluid in which the fluid moves regularly along the streamlines. In nature, we often observe a different type of flow known as turbulent flow in which the apparently steady motion of the fluid is only steady in so far as the temperal mean values of the velocities and pressure are concerned whereas actually both the velocities and pressure are irregularly fluctuating. The mean motions of such flows do not satisfy the Navier-Stokes equations for a viscous fluid. Such turbulent flows occur often at high Reynolds number conditions. The frictional force and heat transfer due to turbulent motion are much larger than those of a laminar flow.

It was Osborne Reynolds who first made systematic investigations of turbulent flow. In his experiments, a filament of colored liquid was introduced into water flowing along a long straight circular tube with a smooth entry. Reynolds defined a nondimensional quantity which is Ud/ν_g, where U is the mean velocity of the water in the tube, d is the diameter of the tube and ν_g is the coefficient of kinematic viscosity of the water. This number is now called Reynolds number in honor of him. At a small Reynolds number, the colored filament is maintained as a straight line with slight increase in width through molecular diffusion. The flow is laminar and the pressure drop along the axis of the tube agrees with that given by the Hagen-Poiseuillé formula based on Navier-Stokes equations. At large Reynolds numbers, the colored filament can no longer maintain its straight line shape but is uniformly mixed with the whole fluid after a short distance from the entrance section. The flow is turbulent and the pressure drop is much larger than that given by the Hagen-Poiseuille formula. Reynolds found that there was a critical Reynolds number R_{ec} that $R_{ec} = 2000$. If the Reynolds number of the test is smaller than this critical value, the flow is laminar; if the Reynolds number of the test is larger than the critical value, the flow is turbulent. However it was found later that this critical Reynolds number depends on the test conditions, particularly the shape of the inlet and the flow conditions of the supply of the water.

If the water is free of disturbance before entering the tube, the critical Reynolds number may be very high, say over 24,000. In practice, we are interested in the lowest critical Reynolds number which is obtained for an arbitrary disturbance in the entrance flow.

Reynolds found that in the turbulent flow in a tube, the pressure drop is proportional to the 1.73 power of U instead of U as in the laminar flow. It is found later that the mean velocity distribution of a turbulent flow in a tube is more uniformly distributed in the central part of the tube with a larger gradient near the wall than the corresponding laminar flow case. Fig. 3.7 shows the mean velocity distribution of the laminar and the turbulent flows in a circular pipe with the same amount of fluid flowing per unit time.

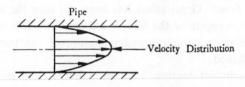

(a)   Laminar flow — parabolic distribution

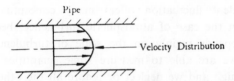

(b)   Turbulent flow — more uniform distribution

Figure 3.7   Velocity distributions in a circular pipe

There is a turbulent boundary layer over a solid body when the Reynolds number is high, as we have discussed in (4). The skin friction of a turbulent boundary layer is much larger than that of a laminar boundary layer of the same body at the same Reynolds number. Furthermore, the resistance to the separation of the boundary layer flow is much higher in the case of a turbulent flow than that of a laminar flow.

One of the most interesting problems in fluid dynamics is the transition of a laminar flow to a turbulent flow. In general, there is a transition region through which the flow is gradually developed from a laminar flow to a fully developed turbulent flow. In the transition region, the turbulent mixing and the laminar motion alternate more or less regularly. The transition is closely related to the basic instability of the laminar flow.[10] It takes some time for the instability of a laminar flow to develop into the flow in a transition region. It was found that the three-dimensional character plays an important role in the transition region even the original laminar flow is two-dimensional. Furthermore, if the outside disturbance is large enough, it may cause transition at a much lower Reynolds number than any disturbance due to the basic instability of a laminar flow. Great effort has been put into the investigation of the transition region of the flow and there are still many points to be settled before a complete understanding of the transition flow region is obtained.

In the turbulent motion, even though the fluid is regarded as a continuum where the average over the molecular motion has been taken, we still have to deal with turbulent fluctuations superimposed on a mean motion. However, such a separation into a mean motion and a turbulent fluctuation offers many conceptual difficulties, particularly in the case of atmospheric turbulence, because such a separation depends mainly on the scale of turbulence. In every observation we are able to measure some quantities only if they are large enough and we neglect or ignore all the others which are smaller than the response corresponding to our actual scale. In other words we shall have a quite different description of the atmosphere if in one case we can get an accurate measurement of the motion of all eddies having a diameter greater than one foot and in the other case only of eddies having diameters larger than 100 miles.

Let us give an illustration to show the influence of scale on turbulence. Consider a jet of smoke in a flow which is superposition of eddies of all sizes. If we take into account of all eddies present (Fig. 3.8a), especially those with diameters smaller than the diameter of the jet, we shall just have the appearance that we call diffusion, i.e., the jet will grow larger and larger. The jet would seem to disappear completely at some distance from the emitting point. On the other hand, if we suppress all the small eddies and consider eddies with diameters much larger than the diameter of the jet, the jet will appear to us as a very smooth ribbon of smoke bending and oscillating gently about a mean position (Fig. 3.8b). Which one of the above two pictures will be obtained depends on the scale used in the average process.

Once the scale of turbulence is chosen, we may write for instance, the instantaneous velocity component u_i as

$$u_i = \bar{u}_i + u'_i \qquad (3.124)$$

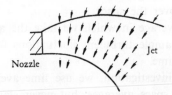

Nozzle

Jet

(a)   All eddies are included

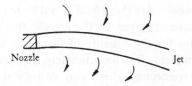

Nozzle

Jet

(b)   Only large eddies are considered

Figure 3.8 A jet of smoke

where u_i is the ith component of the total fluid velocity, $\bar{u}_i$ is the ith mean velocity component and u'_i is the ith component of the turbulent fluctuating velocity. Similarly, any physical quantity Q in a turbulent flow may be decomposed into a mean value $\bar{Q}$ and a turbulent fluctuating part Q'. In the study of a turbulent flow, we have to study not only the mean value of the quantity $\bar{Q}$ but also its fluctuating part Q'.

In taking the average of a turbulent quantity, the result depends not only on the scale used but also on the method of averaging. In practice, four different methods of averaging have been used to obtain the mean value of a turbulent quantity. These methods are:

(1) Time average in which we take the average at a fixed point in space over a long period of time. The period is determined by the scale used in the averaging process.

(2) Space average in which we take the average over all the space at a given time. The size of the space is determined by the scale used in the averaging process.

(3) Space-time average in which we take average over a long period of time and over the space.

(4) Statistical average in which we take the average over the whole collection of sample turbulent fluctuations for a fixed point in space and a fixed time.

In experimental investigation, we use time average almost exclusively and seldom space averages, but never statistical averages. In the theory, we use almost exclusively the statistical averages. The question is: what is the relation between the time average and the statistical average? Are they equal or not? In classical statistical mechanics, the answer is given by the ergodic theorem which states sufficient conditions for the equality of these two kinds of averages for almost all samples. But unfortunately, in fluid mechanics, no ergodic theorem connecting these two entirely different types of averages has yet been proven. Experimenters are very much tempted to translate the theoretical results for quite different types of averages. An ergodic theorem should be proven before a complete rational theory of turbulent flow can be established.

In turbulent flow, we usually assume that the instantaneous velocity components satisfy the Navier-Stokes equations when the fluid may be considered as a continuum. However, in actual measurements, we have only the mean quantities of the flow variables. We have to find the corresponding equations for these mean quantities in a turbulent flow which are known as Reynolds equations.

The following Reynolds rules of average have been used in deriving the mean flow equations of a turbulent flow:

$$\text{R1}: \overline{f+g} = \bar{f} + \bar{g}$$

$$\text{R2}: \overline{cf} = c\bar{f} \tag{3.125}$$

$$\text{R3}: \overline{\bar{f}g} = \bar{f}\bar{g}$$

$$\text{R4}: \overline{\lim f_n} = \lim \overline{f_n} \quad (f_n = \text{sequence of a function})$$

where the bar refers to the average value, and c is a constant.

For simplicity, let us consider the flow of an incompressible fluid. If we substitute the instantaneous velocity components and the pressure into Navier-Stokes equations (3.107) and average the resultant equations according to Reynolds rules (3.125), we have the following Reynolds equations for the mean velocity components and the mean pressure of the turbulent flow:

$$\frac{\partial \bar{u}}{\partial x} + \frac{\partial \bar{v}}{\partial y} + \frac{\partial \bar{w}}{\partial z} = 0 \tag{3.126a}$$

$$\rho \left(\frac{\partial \bar{u}}{\partial t} + \bar{u}\frac{\partial \bar{u}}{\partial x} + \bar{v}\frac{\partial \bar{u}}{\partial y} + \bar{w}\frac{\partial \bar{u}}{\partial z} \right)$$

$$= -\frac{\partial \bar{p}}{\partial x} + \mu \nabla^2 \bar{u} - \frac{\partial \rho \overline{u'^2}}{\partial x} - \frac{\partial \rho \overline{u'v'}}{\partial y} - \frac{\partial \rho \overline{u'w'}}{\partial z} \tag{3.126b}$$

$$\rho \left(\frac{\partial \bar{v}}{\partial t} + \bar{u}\frac{\partial \bar{v}}{\partial x} + \bar{v}\frac{\partial \bar{v}}{\partial y} + \bar{w}\frac{\partial \bar{v}}{\partial z} \right)$$

$$= -\frac{\partial \bar{p}}{\partial y} + \mu \nabla^2 \bar{v} - \frac{\partial \rho \overline{u'v'}}{\partial x} - \frac{\partial \rho \overline{v'^2}}{\partial y} - \frac{\partial \rho \overline{v'w'}}{\partial z} \tag{3.126c}$$

$$\rho \left(\frac{\partial \bar{w}}{\partial t} + \bar{u}\frac{\partial \bar{w}}{\partial x} + \bar{v}\frac{\partial \bar{w}}{\partial y} + \bar{w}\frac{\partial \bar{w}}{\partial z} \right)$$

$$= -\frac{\partial \bar{p}}{\partial z} + \mu \nabla^2 \bar{w} - \frac{\partial \rho \overline{u'w'}}{\partial x} - \frac{\partial \rho \overline{v'w'}}{\partial y} - \frac{\partial \rho \overline{w'^2}}{\partial z} \tag{3.126d}$$

The additional terms due to the turbulent fluctuations over the Navier-Stokes equations are known as Reynolds stresses or eddy stresses. For an incompressible fluid these eddy or turbulent stresses are components of a tensor of second order, i.e.,

ith turbulent normal stress: $-\rho \overline{u'_i}^2$

ijth turbulent shearing stress: $-\rho \overline{u'_i u'_j}$ $(i \neq j)$

These stresses are due to the rate of transfer of momentum across the corresponding surface due to turbulent velocity fluctuations which are usually larger than the viscous stresses (3.52), based on mean velocity gradients.

The solutions of Reynolds equations (3.126) will represent properly the turbulent flow. However there are only four equations in (3.126) for the ten unknowns: the mean pressure, three mean velocity components $\overline{u_i}$, and six Reynolds stresses $-\rho \overline{u'_i u'_j}$. In general, Reynolds equations (3.126) are not sufficient to determine these unknowns. This is one of the main difficulties in the theoretical investigation of turbulent flow.[11] At the present time, it is not possible to solve the Reynolds equations for any practical flow problems. Additional assumptions and hypotheses are necessary to simplify these equations in order to obtain some approximate solution for practical cases.[11]

In a similar manner, the Reynolds equations of motion for the turbulent flow of a compressible fluid may be obtained in which both the fluctuations of the velocity components and the fluctuation of the state variables such as pressure, density, and temperature should be considered. The eddy stresses will be affected by both the fluctuating velocity components and the fluctuating density.

When other physical quantities are important in the fluid dynamics such as electromagnetic fields in the flow of an electrically conducting fluid, we may have contributions due to the fluctuations of electromagnetic fields in the fundamental equations of turbulent flow. We shall discuss some of them in the chapters following.

When physical quantities other than the velocity vector and state variables are important, such as chemical reactions, electromagnetic fields, thermal radiations, etc., we may have a special type of flow depending on the special value of some nondimensional parameters. We shall discuss these cases in following chapters, especially in chapter IV where the various important nondimensional parameters of fluid mechanics are discussed.

14. General Orthogonal Coordinates

In our previous discussions, we have used the simple Cartesian coordinates. For many problems, it may be convenient to use systems of curvilinear coordinates instead of Cartesian coordinates. In fluid mechanics, the most convenient curvilinear coordinates are the orthogonal ones. In this section, we are going to write down some general properties of orthogonal curvilinear coordinates which will be useful in the analysis of fluid mechanics.

Let three independent orthogonal families of surfaces be

$$f_1(x, y, z)=a_1; \quad f_2(x, y, z)=a_2; \quad f_3(x, y, z)=a_3 \quad (3.127)$$

where x, y, and z are the Cartesian coordinates. The surfaces $a_1=$ constant, $a_2=$ constant, and $a_3=$ constant form an orthogonal system. The values of a_1, a_2, and a_3 may be used as coordinates of a point in space. The relations between the two system of coordinates x, y, z and a_1, a_2, a_3 may be given by Eq. (3.127) or by the relations:

$$x=x(a_1, a_2, a_3); \quad y=y(a_1, a_2, a_3); \quad z=z(a_1, a_2, a_3) \quad (3.128)$$

The element of length $ds=(dx^2+dy^2+dz^2)^{1/2}$ may be expressed in terms of the orthogonal coordinates a_1, a_2, a_3 as follows:

$$(ds)^2=h_1^2(da_1)^2+h_2^2(da_2)^2+h_3^2(da_3)^2 \quad (3.129)$$

where

$$h_1=\left[\left(\frac{\partial x}{\partial a_1}\right)^2+\left(\frac{\partial y}{\partial a_1}\right)^2+\left(\frac{\partial z}{\partial a_1}\right)^2\right]^{1/2}$$
$$h_2=\left[\left(\frac{\partial x}{\partial a_2}\right)^2+\left(\frac{\partial y}{\partial a_2}\right)^2+\left(\frac{\partial z}{\partial a_2}\right)^2\right]^{1/2} \quad \Bigg\} \quad (3.130)$$

$$h_3 = \left[\left(\frac{\partial x}{\partial a_3} \right)^2 + \left(\frac{\partial y}{\partial a_3} \right)^2 + \left(\frac{\partial z}{\partial a_3} \right)^2 \right]^{1/2} \Biggr\}$$

From the above expressions, we see that the elements of length at a point (a_1, a_2, a_3) in the direction of increasing a_1, a_2, and a_3 are respectively $h_1 da_1$, $h_2 da_2$, and $h_3 da_3$.

Let q be a vector of components q_{a_1}, q_{a_2}, and q_{a_3} in the direction of a_1, a_2, and a_3 respectively, i.e.,

$$q = i_1 q_{a_1} + i_2 q_{a_2} + i_3 q_{a_3} \tag{3.131}$$

where i_1, i_2, and i_3 are respectively the unit vector in the direction of a_1, a_2, and a_3.

The gradient of a scalar function ϕ in the orthogonal curvilinear coordinates is

$$\nabla \phi = i_1 \frac{1}{h_1} \frac{\partial \phi}{\partial a_1} + i_2 \frac{1}{h_2} \frac{\partial \phi}{\partial a_2} + i_3 \frac{1}{h_3} \frac{\partial \phi}{\partial a_3} \tag{3.132}$$

The divergence of a vector q is

$$\nabla \cdot q = \frac{1}{h_1 h_2 h_3} \left\{ \frac{\partial}{\partial a_1} (q_{a_1} h_2 h_3) + \frac{\partial}{\partial a_2} (q_{a_2} h_3 h_1) + \frac{\partial}{\partial a_3} (q_{a_3} h_1 h_2) \right\} \tag{3.133}$$

The curl of a vector q is

$$\nabla \times q = i_1 \omega_1 + i_2 \omega_2 + i_3 \omega_3 \tag{3.134}$$

where

$$\omega_1 = \frac{1}{h_2 h_3} \left\{ \frac{\partial}{\partial a_2} (q_{a_3} h_3) - \frac{\partial}{\partial a_3} (q_{a_2} h_2) \right\}$$

$$\omega_2 = \frac{1}{h_3 h_1} \left\{ \frac{\partial}{\partial a_3} (q_{a_1} h_1) - \frac{\partial}{\partial a_1} (q_{a_3} h_3) \right\}$$

$$\omega_3 = \frac{1}{h_1 h_2} \left\{ \frac{\partial}{\partial a_1} (q_{a_2} h_2) - \frac{\partial}{\partial a_2} (q_{a_1} h_1) \right\}$$

the Laplacian of a scalar function ϕ is

$$\nabla^2 \phi = \frac{1}{h_1 h_2 h_3} \left\{ \frac{\partial}{\partial a_1} \left(\frac{h_2 h_3}{h_1} \frac{\partial \phi}{\partial a_1} \right) + \frac{\partial}{\partial a_2} \left(\frac{h_3 h_1}{h_2} \frac{\partial \phi}{\partial a_2} \right) + \frac{\partial}{\partial a_3} \left(\frac{h_1 h_2}{h_3} \frac{\partial \phi}{\partial a_3} \right) \right\}$$

(3.135)

The most common curvilinear coordinates are as follows:

(1) Cylindrical coordinates. These coordinates are related with circular cylinders with radius r and angular coordinate θ and axial coordinate z. The relation between the cylindrical coordinates (r, θ, z) and the Cartesian coordinates (x, y, z) are:

$$x = r \cos \theta, \quad y = r \sin \theta, \quad z = z$$ (3.136)

If we take

$$a_1 = r, \quad a_2 = \theta, \quad a_3 = z$$ (3.137)

we have

$$h_1 = 1, \quad h_2 = r, \quad h_3 = 1$$ (3.138)

Substituting Eqs. (3.137) and (3.138) into Eqs. (3.132) to (3.135), we have the gradient of ϕ, divergence of q, curl of q, and Laplacian of ϕ in cylindrical coordinates respectively.

(2) Spherical coordinates. These coordinates are related to spheres with radius r and two angular coordinates θ and ϕ such that

$$x = r \sin \theta \cos \phi, \quad y = r \sin \theta \sin \phi, \quad z = r \cos \theta$$ (3.139)

If we take

$$a_1 = r, \quad a_2 = \theta, \quad a_3 = \phi$$ (3.140)

we have

$$h_1 = 1, \quad h_2 = r, \quad h_3 = r \sin \theta$$ (3.141)

By the help of Eqs. (3.140) and (3.141), we may easily find the fundamental equations in spherical coordinates.

(3) Elliptical coordinates. These coordinates are related with elliptical cylinders such that

$$x = c\xi, \quad y = c \left[(\xi^2 - 1)(1 - \eta^2) \right]^{\frac{1}{2}}, \quad z = z$$ (3.142)

It is easy to show that $\xi =$ constant is a cylinder of elliptical cross section and η=constant is hyperbolic cylinder of two sheets. The factor c is a constant. If we take

$$a_1=\xi, \quad a_2=\eta, \quad a_3=z \tag{3.143}$$

we have

$$h_1 = c\left(\frac{\xi^2-\eta^2}{\xi^2-1}\right)^{\frac{1}{2}}, \quad h_2=c\left(\frac{\xi^2-\eta^2}{1-\eta^2}\right)^{\frac{1}{2}}, \quad h_3 = 1 \tag{3.144}$$

Eqs. (3.143) and (3.144) give the corresponding values for the elliptical coordinates. If we investigate the flow around an elliptical cylinder in a flow field, it is convenient to use the elliptical coordinates because on the boundary of the elliptical cylinder, we have simply ξ=constant.

15. Moving Coordinates

It is sometimes convenient in special problems to employ a system of reference axes which is in motion. Let U, V, W be the x-, y-, and z-component of velocity of the origin of the references axes and ω_x, ω_y, and ω_z, the angular velocities of the frame of the reference axes. Let u, v, w be the absolute velocities of the fluid at a point (x, y, z) rigidly connected to the frame of reference axes and u', v', w', the velocities of the fluid at the same point relative to the frame. We have

$$\left.\begin{aligned} u&=U+u'-y\omega_z+z\omega_y \\ v&=V+v'-z\omega_x+x\omega_z \\ w&=W+w'-x\omega_y+y\omega_x \end{aligned}\right\} \tag{3.145}$$

If we consider the increase in mass in a small rectangular element of volume attached to the frame of reference axes with the center at (x, y, z) where ρ is the density of the fluid, we obtain the equation of continuity from the conservation of mass without any mass source [Eq. (3.32)] as follows:

$$\frac{\partial\rho}{\partial t}+\frac{\partial\rho u'}{\partial x}+\frac{\partial\rho v'}{\partial y}+\frac{\partial\rho w'}{\partial z}=\frac{\partial\rho}{\partial t}+\nabla\cdot(\rho q')=0 \tag{3.146}$$

Hence the equation of continuity has the same form as that for the case where the reference axes are fixed in space.

For the expression of the acceleration [Eq. (3.15)], we shall show that the expression in the moving axes differ from those in the fixed axes.

Let A be the absolute velocity in the direction fixed in space whose direction cosines referred to the moving axes are l_A, m_A, n_A, i.e., $A = l_A u + m_A v + n_A w$. In time δt, the coordinates of the fluid particle at (x, y, z) will increase by an amount $(u', v', w')\, \delta t$ so that to a first approximation, $l_A u$ will become

$$(l_A + \delta l_A)\left\{ u + \left(\frac{\partial u}{\partial t} + u'\frac{\partial u}{\partial x} + v'\frac{\partial u}{\partial y} + w'\frac{\partial u}{\partial z} \right)\delta t \right\}$$

whence we get

$$\frac{DA}{Dt} = \frac{Dl_A}{Dt}u + \frac{Dm_A}{Dt}v + \frac{Dn_A}{Dt}w + l_A\left(\frac{\partial u}{\partial t} + u'\frac{\partial u}{\partial x} + v'\frac{\partial u}{\partial y} \right.$$

$$\left. + w'\frac{\partial u}{\partial z} \right) + m_A\left(\frac{\partial v}{\partial t} + u'\frac{\partial v}{\partial x} + v'\frac{\partial v}{\partial y} + w'\frac{\partial v}{\partial z} \right)$$

$$+ n_A\left(\frac{\partial w}{\partial t} + u'\frac{\partial w}{\partial x} + v'\frac{\partial w}{\partial y} + w'\frac{\partial w}{\partial z} \right) \tag{3.147}$$

But since l_A, m_A, n_A are direction cosines referred to the moving axes of a line in space, therefore

$$\left.\begin{array}{l} \dfrac{Dl_A}{Dt} - m_A\omega_z + n_A\omega_y = 0 \\[2ex] \dfrac{Dm_A}{Dt} - n_A\omega_x + l_A\omega_z = 0 \\[2ex] \dfrac{Dn_A}{Dt} - l_A\omega_y + m_A\omega_x = 0 \end{array}\right\} \tag{3.148}$$

and substituting Eq. (3.148) into Eq. (3.147), we have

$$\frac{DA}{Dt} = l_A\left(\frac{\partial u}{\partial t} - v\omega_z + w\omega_y + u'\frac{\partial u}{\partial x} + v'\frac{\partial u}{\partial y} + w'\frac{\partial u}{\partial z} \right)$$

$$+m_A\left(\frac{\partial v}{\partial t}-w\omega_x+u\omega_z+u'\frac{\partial v}{\partial x}+v'\frac{\partial v}{\partial y}+w'\frac{\partial v}{\partial z}\right)$$

$$+n_A\left(\frac{\partial w}{\partial t}-u\omega_y+v\omega_x+u'\frac{\partial w}{\partial x}+v'\frac{\partial u}{\partial y}+w'\frac{\partial w}{\partial z}\right)\quad(3.149)$$

In general curvilinear coordinates such as cylindrical or spherical coordinates, at every instance we consider that the origin of the moving axes is coincident with that of the fixed axes and that the moving axes have no linear velocities but only angular velocities. Hence

$$x=y=z=U=V=W=0,\quad u=u',\quad v=v',\quad w=w'\quad(3.150)$$

From Eq. (3.150), we see that the absolute accelerations of the fluid particle which should be used in the equations of motion in the moving coordinates are as follows:

$$\left.\begin{aligned}\left(\frac{DA}{Dt}\right)_x &=\frac{Du'}{Dt}-v'\omega_z+w'\omega_y\\[2mm]\left(\frac{DA}{Dt}\right)_y &=\frac{Dv'}{Dt}-w'\omega_x+u'\omega_z\\[2mm]\left(\frac{DA}{Dt}\right)_z &=\frac{Dw'}{Dt}-u'\omega_y+v'\omega_x\end{aligned}\right\}\quad(3.151)$$

where

$$\frac{D}{Dt}=\frac{\partial}{\partial t}+u'\frac{\partial}{\partial x}+v'\frac{\partial}{\partial y}+w'\frac{\partial}{\partial z}$$

In cylindrical coordinates, we have

$$u'=q_r,\quad v'=q_\theta,\quad w'=q_z,\quad \omega_x=\omega_y=0,$$

$$\omega_z=\frac{d\theta}{dt}=\frac{q_\theta}{r},\,dx=dr,\quad dy=rd\theta,\quad dz=dz\quad(3.152)$$

Substituting Eq. (3.152) into Eq. (3.151), we have the acceleration terms in the cylindrical coordinates as follows:

$$\left.\begin{array}{l} \text{The radial acceleration} = \dfrac{Dq_r}{Dt} - \dfrac{q_\theta^2}{r} \\[3mm] \text{The tangential acceleration} = \dfrac{Dq_\theta}{Dt} + \dfrac{q_r q_\theta}{r} \\[3mm] \text{The axial acceleration} = \dfrac{Dq_z}{Dt} \end{array}\right\} \quad (3.153)$$

where

$$\frac{D}{Dt} = \frac{\partial}{\partial t} + q_r \frac{\partial}{\partial r} + \frac{q_\theta}{r}\frac{\partial}{\partial \theta} + q_z \frac{\partial}{\partial z}$$

Similar expressions may be obtained for the general orthogonal coordinates. The corresponding expressions for the angular velocities are

$$\omega_x = \frac{1}{h_2}\left(\frac{\partial h_3}{\partial a_2}\right)\frac{da_3}{dt} - \frac{1}{h_3}\left(\frac{\partial h_2}{\partial a_3}\right)\frac{da_2}{dt}$$

$$\omega_y = \frac{1}{h_3}\left(\frac{\partial h_1}{\partial a_3}\right)\frac{da_1}{dt} - \frac{1}{h_1}\left(\frac{\partial h_3}{\partial a_1}\right)\frac{da_3}{dt}$$

$$(3.154)$$

$$\omega_z = \frac{1}{h_1}\left(\frac{\partial h_2}{\partial a_1}\right)\frac{da_2}{dt} - \frac{1}{h_2}\left(\frac{\partial h_1}{\partial a_2}\right)\frac{da_1}{dt}$$

$$u' = q_{a_1}, \ v' = q_{a_2}, \ w' = q_{a_3}, \ dx = h_1 da_1, \ dy = h_2 da_2, \ dz = h_3 da_3$$

PROBLEMS

1. Find the expression of acceleration at a point in the flow field in the following coordinates systems:
   (a) spherical coordinates,
   (b) elliptical coordinates.
2. Find the equation of continuity in the following coordinates:
   (a) cylindrical coordinates,
   (b) spherical coordinates.
3. Find the components of the strain tensor (3.20) in the following coordinates:

(a) cylindrical coordinates,
(b) spherical coordinates.

4. The parabolic coordinates are $a_1 = \xi$, $a_2 = \eta$, and $a_3 = -z$, where

$$\xi = (2r)^{\frac{1}{2}} \sin\left(\frac{1}{2}\,\theta\right), \quad \eta = (2r)^{\frac{1}{2}} \cos\left(\frac{1}{2}\,\theta\right)$$

where r, θ, and z are the cylindrical coordinates. Find the Laplace equation in parabolic coordinates.

5. Prove the Stokes theorem (3.22).

$$\oint_C q \cdot ds = \oint_A \omega \cdot dA$$

6. Define two streamfunctions [Eq. (3.44)] for an unsteady three-dimensional flow in cylindrical coordinates.

7. Derive the dissipation function Φ of Eq. (3.53) from the equations of motion.

8. By means of Bernoulli's equation (3.61), find the relation of the flow velocity and the difference of stagnation pressure and the static pressure in terms of the square of Mach number M up to M^6, where $M = q/a$, q is the magnitude of the local velocity and a is the local sound speed corresponding to q in a steady isentropic flow.

9. Prove Gauss's theorem [cf. Eq. (3.77)]

$$\oint_V (\nabla \cdot q)\, dV = \oiint_A q \cdot dA \qquad (3.155)$$

10. Consider the flow of a compressible fluid through a curved channel shown in Fig. 3.9 at the entrance section 1, the velocity is q_1 and the pressure is p_1 and the temperature is T_1 and at the exit section 2, the velocity is q_2 and the pressure is p_2. Find the resultant force F exerted on the wall of this curved channel by assuming that the flow is isentropic.

11. Consider the case of a liquid of density ρ between two con-

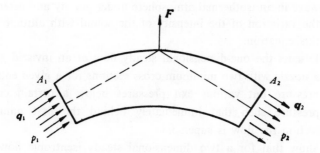

Figure 3.9   Flow through a curved pipe

centric spheres of radii R_1 and R_2 respectively with $R_2 > R_1$. If the inside sphere R_1 suddenly shrinks into another concentric sphere, calculate the impulsive pressure which is required to produce this motion.

12. Neglecting the gravitational force, derive the Euler's equations of motion for (i) cylindrical coordinates and (ii) spherical coordinates.

13. Show that in an inviscid fluid flow without any body force, Kelvin's theorem always holds when the flow is subsonic and starts with a uniform flow and that Kelvin's theorem may not be true when the flow is supersonic and with a uniform free-stream.

14. Define a streamfunction ψ for the one dimensional unsteady flow of a compressible fluid so that the equation of continuity is automatically satisfied. Derive the differential equation for this streamfunction for an isentropic flow. Find the physical characteristics of this flow and show that certain quantities known as Reimann invariants are constant along these characteristics.

15. Consider a very weak shock traveling in an ideal gas toward a gaseous surface of another ideal gas. Discuss the conditions of the reflected and the refracted shock waves at the gaseous surface.

16. Derive the differential equation of a one-dimensional sound

wave in an isothermal atmosphere under gravity and determine the variation of the intensity of the sound with altitude from this equation.

17. Discuss the one-dimensional steady flow of an inviscid gas in a nozzle with two minimum cross sections for a given entrance pressure but various exit pressures below a certain critical pressure such that immediately behind the first minimum section the flow is supersonic.

18. Show that for a two dimensional steady isentropic flow, the streamfunction ψ for a transonic flow, i.e., the Mach number is close to unity, satisfies the Tricomi equation

$$\sigma \frac{\partial^2 \psi}{\partial \theta^2} + \frac{\partial^2 \psi}{\partial \tau^2} = 0 \qquad (3.156)$$

where θ is the angle of the streamline with respect to the x-axis and

$$\sigma = -\int_{a*}^{q} \left(\frac{\rho}{\rho_0} \right) \frac{dq}{q} \qquad (3.157)$$

$a*$ is the critical sound speed when $q=a$.

19. Find the general solutions of Eq. (3.157).

20. Calculate (i) the increase of entropy across an oblique shock with a shock angle α and the Mach number in front of the shock M_1, (ii) the rate of change of entropy with respect to the shock angle α for a given M_1, and (iii) the rate of change of entropy with respect to M_1 for a given shock angle α.

21. Consider a vertical column of gas of constant temperature and take the spatial coordinate x as the vertical distance positive upward. If a piston is moving upward at a speed of one half of sound speed in the gas, calculate the flow field in front of the piston and determine the shock wave produced by the piston. What would be the temperature behind this shock wave?

22. Derive the Lagrangian equation of continuity.

23. A flame front is a surface of discontinuity in combustion which

has a very small velocity of propagation. If r is the ratio of the stagnation temperature after combustion to that before combustion, find the relations between velocities, pressures, and temperatures across a flame front in term of r and the Mach number M_1 in front of this flame front which is assumed to be a very small quantity.

24. From the linearized equation of a viscous, heat conducting, and compressible fluid, show that the disturbance may be divided into transverse waves and longitudinal waves. Find the dispersion relations of these two types of waves and discuss the results in detail.

25. Derive the energy equation of a turbulent flow of a compressible fluid with the help of Reynolds rules of averaging (3.125) and the energy equation of a compressible fluid (3.95).

26. Derive the three dimensional boundary layer equations of a finite plate in a uniform stream of an incompressible and viscous fluid. Show that the principle of independence holds if the plate is a yawed infinite plate, i.e., all the variables are independent of the spanwise coordinate. By the principle of independence, we mean that the solutions of the longitudinal velocity u and vertical velocity v are independent of the spanwise velocity w.

27. Derive the three-dimensional boundary layer equations on a body of revolution with a spin at zero angle of attack in a uniform stream of a compressible and viscous fluid for the following two cases:
    (a) the radius of the body is much larger than the boundary layer thickness; and
    (b) the radius of the body is of the same order of magnitude of the boundary layer thickness.

28. Derive the equation of vorticity in a viscous and incompressible fluid.

REFERENCES

[1] Burgers, J. M. "The bridge between particle mechanics and continuum mechanics", Proceedings of the International Symposium on Plasma Dyna-

mics. Addison-Wesley Publishing Co., 1960, pp. 119—186.

[2] Emmons, H. G. (ed.) Fundamentals of Gas Dynamics. Vol. 3 of High Speed Aerodynamics and Jet Propulsion. Princeton University Press, 1958.

[3] Goldstein, S. (ed.) Modern Developments in Fluid Dynamics. Vols. I & II. Oxford University Press, 1938.

[4] Hayes, W. D. and Probstein, R. F. Hypersonic Flow Theory. Academic Press, 1959.

[5] Hinze, J. O. Turbulence: An Introduction to its Mechanism and Theory. McGraw-Hill, 1959. 2nd ed. 1975.

[6] Lamb, H. Hydrodynamics. (6th ed.) Cambridge University Press, 1932.

[7] Landau, L. D. and Lifshitz, E. M. Fluid Dynamics. Pergamon Press, 1959.

[8] Lin C. C. (ed.) Turbulent Flows and Heat Transfer. Vol. 5 of High Speed Aerodynamics and Jet Propulsion. Princeton University Press, 1959.

[9] Love, A. E. A. A Treatise on the Mathematical Theory of Elasticity. Dover Publications, 1944.

[10] Pai, S.-I. Viscous Flow Theory. Vol. I. Laminar Flow. D. Van Nostrand, 1956.

[11] Pai, S.-I. Viscous Flow Theory. Vol. II. Turbulent Flow. D. Van Nostrand, 1957.

[12] Pai, S.-I. Introduction to the Theory of Compressible Flow. D. Van Nostrand, 1959.

[13] Prandtl, L. Fluid Dynamics. Hafner Publishing Co., 1952.

[14] Sandri, G. "The physical foundations of modern kinetic theory", in Dynamics of Fluids and Plasmas. S. I. Pai, et al. (eds.) Academic Press, 1966, pp. 341—398.

[15] Schlichting, H. Boundary Layer Theory. (6th ed.) McGraw-Hill, 1968.

[16] Sears, W. R. (ed.) General Theory of High Speed Aerodynamics. Vol. 6 of High Speed Aerodynamics and Jet Propulsion. Princeton University Press, 1954.

[17] Streeter, V. L. (ed.) Handbook of Fluid Dynamics. McGraw-Hill, 1961.

[18] von Mises, R. Mathematical Theory of Compressible Fluid Flow. Academic Press, 1958.

DIMENSIONAL ANALYSIS
AND DYNAMIC SIMILARITY

1. Introduction

In chapter III, we have shown that the fluid flows are usually governed by a system of nonlinear partial differential equations, such as the Navier-Stokes equations (3.120) for a viscous and incompressible fluid, Euler's equations (3.110) for an inviscid and compressible fluid, etc. Even though these equations are among the simple cases in comparison with what we are going to discuss in later chapters, there is no general method of finding the solution of these nonlinear equations. Hence there is less chance to solve more complicated equations when the interaction between fluid mechanics and other physical phenomena becomes important. In order to get some understandings of these complicated flow phenomena, two approaches have been used: one is to simplify these fundamental equations according to some physical considerations so that the resultant equations may be solved; and the second is experimental investigation of the flow problem under conditions which are similar to the actual case but under proper control. In order to achieve these approaches, it is desirable to find out the important parameters in a given flow problem. There are two methods to find these important parameters: one is known as inspection analysis and the other is known as dimensional analysis. However, the basic idea of these two methods is the dimensional analysis which will be discussed in section 2.

In experimental investigations, it frequently happens that the test model is of a different size from the actual body, that the test fluid is the same as the actual fluid but in a different thermodynamic state, or that the test fluid is different from the one under actual conditions. These circumstances usually are the results of an attempt either to reduce the cost of the tests or to have better control over the testing conditions or both. Also in experimental investigations, one occasionally desires to compare experimental results on the same or geometrically similar models tested under different conditions. Also one might want to apply one series of tested results to similar cases. When the test conditions are different from the actual conditions, or when geometrically similar models are tested under different sets of conditions, it is of course necessary to find the relations that exist between the two sets of conditions. These relations depend essentially on some important parameters of the problem. If we know these parameters, it will be easy to correlate the experimental results. We shall discuss such model similarity in section 7.

From chapter III section 13, we see that the flow field may be divided into different types. In each type of flow, one or a few of the nondimensional parameters are important. If we know the range of values of these important parameters, the fundamental equations of fluid dynamics may be greatly simplified. For instance, the boundary layer equations may be derived from the Navier-Stokes equations when the Reynolds number of the flow is high. Furthermore, by dimensional analysis, we may obtain the similar solutions for various fluid dynamic problems. By the concept of these similar solutions, it is often possible to reduce a set of partial differential equations into a set of total differential equations which are much easier to handle. These similar solutions would give us significant insight of the complicated phenomena of fluid dynamics. We shall discuss some of these similar solutions in section 8 which are a powerful method of analysis of fluid dynamic problems.

By dimensional analysis, we may find out many characteristic velocities, characteristic frequencies, and characteristic lengths which have important meanings in special problems. It helps to understand the real phenomena in a flow field. We shall discuss some of these characteristic velocities, frequencies, and lengths in section 3 as well as in later chapters.

2. Theory of Dimensions

Every physical quantity has certain dimensions. These dimensions are closely associated with the units of measurements. But there are a few basic dimensions or units by which the dimensions of all the other physical quantities may be expressed. Furthermore, all the physical quantities may be divided into two classes: one is dimensional, and the other is dimensionless or nondimensional. All the nondimensional quantities are independent of the scale of measurement. It would be desirable to present the results of a study in nondimensional quantities which are more general than those in dimensional quantities depending on the scale of measurement. However, in actual life, most of the quantities used are dimensional. We should find an efficient way to express dimensional quantities in terms of nondimensional quantities which will be discussed in section 6.

In classical mechanics which includes fluid mechanics, there are only three basic dimensions which are usually taken as the length L, the mass m, and the time t. If we establish the scales or units of these three basic dimensions, all the other physical quantities in mechanics can be expressed by these three basic dimensions. There are several systems of units of measurement, which have been used in the world. The most common ones are (i) CGS systems of units in which the length is expressed in centimeters, the mass in grams, and the time in seconds, (ii) MKS system in which the length is in meters, the mass in kilograms, and the time in seconds, and (iii) Ft-lb-sec system in which the length is in feet, the mass in pound, and the time in seconds. There are definite conversion factors between these systems. For instance, one meter is equal to 100 cm. We assume that a definite system is used and we will not discuss them anymore.

If we establish the system of units of the basic dimensions L, m, and t, the dimensions of all other physical quantities in mechanics can be expressed by these three basic dimensions. A few examples are given below

(1) Velocity V. The dimensions of a velocity are

$$V = \text{velocity} = \frac{\text{length}}{\text{time}} = \frac{L}{t} \qquad (4.1)$$

(2) Acceleration a_c. The dimensions of an acceleration are

$$a_c = \text{acceleration} = \frac{\text{velocity}}{\text{time}} = \frac{L}{t^2} \tag{4.2}$$

(3) Force F. The dimensions of a force are

$$F = \text{force} = \text{mass} \times \text{acceleration} = m\frac{L}{t^2} \tag{4.3}$$

(4) Work W. The dimensions of the work are

$$W = \text{work} = \text{force} \times \text{distance} = \frac{mL^2}{t^2} \tag{4.4}$$

(5) Coefficient of viscosity μ. The dimensions of the coefficient of viscosity are

$$\mu = \frac{\text{stress}}{\text{velocity gradient}} = \frac{\text{force/area}}{(L/t)/L} = \frac{(mL/t^2)/L^2}{1/t} = \frac{m}{Lt} \tag{4.5}$$

If we study the classical mechanics, it is sufficient to use three basic dimensions such as L, m, and t. However the choice is not unique. We may use other combinations of physical quantities so that there are three independent dimensions. For instance, we may use force F, length L, and time t as three basic dimensions. The application of dimensional analysis is independent of the choice of the basic dimensions, because the ratio of two numerical values of any physical quantity must be independent of the choice of the scale for the fundamental units. For instance, the ratio of two different volumes measured in cubic meters must be the same as the ratio of the same volumes measured in cubic feet. Because of this fact, we can show that all the physical quantities Q in mechanics may be expressed in the form of a monomial power:

$$Q = L^a m^b t^c \tag{4.6}$$

(see problem 1 and reference [12]). We see that the formulas (4.1) to (4.5) belong to the form of Eq. (4.6). Actually, if there are n basic dimensions in a physical problem, i.e., Q_1, Q_2, $\cdots$, Q_n, any physical quantity in this problem may be expressed in the form:

$$Q = Q_1^a \, Q_2^b \cdots Q_n^c \tag{4.7}$$

Eq. (4.7) is very useful in dimensional analysis.

If we study the fluid mechanics of a compressible fluid, we have to add another basic dimension to the above three dimensions because besides the mechanics, we have to study heat transfer. It is usually to use the temperature T as the fourth dimension. Of course, we have various scales for temperature such as degrees Kelvin °K, degrees Rankine °R, degrees centigrade °C, and degrees Fahrenheit °F. The physical quantities in gasdynamics may be expressed in the four basic dimensions: L, m, t, and T. For instance:

(6) The specific heat at constant pressure c_p. The dimensions of c_p are

$$c_p = \frac{\text{(quantity of heat)}/\text{mass}}{\text{temperature}} = \frac{mL^2}{t^2}\frac{1}{m}\frac{1}{T} = \frac{L^2}{t^2}\frac{1}{T} \tag{4.8}$$

(7) Thermal conductivity κ. The dimensions of the coefficient of thermal conductivity or heat conductivity κ are

$$\kappa = \frac{\text{quantity of heat}}{\text{area} \times \text{time}} \times \frac{1}{\text{temperature gradient}}$$

$$= \frac{mL^2}{t^2}\frac{1}{L^2 t}\frac{1}{T/L} = \frac{mL}{t^3 T} \tag{4.9}$$

If we study the fluid mechanics of a plasma, we have to add one more basic dimension to the above four basic dimensions. We may take the absolute electric charge e as the fifth dimension. Some of the physical quantities which have the dimensions including the electric charge e are given below:

(8) The excess electric charge of the fluid ρ_e. The dimensions of ρ_e are

$$\rho_e = \frac{\text{charge}}{\text{volume}} = \frac{e}{L^3} \tag{4.10}$$

(9) The electric current density J. The dimensions of J are

$$J = \frac{\text{current}}{\text{area}} = \frac{\text{charge}}{\text{time}} \times \frac{1}{\text{area}} = \frac{e}{tL^2} \tag{4.11}$$

(10)   The magnetic field strength H.   The dimensions of H are

$$H = \text{current} \times \text{length} = \frac{e}{tL} \qquad (4.12)$$

(11)   The electric field strength E.   The dimensions of E are

$$E = \frac{\text{force/volume}}{\text{charge}} = \frac{mL}{t^2 e} \qquad (4.13)$$

The formulas from which Eq. (4.10) to (4.13) are derived will be discussed in chapter VI.

One of the basic principles of the theory of dimensions is that all the terms in an equation must have the same dimensions. For instance, in the Euler's equation of motion (3.57), we have the following dimensions for the various terms:

$$\text{The inertial terms: } \rho \frac{Dq}{Dt} = \frac{m}{L^3} \frac{L/t}{t} = \frac{m}{L^2 t^2}$$

$$\text{The gravitational force: } \rho g = \frac{m}{L^3} \frac{L}{t^2} = \frac{m}{L^2 t^2}$$

$$\text{The pressure gradient: } \nabla p = \frac{\text{force/area}}{\text{length}} = \frac{mL}{t^2} \frac{1}{L^2} \frac{1}{L} = \frac{m}{L^2 t^2}$$

All the three terms have the dimensions $m/(L^2 t^2)$.

The number associated with a given physical quantity depends on the units used. For instance, a length of 1 yard is equal to 3 feet. In one case we have the number 3 and in the other case we have the number 1. In order to avoid the possible confusion due to the choice of units, it is convenient to use nondimensional quantities which are independent of the units used. For example, we may express all the physical quantities in Euler's equation (3.57) as follows:

$$x_i^* = x_i/L, \ \nabla^* = L\nabla, \ q^* = q/U, \ \rho^* = \rho/\rho_0, \ t^* = tU/L,$$
$$p^* = p/(\rho_0 U^2), \ g^* = gL/U^2 \qquad (4.14)$$

where star refers to the nondimensional quantities; L, U, and ρ_0 are respectively the reference length, velocity, and density. Now if we introduce the nondimensional expressions of (4.14) into Eq. (3.57), we have

$$\rho^* \frac{Dq^*}{Dt^*} = \rho^* g^* - \nabla^* p^* \qquad (4.15)$$

Eq. (4.15) is the nondimensional form of Euler's equation of motion (3.57). Eq. (4.15) will give the same number no matter what units are used. We shall show later that all important parameters in fluid mechanics are nondimensional quantities.

3. Some Characteristic Velocities, Characteristic Frequencies, and Characteristic Lengths in Fluid Mechanics

In Eq. (4.14), we show that a combination of several physical quantities may result in nondimensional quantity. Similarly, a combination of several quantities may result a new quantity which has the dimensions of a velocity, a frequency, or a length. In many problems of fluid mechanics, these new quantities have special physical significance. We shall call them characteristic velocity, characteristic frequency, or characteristic length depending on their final dimensions. We shall discuss their physical significance in detail when we study the corresponding fluid mechanics problems in which they are important in later chapters. Now we list a few most important characteristic velocities, characteristic frequencies, and characteristic lengths in this section and state briefly their physical significance for future reference.

(A) Characteristic Velocities

(1) Velocity of sound a. It is defined as follows:

$$a = \sqrt{\left(\frac{\partial p}{\partial \rho}\right)_s} = (\gamma RT)^{1/2} \qquad (4.16)$$

Since the pressure may be considered as a function of density ρ and entropy S, the partial differentiation is performed at constant entropy. In Eq. (4.16), the perfect gas law is used. The dimensions of the velocity of sound or sound speed a are

$$a = \left(\frac{m}{Lt^2} \frac{L^3}{m}\right)^{1/2} = \frac{L}{t} = \text{velocity} \qquad (4.16a)$$

The sound speed is in the speed of propagation of infinitesimal disturbance in a gas with gas constant R and ratio of specific heats γ at a temperature T. At standard atmosphere, the sound speed at sea level is about 340 meters/sec. The sound speed of the fluid is one of the most important characteristic velocities in the study of the flow field of a compressible fluid [see Eq. (3.115)].

(2) Velocity of propagation of a surface wave v_g. It is defined as

$$v_g = (gh_0)^{1/2} \tag{4.17}$$

where h_0 is the mean height of a surface of water. Eq. (4.17) represents the speed of propagation of small disturbance on the surface of the water. The dimensions of v_g are

$$v_g = \left(\frac{L}{t^2}L\right)^{1/2} = \frac{L}{t} = \text{velocity} \tag{4.17a}$$

The velocity v_g is important when the gravitational force is important such as the free surface effect on a hydrofoil and other similar problems.

(3) Velocity of Alfven's wave[1] V_H. It is defined as

$$V_H = H(\mu_e/\rho)^{1/2} \tag{4.18}$$

Eq. (4.18) gives the speed of propagation of a wave in an incompressible and inviscid fluid of density ρ and of infinite electric conductivity under a homogeneous magnetic field H with μ_e as the magnetic permeability. Such a wave was first predicted by Alfven.[1] Hence we call it Alfven's wave. It is one of the basic waves in magnetogasdynamics as we shall discuss in chapter VI. If we take $B = \mu_e H = 1000$ gausses and $\rho = 1.16 \times 10^3$ kg/m³, we have $V_H = 0.77$ m/sec, which is a typical Alfven's wave speed in mercury. It is easy to show that the dimensions of V_H are the same as those of a velocity. (See problem 1.)

(4) Velocity of light c. It is defined as

$$c = (\epsilon\mu_e)^{-1/2} \tag{4.19}$$

where ϵ is the inductive capacity, sometimes referred to as the dielectric constant. The velocity of light c is the speed of propagation of electromagnetic waves. In free space, $c = 3.0 \times 10^8$ m/sec. The velocity of light c is important in the wave motion in a plasma and it is also important in relativistic fluid mechanics. (See chapter IX.)

(B) Characteristic Frequencies[6]

(1) Plasma frequency ω_p. It is defined as

$$\omega_p = e\left(\frac{n}{m\epsilon}\right)^{\frac{1}{2}} \tag{4.20}$$

where n is the number density of the charged particles, electrons or ions, and m is the mass of a charged particle. For electrons with $n = 10^{22}/m^3$, $\omega_{p_e} = 5.62 \times 10^{12}$ sec^{-1}. The dimensions of ω_p is $1/t$ which is the same as that of a frequency. Eq. (4.20) gives the frequency of oscillation of a plasma due to the electric field alone as we shall discuss in chapters VI and IX.

(2) Cyclotron frequency ω_c or Larmor frequency. It is defined as

$$\omega_c = \frac{eB}{m} \tag{4.21}$$

where B is the magnetic induction and m is the mass of a charged particle. Eq. (4.21) gives the angular frequency with which a particle of mass m and charge e gyrates in a cyclotron under the magnetic induction B. It is one of the most important quantities in the study of plasma dynamics as we shall discuss in chapters VI and IX. If B is 1000 gausses, the electron cyclotron frequency will be $\omega_{ec} = 1.76 \times 10^{10}$ sec^{-1}.

(C) Characteristic Lengths

(1) Mean free path of gas L_f. It is defined as

$$L_f = 1.255\sqrt{\gamma}\left(\frac{\mu}{a\rho}\right) \tag{4.22}$$

It is a measure of the mean distance traveled between collisions of particles in a gas, as we shall discuss in chapter VIII. When L_f is much smaller than the typical length in a flow field, we may treat the fluid as a continuum. Otherwise the discrete character of the fluid should be taken into account. For air, $\gamma = 1.4$. In standard atmosphere, we may take $\mu = 1.8 \times 10^{-4}$ gr/cm·sec, $\rho = 1.23 \times 10^{-3}$ gr/cm and $a = 340$ m/sec, we have then

$$L_f = 0.64 \times 10^{-5} \text{ cm}$$

Hence the mean free path L_f of a gas is usually very small. As a result, the continuum theory of gasdynamics is very good for most engineering problems. But the mean free path increases with the altitude of the atmosphere. At an altitude of 100 km, we have $\mu = 2 \times 10^{-4}$ gr/cm·sec, $\rho = 2 \times 10^{-9}$ gr/cc, and $a = 400$ m/sec. Then $L_f = 3.7$ cm. At very high altitude, the mean free path L_f of the air may not be negligible in comparison with the typical length of the flow problem. (See chapter VIII.)

(2) Mean free path of radiation L_{R_ν}. When we study the thermal radiation effects (see chapter VII), the linear absorption coefficient ρk_ν is important which is defined by the following formula:

$$dI_\nu = -\rho k_\nu I_\nu ds \qquad (4.23)$$

The loss of specific intensity of radiation I_ν along the ray of radiation with frequency ν over a distance ds in a medium of density ρ and absorption coefficient k_ν is given by Eq. (4.23). Integration of Eq. (4.23) gives

$$I_\nu(s) = I_\nu(s_0) \exp\left(-\int_{s_0}^{s} \rho k_\nu ds\right) = I_\nu(s_0) \exp\left(-\tau_\nu\right) \qquad (4.24)$$

where τ_ν is known as the optical thickness of radiation of the layer $(s - s_0)$ and s_0 is a reference point where the specific intensity $I_\nu(s_0)$ is given and

$$L_{R_\nu} = \frac{1}{\rho k_\nu} = \text{mean free path of radiation} \qquad (4.25)$$

The mean free path of radiation represents the mean distance traveled by photons between absorption. The optical thickness is a non-dimensional distance which shows the effective length in absorption of radiation. For a given physical length $(s - s_0)$, if τ_ν is large, the medium is said to be optically thick; while if τ_ν is small, the medium is said to be optically thin. Proper approximations may be made according to whether τ_ν is very small or very large as we shall discuss in chapter VII.

(3) Debye length L_D. It is defined as

$$L_D = \left(\frac{RT\epsilon}{e^2 n} \right)^{1/2} \tag{4.26}$$

In the analysis of an electrolyte, the Debye length was originally obtained as the distance that the electric field of a point charge is shielded by a local increase of concentration of charges of the opposite sign. In a plasma, it is a measure of the distance over which the excess electric charge may be different appreciably from zero, as we shall discuss in chapters VI and IX. If we take $T = 10^4$ °K and $n = 10^{22}/m^3$, we have $L_D = 4.75 \times 10^{-3}$ mm. Hence the Debye length in a plasma is usually very small and thus the plasma has the tendency toward electrical neutrality, i.e., without any excess charge.

(4) Larmor radius L_L. It is defined as

$$L_L = \frac{a}{\omega_c} = (\gamma RT)^{1/2} \frac{m}{eB} \tag{4.27}$$

In Eq. (4.27), we use the sound speed a to represent the typical velocity of a charged particle in a plasma. The Larmor radius is a measure of the radius of the helical path of a charged particle of mass m and charge e in a magnetic induction B. The Larmor radius has a similar position in plasma dynamics as the mean free path of a gas in ordinary gasdynamics. If we take $\omega_c = 1.76 \times 10^{10}$ sec^{-1} and $a = 1.18 \times 10^4$ m/sec, we have $L_L = 6.7 \times 10^{-3}$ mm. If we consider the charged particles as singly charged hydrogen ions, we have for the same magnetic induction and temperature $L_{Li} = 1.23$ mm.

4. Reynolds Laws of Similarity and Inspection Analysis

Osborne Reynolds[11] was the first one to consider the laws of similarity from which very important deductions concerning the flow of fluid can be drawn from the fundamental equations of fluid mechanics. The problem is: under what conditions are the forms of the flows of any fluid around geometrically similar bodies themselves geometrically similar? Such similarity is known as dynamical similarity.

What information can the fundamental equations of fluid mechanics supply as to the conditions for this dynamical similarity? The answer is: we shall have dynamical similarity when alternations of the units of length, time, mass, temperature, and electric charge transform the fundamental equations of the fluid mechanics and the boundary conditions in one case into those of the other case so that the equations completely coincide. Assuming that we have a unique solution of the system of equations, this system of equations will furnish a direct description for one case and a corresponding description for the other case by suitable alternations of the units. Obviously the dynamical similarity may be attended if the forces acting at points of similar position in the two flow fields on volume elements at these points have the same ratio. Thus, various laws of similarity will result from this requirement, depending on what kinds of forces are in effect. For instance, if the important forces are only the gravitational force and the inertial forces as shown in Euler's equation (3.57) or (4.15), for dynamical similarity, the ratio of the inertial force to the gravitational force must be the same for all points in the flow field. We now inquire as to the variations of these forces with the variations in the characteristic quantities of the flow fields: the free stream velocity U, the characteristic length L, the density of the fluid ρ, and the gravitational acceleration g. We have

$$\frac{\text{inertial force}}{\text{gravitational force}} = \frac{\rho U^2 L^2}{\rho g L^3} = \frac{U^2}{gL} = \frac{U^2}{v_g^2} = F_r \qquad (4.28)$$

where F_r is known as the Froude number which is a measure of the effect of the gravitational force in the flow field. It is especially important when there is a free surface such as the flow over a ship on a water surface.

In Reynold's original experiment[11], his main interest was in the flow of a viscous fluid in which the inertial and the viscous forces are the main forces. The condition for dynamical similarity in his case is the ratio of the inertial force to the viscous force, i.e.,

$$\frac{\text{inertial force}}{\text{viscous force}} = \frac{\rho U^2 L^2}{\mu UL} = \frac{\rho UL}{\mu} = \frac{UL}{v_g} = R_e \qquad (4.29)$$

where R_e is known as the Reynolds number which is the most important parameter for the flow of a viscous fluid.

In order to get better informations, we must investigate the differential equations which govern the flows by inspection analysis. Now we consider two special cases to illustrate the principle of inspection analysis:

(1) Euler's equations for an isentropic flow. The fundamental equations are:

$$\frac{\partial u_i}{\partial t} + u_j \frac{\partial u_i}{\partial x_j} = g_i - \frac{1}{\rho} \frac{\partial p}{\partial x_i} = g_i - \frac{a^2}{\rho} \frac{\partial \rho}{\partial x_i} \qquad (4.30a)$$

$$\frac{\partial \rho}{\partial t} + u_j \frac{\partial \rho}{\partial x_j} + \rho \frac{\partial u_i}{\partial x_i} = 0 \qquad (4.30b)$$

$$\frac{p}{p_0} = \left(\frac{\rho}{\rho_0}\right)^\gamma \qquad (4.30c)$$

where $a^2 = dp/d\rho = \gamma\, p/\rho$ and subscript 0 refers to some reference values.

Now we consider two flow fields with geometrically similar boundaries in which a length L may characterize the size of the flow field. If L_1 is the characteristic length of the first flow field and L_2, the corresponding length in the second, we may put $L_2 = c_1 L_1$. This same relation holds between all other pairs of corresponding points in the two fields. Furthermore, there are definite relations between all other physical quantities of these two flow fields. In fact, we may write

$$x_{i2} = c_1 x_{i1}, \quad t_2 = c_2 t_1, \quad u_{i2} = c_3 u_{i1}, \quad \rho_2 = c_4 \rho_1,$$

$$p_2 = c_5 p_1, \quad g_{i2} = c_6 g_{i1}, \quad a_2 = c_7 a_1$$

where c_i's are the proportional constants. We express the magnitudes of the physical quantities of the second flow field in terms of those of the first flow field. Now the first term on the left hand side of Eq. (4.30a) gives

$$\frac{\partial u_{i_2}}{\partial t_2'} = \frac{c_3}{c_2} \frac{\partial u_{i_1}}{\partial t_1} \qquad (4.31\text{a})$$

The second term on the left hand side of Eq. (4.30a) gives

$$u_{j_2} \frac{\partial u_{i_2}}{\partial x_{j_2}} = \frac{c_3^2}{c_1} u_{j_1} \frac{\partial u_{i_1}}{\partial x_{j_1}} \qquad (4.31\text{b})$$

The first term on the right hand side of Eq. (4.30a) gives

$$g_{i_2} = c_6 g_{i_1} \qquad (4.31\text{c})$$

The last term on the right hand side of Eq. (4.30a) gives

$$\frac{a_2^2}{\rho_2} \frac{\partial \rho_2}{\partial x_{i_2}} = \frac{c_7^2 a_1^2}{c_1 \rho_1} \frac{\partial \rho_1}{\partial x_{i_1}} \qquad (4.31\text{d})$$

The last term on the right hand side of Eq. (4.30a) may also be written as

$$\frac{1}{\rho_2} \frac{\partial p_2}{\partial x_{i_2}} = \frac{c_5}{c_1 c_4} \frac{1}{\rho_1} \frac{\partial p_1}{\partial x_{i_1}} \qquad (4.31\text{e})$$

In order that the equations of motion (4.30a) for the two flow fields may be identical, the factors for all the terms in Eq. (4.31) must be all equal, i.e.,

$$\frac{c_3}{c_2} = \frac{c_3^2}{c_1} = c_6 = \frac{c_7^2}{c_1} = \frac{c_5}{c_1 c_4} \qquad (4.32)$$

Eq. (4.32) gives various nondimensional parameters which should be equal in order to get dynamical similarity and which characterize the present flow problem. The equality of the first two terms of Eq. (4.32) means

$$\frac{U_2 t_2}{L_2} = \frac{U_1 t_1}{L_1} = \frac{Ut}{L} = R_t = \text{time parameter} \qquad (4.33)$$

where U is the representative velocity of the flow field. The parameter R_t, which may be called the time parameter, characterizes the unsteady property of the flow field. $R_f = 1/R_t$ is sometime used, which may be called the frequency parameter. When R_t of the two flow fields are equal, the properties of unsteadiness of the two flow fields are the same. When R_t is very small, there will be high frequency phenomena in the flow field. Ordinarily, we take R_t of the order of unity in many flow problems.

The equality of the second and the third terms of Eq. (4.32) gives

$$\frac{U_2^2}{g_2 L_2} = \frac{U_1^2}{g_1 L_1} = \frac{U^2}{gL} = \frac{U^2}{v_g^2} = F_r = \text{Froude number} \qquad (4.34)$$

Eq. (4.34) gives the Froude number which has been discussed in Eq. (4.28), and which shows that if the Froude numbers of the two flow fields are equal the effects of the gravitational forces with respect to inertial forces are the same.

The equality of the second and the fourth terms of Eq. (4.32) gives

$$\frac{U_2}{a_2} = \frac{U_1}{a_1} = \frac{U}{a} = M = \text{Mach number} \qquad (4.35)$$

where M is known as the Mach number which is a measure of the compressibility of the fluid due to the velocity of the flow.

The equality of the second and the fifth terms of Eq. (4.32) gives

$$\frac{p_2}{\rho_2 U_2^2} = \frac{p_1}{\rho_1 U_1^2} = \frac{1}{2} \frac{p}{\frac{1}{2} \rho U^2} = \frac{1}{2} C_p \qquad (4.36)$$

where C_p is known as the pressure coefficient of the flow which is the commonly used nondimensional expression for the pressure p.

We find that there are four nondimensional parameters R_t, F_r, M, and C_p which characterize the flow field of Euler's equations under isentropic process. Since there are seven physical quantities (c_1 to c_7), there are only four independent nondimensional parameters in the present problem as we shall show in the next section. We may form other nondimensional parameters but they are simply combinations of these four parameters.

From Eq. (4.30b), we will not find any new parameter. From Eq. (4.30c), we have a new nondimensional parameter γ which is the ratio of specific heats and which characterizes the relative internal complexity of the molecules as we have shown in Eq. (1.41).

(2) Navier-Stokes equations for an incompressible and viscous fluid. The fundamental equations for this case are Eq. (3.107), i.e.,

$$\frac{\partial u_i}{\partial x_i} = 0 \qquad (4.37a)$$

$$\frac{\partial u_i}{\partial t} + u_j \frac{\partial u_i}{\partial x_j} = -\frac{1}{\rho}\frac{\partial p}{\partial x_i} + \frac{\mu}{\rho}\frac{\partial^2 u_i}{\partial x_j^2} \qquad (4.37b)$$

The only new term of Eq. (4.37) from those in Eq. (4.30) is the viscous force. We may write $\mu_2 = c_8\mu_1$. The relation between the two viscous forces of the two flow fields is

$$\frac{\mu_2}{\rho_2}\frac{\partial^2 u_{i_2}}{\partial x_{j_2}^2} = \frac{c_8}{c_4}\frac{c_3}{c_1^2}\frac{\mu_1}{\rho_1}\frac{\partial^2 u_{i_1}}{\partial x_{j_1}^2} \qquad (4.38)$$

We may add the factor $c_8 c_3/(c_4 c_1^2)$ in Eq. (4.32). The equality of $c_3^2/c_1 = c_8 c_3/(c_4 c_1^2)$ gives

$$\frac{\rho_2 U_2 L_2}{\mu_2} = \frac{\rho_1 U_1 L_1}{\mu_1} = \frac{\rho\, UL}{\mu} = R_e = \text{Reynolds number} \qquad (4.39)$$

For the flow of a viscous fluid, the Reynolds number is one of the most important parameters.

5. Dimensional Analysis. π-Theorem

Instead of inspecting the fundamental equations, we may obtain the important parameters from the dimensional analysis of the physical quantities involved in a given flow problem. The basic principle is the homogeneity of dimensions in any equation of a physical problem as we have mentioned in section 2. The basic theorem upon which the application of dimensional analysis for finding the nondimensional quantities characterizing the dynamic similarity of the flow field rests is known as the π-theorem. This important theorem may be stated as follows:[3]

If there are n variables, Q_1, Q_2, $\cdots$, Q_n which are important in a physical problem and there are m independent fundamental units in this system, the complete equation for this problem is

$$\phi(Q_1, Q_2, \cdots, Q_n) = 0 \qquad (4.40)$$

the final solution of the problem may be written in the form:

$$f(\pi_1, \pi_2, \cdots, \pi_{n-m}) = 0 \qquad (4.41)$$

where $\pi_1, \pi_2, \cdots, \pi_{n-m}$ are nondimensional parameters formed by those Q_i's.

Eq. (4.41) may also be written as

$$\pi_1 = F(\pi_2, \pi_3, \cdots, \pi_{n-m}) \qquad (4.42)$$

It should be noted that information as to the nature of the function $f(X)$ or $F(X)$ can naturally be obtained only through experiments or actual theory of the phenomena in question.

The π-theorem may be proved as follows:

The equation (4.40) may be expressed in the form of series of the physical quantities Q_i's as follows:

$$C_1 Q_1^{a_1} Q_2^{a_2} \cdots Q_n^{a_n} + C_2 Q_1^{b_1} Q_2^{b_2} \cdots Q_n^{b_n} + \cdots = 0 \qquad (4.43)$$

where C_i, a_i, and b_i are constants. By dimensional homogeneity, the dimensions of all terms in Eq. (4.43) are the same. In order to make all the terms dimensionless, we divide Eq. (4.43) by $Q_1^{a_1}$, $Q_2^{a_2}$, $\cdots$, $Q_n^{a_n}$ and obtain

$$C_1 + C_2 Q_1^{\alpha_1} Q_2^{\alpha_2} \cdots Q_n^{\alpha_n} + \cdots = 0 \qquad (4.44)$$

where $\alpha_i = b_i - a_i$. Since there are m independent fundamental dimensions, we may express the dimensions of Q_i in terms of these fundamental dimensions. Substituting the dimensions of Q_i in each term of Eq. (4.44), we have m relations between α_i because the powers of m fundamental units must be zero for dimensionless quantities. Since there are only m relations between n unknowns $(n > m)$, we have $n-m$ arbitrary values of α_i, which may be denoted as $\beta_1, \beta_2, \cdots, \beta_{n-m}$. As a result, Eq. (4.44) may be written as:

$$C_1 + C_2 Q_1^{\alpha_1} Q_2^{\alpha_2} Q_3^{\alpha_3} \cdots Q_m^{\alpha_m} \pi_1^{\beta_1} \pi_2^{\beta_2} \cdots \pi_{n-m}^{\beta_{n-m}} + \cdots = 0 \qquad (4.45)$$

Eq. (4.45) may be written in the form of Eq. (4.41), because the values of $\beta_1, \beta_2, \cdots, \beta_{n-m}$ are still arbitrary constants. The values of $\alpha_1, \alpha_2, \cdots, \alpha_m$ in Eq. (4.45) are known in a given problem.

Thus we prove the π-theorem.

It is clearer if we treat a simple problem by the means of π-theorem. We shall consider the case of classical hydrodynamics in which there are only three fundamental dimensions which are the mass m, the length L, and the time t. If there are five physical quantities in a given problem which we are going to study, we call them Q_1, Q_2, Q_3, Q_4, and Q_5. We take Q_1, Q_2, and Q_3 as the fundamental units and express Q_4 and Q_5 in terms of Q_1, Q_2, and Q_3 in the forms of π_1 and π_2. We write

$$Q_4 = \text{constant } Q_1^{\alpha_1} Q_2^{\alpha_2} Q_3^{\alpha_3} \qquad (4.46)$$

where α_1, α_2, and α_3 are integers which are determined by the conditions of dimensional homogeneity as follows:

Let the exponents of the length L, time t, and mass m in the dimensional expressions for Q_1, Q_2, Q_3, and Q_4 be given below:

	L	t	m
Q_1	n_1	n_2	n_3
Q_2	p_1	p_2	p_3
Q_3	q_1	q_2	q_3
Q_4	r_1	r_2	r_3

Because the dimensions in L, m, and t on both sides of Eq. (4.46) must be the same, we have

$$\left. \begin{array}{l} r_1 = \alpha_1 n_1 + \alpha_2 p_1 + \alpha_3 q_1 \\ r_2 = \alpha_1 n_2 + \alpha_2 p_2 + \alpha_3 q_2 \\ r_3 = \alpha_1 n_3 + \alpha_2 p_3 + \alpha_3 q_3 \end{array} \right\} \qquad (4.47)$$

We may solve α_1, α_2, and α_3 from Eq. (4.47). Hence we have the nondimensional quantity π_1 as follows:

$$\pi_1 = \frac{Q_4}{Q_1^{\alpha_1} Q_2^{\alpha_2} Q_3^{\alpha_3}} \qquad (4.48)$$

Similarly, we have the second nondimensional quantity π_2 as follows:

$$\pi_2 = \frac{Q_5}{Q_1^{\alpha_4} Q_2^{\alpha_5} Q_3^{\alpha_6}} \qquad (4.49)$$

where α_4, α_5, and α_6 are also integers.

By the π-theorem (4.42), we have

$$\pi_1 = F(\pi_2) \tag{4.50}$$

Let us apply the π-theorem to the isentropic flow governed by the Euler's equations discussed before. In this case, we have seven important physical quantities as follows:

$Q_1 = L =$ typical length of a body in the flow field,
$Q_2 = U =$ typical velocity of the flow,
$Q_3 = \rho =$ density of the fluid,
$Q_4 = p =$ pressure of the fluid,
$Q_5 = t =$ typical time scale of the flow field,
$Q_6 = g =$ gravitational acceleration,
$Q_7 = a =$ sound speed of the fluid.

Since there are seven Q_i's and three fundamental units, we may form four nondimensional quantities π_1 to π_4. Now we take the first three Q_i's as the fundamental units. The dimensions of the first four variables are

	L	t	m
L	1	0	0
U	1	-1	0
ρ	-3	0	1
p	-1	-2	1

Eq. (4.47) for the present case becomes

$$-1 = \alpha_1 + \alpha_2 - 3\alpha_3$$

$$-2 = -\alpha_2$$

$$1 = \alpha_3$$

or

$$\alpha_1 = 0, \ \alpha_2 = 2, \ \alpha_3 = 1$$

or

$$\pi_1 = \frac{p}{\rho U^2} = \frac{1}{2} C_p$$

Similarly for $Q_5 = t$, we have

$$\pi_2 = \frac{Ut}{L} = R_t$$

For $Q_6 = g$, we have

$$\pi_3 = \frac{U^2}{gL} = F_r$$

For $Q_7 = a$, we have

$$\pi_4 = \frac{U}{a} = M$$

The π-theorem gives

$$\pi_1 = F(\pi_2, \ \pi_3, \ \pi_4)$$

or

$$C_P = f(R_t, \ F_r, \ M) \tag{4.51}$$

We have that the pressure coefficient C_p on the body in the flow field will be a function of the Mach number M, Froude number F_r, and time parameter R_t.

Thus we see that the dimensional analysis lacks the pictorial quality of the dynamical similarity consideration of last section, but it has the advantages of not using the knowledge of the equations governing the problem. In many very complicated problems, we may not have the correct fundamental equations available but we must know the important physical quantities in such problems. Thus we may find the important parameters from these physical quantities first. After we know the important parameters and their range of values, we may formulate the proper theory and then write down the fundamental equations.

6.   Important Nondimensional Quantities in Fluid Mechanics

In fluid mechanics, we desire to find the forces, moments, heat transfers, and other physical quantities in the flow field under certain given conditions. It is usually desirable to express these quantities in nondimensional coefficients so that the results may be applied to other geometrically and dynamically similar systems. The applications to geometrically similar systems are self-evident. But for the applications to dynamically similar systems, we have to find out the relations of these nondimensional coefficients with those important nondimensional parameters which characterize the dynamical similarity of the important terms for a given problem. It is one of the main objects of fluid mechanics to find such relations. We shall give the definitions of those nondimensional coefficients and nondimensional parameters which occur in various problems of fluid mechanics in this section.

(A) Classical Gasdynamics

First we consider the flow problems of a viscous, heatconducting, nonelectrically conducting and compressible fluid without any body force. These are problems treated in ordinary gasdynamics textbook.[8] In these problems, we would like to know the force, pressure, and moment on a body in the flow field and the heat transfer rate on the body. Hence the following nondimensional coefficients are often used:

(1) Force coefficient C_F. The nondimensional coefficient of the force F on a body may be expressed as follows:

$$C_F = \frac{F}{\frac{1}{2}\rho_0 U^2 L^2} \tag{4.52}$$

The force F may be lift, drag, thrust, skin friction, or other force. In problems where there is a uniform free stream of velocity U, we may take this velocity as our reference velocity U in Eq. (4.52). In problems where the essential motion is rotational, we may take $U = \Omega L$, where Ω is the angular velocity. The value of L is the characteristic length of the system. It is often to use certain area S to replace L^2 in Eq. (4.52). For instance, if we calculate the lift coefficient of an airplane, we may use the wing area to replace L^2. The value of ρ_0 is a certain reference density of the fluid. For incompressible fluid, the density is a constant which may be used in Eq. (4.52). For a compressible fluid, the density is not a constant in the whole flow field. Hence we should use some certain known density in the problem as ρ_0. For instance, we may use the density of the free stream as ρ_0.

Eq. (4.52) is used for most engineering problems. It is particularly good for problems of large Reynolds numbers where the inertial force is much larger than the viscous force. In certain problems where the viscous force is larger than the inertial force, that is for the case of small Reynolds numbers, the proper nondimensional coefficient of force will be

$$C'_F = \frac{F}{\mu UL} \tag{4.53}$$

where μ is certain reference value of the coefficient of viscosity of the fluid. It is used only for very slow motion problems such as in lubrication problems. The relation between these two coefficients of force is

$$C_F = 2C'_F \frac{\mu}{\rho_0 UL} = \frac{2C'_F}{R_e} \tag{4.54}$$

where R_e is the Reynolds number of the flow field.

(2) Moment coefficient C_M. The nondimensional coefficient of moment M_0 may be expressed as follows:

$$C_M = \frac{M_0}{\frac{1}{2}\rho_0 U^2 SL} \tag{4.55}$$

The quantities U, L, S, and ρ_0 are similar to those discussed for C_F.

(3) Pressure coefficient C_p. The nondimensional coefficient of pressure p is

$$C_p = \frac{p}{\frac{1}{2}\rho_0 U^2} \tag{4.56}$$

(4) Coefficient of heat transfer. In ordinary gasdynamics, the exchange of heat between the fluid and solid surface is mainly due to heat conduction. Hence ordinarily the coefficient of heat transfer is associated with heat conduction. For engineering practice, we often use a film coefficient $h(x)$ to express the heat transfer rate:

$$h(x) = \frac{q(x)}{T_w - T_{wi}} \tag{4.57}$$

where $q(x)$ is the quantity of heat transferred through unit area per unit time which is in general a function of position x along the solid surface. T_w is the temperature of the surface of the solid wall considered. T_{wi} is the temperature of the solid surface in the case of no heat transfer, i.e., for insulated wall. Some authors use other reference temperature T_r of the fluid such as the free stream temperature T_∞ instead of T_{wi} in Eq. (4.57) to define the film coefficient. By heat conduction Eq. (1.55), the magnitude of $q(x)$ is

$$q(x) = -\kappa \left(\frac{\partial T}{\partial n} \right)_{n=0} \tag{4.58}$$

where n is the normal to the surface of the solid wall. From Eq. (4.57) and Eq. (4.58), we may define a nondimensional coefficient of heat transfer which is known as the Nusselt number as follows:

$$N_u = \frac{h(x) \cdot L}{\kappa} = \frac{L}{T_{wi} - T_w} \left(\frac{\partial T}{\partial n} \right)_{n=0} \tag{4.59}$$

where L is a characteristic length of the problem.

In general these nondimensional coefficients are functions of some nondimensional parameters. The following are some important nondimensional parameters in fluid dynamics.

(1) Mach number M. Mach number M is a measure of the compressibility of the fluid due to high speed of the flow and is defined as

$$M = \frac{\text{velocity}}{\text{speed of sound}} = \frac{U}{a} \tag{4.60}$$

where sound speed a for a perfect gas is $(\gamma R T)^{1/2}$. It is easy to show that the ratio of the variation of density of a gas to the variation of flow velocity is, to a first approximation, proportional to the square of the Mach number of the flow. Hence for very small Mach number, the variation of density of a gas in the flow field is negligible. For large Mach number, the effect of compressibility, i.e., the variation of density due to the variation of velocity, is important. When M is smaller than unity, the flow is said to be subsonic and when M is larger than unity, the flow is supersonic. The flow field of a subsonic flow differs greatly from that of a supersonic flow. For instance, for a steady flow of an inviscid fluid, the fundamental equation for the flow field is elliptical for subsonic flow while that for supersonic flow is hyperbolic. Because of the variation of local Mach number in the flow field, the flow may be subsonic in one part of the flow field and supersonic in another part of the flow field with a transonic flow field ($M \cong 1$) between them. The fundamental equation is of a mixed type which changes from hyperbolic type in the supersonic part to the elliptical type in the subsonic part and with the parabolic type at $M = 1$. Shock wave occurs only in the supersonic flow field for the steady case but never in the corresponding subsonic case. When Mach number M is of the order of unity, the flow is considered as transonic while Mach number M is much larger than unity, the flow is considered as hypersonic. Many new features of the flow field in the regions of transonic flow and hypersonic flow differ from those of ordinary subsonic and supersonic flows.[8]

The sound speed of a gas, i.e., the speed of propagation of infinitesimal wave in an ideal gas, will be affected by other physical phenomena, such as chemical reactions (chapter V), electromagnetic fields (chapter VI), and thermal radiation (chapter VII). Sometimes, the magnitude of the sound speed will be changed by these effects and sometimes new modes of sound waves will be introduced by these effects and new phenomena occur. We shall discuss them in detail later.

(2) The ratio of specific heats γ. This is the ratio of specific heat at constant pressure to that at constant volume [see Eq. (1.41)]. It is a measure of the relative complexity of the molecules of a gas and has been discussed in chapter I section 11.

(3) Reynolds number R_e.

$$R_e = \frac{\text{inertial force}}{\text{viscous force}} = \frac{\rho\, UL}{\mu} = \frac{UL}{\nu_g} \tag{4.61}$$

This is the most important parameter for the fluid dynamics of a viscous fluid. When the Reynolds number is small, the viscous force is predominant and the effect of viscosity is important in the whole flow field.   When the Reynolds number is large, the inertial force is predominant and the effect of viscosity is important only in a narrow region near a solid boundary or other restricted region which is known as the boundary layer region or transition region. Outside these transition or boundary layer regions, the flow may be considered as inviscid.   If the Reynolds number is enormously large, the flow becomes turbulent.

The value of the coefficient of kinematic viscosity $\nu_g = \mu/\rho$ shows the effects of viscosity of a fluid.   If other things are the same, the smaller the value of the kinematic viscosity is, the narrower the region which is affected by viscosity will be.   This viscous region is generally known as the boundary layer of the velocity field when the kinematic viscosity is very small.   The thickness of the boundary layer of the velocity field of a laminar flow is proportional to the square root of the kinematic viscosity ν_g.   Thus the kinematic viscosity shows the momentum diffusivity due to the viscous effect.

The property of the boundary layer may be affected by other physical phenomena so that the thickness of the boundary layer may be changed or a new boundary layer or other transition region may occur in the flow field, which will be discussed in later chapters.

(4)   Prandtl number P_r.

$$P_r = \frac{\text{kinematic viscosity}}{\text{thermal diffusivity}} = \frac{\nu_g}{\kappa/(c_p\rho)} = \frac{c_p\mu}{\kappa} \qquad (4.62)$$

The value of $\kappa/(c_p\rho)$ shows the thermal diffusivity due to the heat conductivity.   The smaller the value of thermal diffusivity is, the narrower will be the region which is affected by the heat conductivity and which is known as thermal boundary layer when the thermal diffusivity $\kappa/(c_p\rho)$ is very small.   Thus the thickness of the thermal boundary layer is proportional to the square root of thermal diffusivity $[\kappa/(c_p\rho)]^{1/2}$.   The Prandtl number P_r shows the relative importance of the heat conductivity and viscosity of a fluid.   As a first approximation, the Prandtl number gives the square of the ratio of the thickness of the velocity boundary layer to that of the thermal boundary layer.   The Prandtl number depends only on the physical properties of the fluid but not on the flow conditions such as the velocity and/or the size of the flow field. For ordinary gas, the Prandtl number is of the order of unity and hence the thickness of the velocity boundary layer is on the same order of magnitude as that of the thermal boundary layer. If the value of the Prandtl number of a fluid differs greatly from unity, the thickness of a velocity boundary layer will also differ greatly from that of a thermal boundary layer.

Since the physical properties of a fluid may be affected by other physical phenomena, the Prandtl number may be changed by these physical phenomena. For instance, the thermal diffusivity of a gas depends on both thermal conductivity and thermal radiation. If we include the thermal radiation effects (chapter VII) in the flow field, we should use an effective Prandtl number in our analysis instead of that defined by Eq. (4.62) when we study the thermal boundary layer of a radiating gas. In some physical problems, such as magnetofluid dynamics, we may obtain a certain physical quantity which has a similar influence as an ordinary Prandtl number and which may be called a magnetic Prandtl number or another suitable name as we will discuss later.

(5)   The time or frequency parameter R_t.

$$R_t = \frac{U t_0}{L} = \frac{U}{Lf} = \frac{1}{R_f} \tag{4.63}$$

This parameter characterizes the time scale of the flow field with respect to the flow velocity and the dimension of the flow field. The quantity (L/U) may be considered as a characteristic time of the flow field. In the steady flow problem, the typical time t_0 is infinite and R_t is infinite so that the unsteady term $(\partial Q/\partial t)$ is zero. However for convective terms the value of R_t should be taken as unity.

In many unsteady flow problems, we may use a typical frequency $f = 1/t_0$ to represent the time scale and we may use the reduced frequency R_f to replace the time parameter R_t. For high frequency phenomena such as in a flutter problem or a high frequency wave motion, we have a large value of reduced frequency R_f and a small value of the time parameter R_t.

(B) *Flow Field Including Gravitational Force*

The second problem which we are going to discuss is the flow field including gravitational force which is important in dynamical meteorology and in the free surface effects on hydrofoils and many other problems. Two new physical phenomena occur in the present case in addition to what we have already discussed in (A). One is that we introduce the gravitational force ρg in our flow field and the other is the bouyancy force of free convection on heated fluid of variable density. Hence we have two new parameters:

(1) Froude number F_r.

$$F_r = \frac{U^2}{gL} \qquad (4.64)$$

The Froude number is important when there is a free surface and it is also important when the variation of the height in the flow field is large, such as in the problem of dynamical meteorology (see chapter III section 6).

(2) Grashoff number G_r.

$$G_r = \frac{L^3 \beta g (T_1 - T_0)}{\nu_g^2} \qquad (4.65)$$

By free convection, we mean flows in which the motion is caused by the effect of gravity on heated fluids of variable density. In free convection, we have a body force

$$F_c = - g\rho\beta(T - T_0) \qquad (4.66)$$

where β is the coefficient of thermal expansion and $T - T_0$ is the excess temperature of the heated parts of the fluid over the parts which remain cold. The nondimensional parameter which characterizes the free convection is the Grashoff number defined by Eq. (4.66), where T_1 and T_0 are two representative temperatures.

(C) *Flow Field Including Chemical Reactions* (*Chapter V*)

The third problem which we are going to discuss in the flow problem including chemical reactions which is important in combustion problems and other high temperature flow such as reentry of space vehicles. This problem is sometimes referred to as aerothermochemistry which was suggested by the late Professor Theodore von Kármán and in which we have to study aerodynamics, thermodynamics, and chemical reactions. Besides the physical quantities of ordinary gasdynamics or aerodynamics discussed in (*A*), we have to study the diffusivity of matter of various species in a mixture in which chemical reaction takes place. We have one new parameter for the diffusion of matter and two new parameters for the change of matter and energy due to chemical reactions.

(1) Schmidt number S_c. The Schmidt number is a measure of the diffusion effect which is defined as:

$$S_c = \frac{\mu}{\rho D} \tag{4.67}$$

where D is the coefficient of diffusion of the fluid [see Eq. (1.72) or (3.39)]. It is interesting to notice that there are at least three German professors named Schmidt who have done important work in diffusion and each of them may be honored by this number.

(2) Damkohler numbers D_{am_1} and D_{am_2}. According to Damkohler, the following two nondimensional parameters are important in aerothermochemistry. For the change of matter, we use the first nondimensional parameter of Damkohler which is defined as

$$D_{am_1} = \frac{U_i L}{U} \tag{4.68}$$

where L/U may be called the residence time here while $1/U_i$ is the chemical time defined by the equation:

$$U_i = \text{reaction frequency} = \frac{1}{\rho k_i} \frac{d}{dt} (\rho k_i) \tag{4.69}$$

where k_i is the mass concentration of ith component in the mixture. The residence time is the time available for chemical reaction. We may take L as the length of a combustion chamber and U as the average velocity of the combustible mixture. The residence time is then the time at which the mixture is in the combustion chamber.

We may also choose L as the mean free path of the gas molecule and U as the average molecular velocity. In this case L/U is the collision time of the molecules. Hence the chemical reactions depend on the residence time, the collision time, and the chemical time.

For the change of energy, we use the second nondimensional parameter of Damkohler which is defined as

$$D_{am_2} = \frac{Q}{c_p T_0} \tag{4.70}$$

where Q is the reaction heat and $c_p T_0$ is the initial enthalpy of the flow field before the chemical reaction takes place.

(D) Magnetofluid Dynamics (Chapter VI)

The fourth problem which we are going to discuss is the flow problem of an electrically conducting fluid. Besides ordinary gas-dynamic forces, we have to consider electromagnetic forces. The electromagnetic properties of the medium will introduce many new parameters. Here we shall discuss two new parameters which are important in magnetofluid dynamics. The other parameters of magnetofluid dynamics will be discussed in chapters VI and IX.

(1) Magnetic pressure number R_H and magnetic Mach number M_m.

$$R_H = \frac{\text{magnetic pressure}}{\text{dynamic pressure}} = \frac{\frac{1}{2}\mu_e H^2}{\frac{1}{2}\rho_0 U^2} = \frac{V_H^2}{U^2} = \frac{1}{M_m^2} \tag{4.71}$$

The magnetic pressure number is the ratio of the magnetic pressure to the dynamic pressure. Only when R_H is of the order of unity or larger, will the fluid flow be affected noticeably by the magnetic field. If R_H is much smaller than unity, the terms due to the magnetic field in both the equations of motion and the equation of energy may be neglected and the gas motion will not be affected noticeably by the magnetic field.

The magnetic Mach number is the ratio of the flow velocity to the speed of the Alfven wave, i.e., $M_m = U/V_H$, and it is closely related to the magnetic pressure number by Eq. (4.71). Since the

Alfven's wave is one of the basic wave motions in a plasma, it will affect the effective sound speed in a plasma and it will introduce new modes of sound wave in a plasma. The flow pattern of a plasma will differ when M_m is greater or less than unity as we shall see in chapter VI. When M_m is larger than unity, the flow is said to be super-Alfven flow and when M_m is smaller than unity, the flow is said to be sub-Alfven flow.

(2) Magnetic Reynolds number R_σ.

$$R_\sigma = \mu_e \sigma_e UL = \frac{UL}{\nu_H} \tag{4.72}$$

where σ_e is the electrical conductivity of the fluid and $\nu_H = 1/(\mu_e \sigma_e)$ is known as magnetic diffusivity. The magnetic Reynolds number R_σ determines the diffusion phenomena of the magnetic field along streamlines in a similar manner as the ordinary Reynolds number determines the diffusion phenomena of vorticity along streamlines. Hence the magnetic Reynolds number determines the effect of the flow field on the magnetic field. If R_σ is negligibly small, the magnetic field is practically unaffected by the flow field. On the other hand, if R_σ is very large, the magnetic field will stay with the flow, the so-called frozen-in fields, and it will be greatly influenced by the motion of the fluid. When R_σ is very large, we have the magnetic boundary layer and the thickness of the magnetic boundary layer is proportional to the square root of magnetic diffusivity $(\nu_H)^{1/2}$.

(E) Radiation Gasdynamics (Chapter VII)

The fifth problem which we are going to discuss is the flow problem including the effects of thermal radiation. When the temperature of a gas is very high and its density is very low, thermal radiation has a significant influence on the flow field. We have to study simultaneously the gasdynamic field and radiation field and the term Radiation Gasdynamics has been used for this new branch of fluid dynamics, which will be discussed in chapter VII. The following are three new parameters for radiation gasdynamics:

(1) Radiation pressure number R_p.

$$R_p = \frac{\text{radiation pressure}}{\text{gas pressure}} = \frac{a_R T^4}{3p} \qquad (4.73)$$

where $a_R = 7.67 \times 10^{-15}$ erg·cm^{-3}·°K^{-4} is known as the Stefan-Boltzmann constant. Since a_R is an extremely small number, the radiation pressure $p_R = a_R T^4/3$ becomes comparable with the gas pressure p only when the temperature is very large. Fig. 4.1 shows the values of R_p for various pressures and temperatures. When R_p is not negligibly small, we have to consider the radiation energy density and radiation pressure and radiation stresses in the flow field.

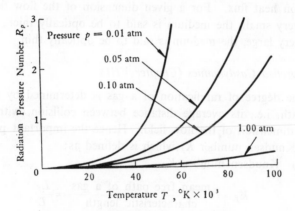

Figure 4.1   Radiation pressure number

(2)   Relativistic parameter R_r.

$$R_r = \frac{\text{flow velocity}}{\text{velocity of light}} = \frac{U}{c} \qquad (4.74)$$

Since the thermal radiation effect is due to the motion of photons at a speed of light c, the relativistic parameter plays an important role in the heat flux of radiation. In fact, the radiative heat flux in comparison with the convective heat flux is proportional to R_p and inversely proportional to R_r. Usually both R_p and R_r are extremely small quantities. Hence even when R_p is negligibly small, the radiative heat flux may not be negligible. We shall discuss this point in detail later.

(3)   Knudsen number of radiation K_r.

$$K_r = \frac{\text{mean free path of radiation}}{\text{characteristic length of flow}} = \frac{L_R}{L} = L_R^* = \frac{1}{\rho K_R L} \quad (4.75)$$

where ρK_R is certain average linear absorption coefficient of radiation of the medium and $L_R = 1/(\rho K_R)$ is the mean free path of radiation which represents the average distance traveled by photons before they are absorbed by the gas particles. We shall discuss these terms in detail in chapter VII. The mean free path of radiation is one of the most important physical quantities to determine the radiation heat flux. For a given dimension of the flow field L, if K_r is very small, the medium is said to be optically thick, while if K_r is very large, the medium is said to be optically thin.

(F)   *Rarefied Gasdynamics (Chapter VIII)*

The degree of rarefication of a gas is determined by its mean free path, i.e., its average distance between collision, with respect to the dimension of the flow field. Hence the important parameter is the Knudsen number K_f which is defined as:

(1)   Knudsen number K_f.

$$K_f = \frac{\text{mean free path of a gas}}{\text{characteristic length}} = \frac{L_f}{L} \quad (4.76)$$

If the Knudsen number K_f is not small, the discrete character of the gas must be taken into account. For the large Knudsen number, we have the slip flows, free molecule flows and other special transition flows which will be discussed in chapter VIII. In most of the flow problems discussed in this book, except those in chapter VIII, we consider the cases that the Knudsen number is very small so that the gas or plasma may be considered as a continuum. As we shall show later, there are many similarities of the gasdynamic terms and radiation terms according to the values of two Knudsen numbers K_f and K_r. For small Knudsen numbers, the radiative heat transfer terms and the gasdynamic transfer terms such as viscous terms, heat conduction terms may be expressed in terms of gradients of local variables, such as the Rosseland approximation for radiative heat flux, Navier-Stokes relations for viscous stresses, and the Fourier relation for heat conduction. When the Knudsen numbers are large, the radiative as well as the gasdynamic transfer terms should be expressed in terms of integral forms or other complicated forms such as some types of differential equations as we shall discuss later.

(G) *Combinations of Parameters of Fluid Mechanics*

For a given problem of fluid mechanics, there are a definite number of important parameters which characterize the flow field. But these parameters are not unique. We may obtain other important parameters from various combinations of these parameters. These new parameters may be more appropriate for special problems than those described above. We are going to discuss a few of such important parameters below:

(1) Peclet number P_e.

$$P_e = \frac{UL}{\kappa/(c_p\rho)} = P_r R_e \qquad (4.77)$$

The Peclet number which is the product of the Prandtl number and the Reynolds number plays the same role in the thermal boundary layer as the Reynolds number in the velocity boundary layer.

(2) Stanton number S_t.

$$S_t = \frac{h(x)}{Uc_p\rho} = \frac{N_u}{P_e} \qquad (4.78)$$

The Stanton number is another way to express the heat transfer rate which may be considered as the total heat transfer to a typical convective heat transfer.

(3) Lewis-Semenov number L_e.

$$L_e = \frac{\kappa}{\rho c_p D} = \frac{S_c}{P_r} \qquad (4.79)$$

Lewis-Semenov number shows the ratio of thermal diffusivity to matter diffusivity. In the U.S. this number is usually known as the Lewis number, while in the U.S.S.R. it is known as the Semenov number. We shall call it the Lewis-Semenov number which is the ratio of the Schmidt number and the Prandtl number. These three numbers, S_c, P_r, and L_e represent the relative importance between momentum diffusivity ν_g, thermal diffusivity $\kappa/(c_p\rho)$, and matter diffusivity D.

(4) Magnetic Prandtl number P_m.

$$P_m = \frac{\mu\mu_e\sigma_e}{\rho} = \frac{\nu_g}{\nu_H} = \frac{R_\sigma}{R_e} \qquad (4.80)$$

The magnetic Prandtl number is a measure of the relative magnitude of the thickness of the velocity boundary layer to that of the magnetic boundary layer. Hence the physical significance of P_m is similar to that of the Prandtl number P_r.

(5) Magnetic number R_m.

$$R_m = \left(\frac{\text{magnetic force}}{\text{inertial force}}\right)^{\frac{1}{2}} = B\left(\frac{\sigma_e L}{\rho U}\right)^{\frac{1}{2}} = (R_H R_\sigma)^{\frac{1}{2}} \qquad (4.81)$$

When the magnetic Reynolds number R_σ is very large, we use R_H to show the effect on the flow field by a magnetic field because at a large magnetic Reynolds number, the electric current is determined essentially by the magnetic field. When the magnetic Reynolds number is small, we should use R_m to show the effect on the flow field by a magnetic field because at a small magnetic Reynolds number, the electric current depends greatly on the electrical conductivity and the magnetic field.

(6) Hartmann number R_h.

$$R_h = \frac{\text{magnetic force}}{\text{viscous force}} = BL\left(\frac{\sigma_e}{\mu}\right)^{\frac{1}{2}} = (R_e R_H R_\sigma)^{\frac{1}{2}} \qquad (4.82)$$

This number R_h was first used by Hartmann in his classical experiment of channel flow of magnetohydrodynamics in which the important forces are the magnetic force and the viscous force.[6]

(7) Knudsen number K_f. The mean free path of a gas L_f may be taken as follows:

$$L_f = 1.255\sqrt{\gamma}\,\frac{\nu_g}{a} \qquad (4.83)$$

With the expression of Eq. (4.83), the Knudsen number K_f of Eq. (4.76) may be written as follows:

$$K_f = \frac{L_f}{L} = 1.255\sqrt{\gamma}\ \frac{M}{R_e} \tag{4.84}$$

The Knudsen number will be large for a high Mach number M and a low Reynolds number.

(8) Radiative flux number R_F.

$$R_F = \frac{\text{radiative heat flux}}{\text{heat conduction flux}} \tag{4.85}$$

The exact expression of radiative heat flux depends on the optical thickness of the medium as we shall discuss in chapter VII. For an optically thick medium, we have:

$$\text{radiative heat flux} = 4D_R a_R T^3 \nabla T = \kappa_R \nabla T \tag{4.86}$$

where $D_R = cL_R/3$ is the Rosseland diffusion coefficient of radiation (see chapter VII) and then the radiative flux number becomes

$$R_{F_1} = \frac{\kappa_R}{\kappa} = 4\left(\frac{\gamma-1}{\gamma}\right) P_e K_r R_p / R_r \tag{4.87}$$

For an optically thin medium, the radiative heat flux is proportional to $ca_R T^4/K_r$ and then the radiative flux number becomes

$$R_{F_2} = \frac{ca_R T^4 L}{K_r \kappa T} = 3\left(\frac{\gamma-1}{\gamma}\right) P_e\ R_p / (K_r\ R_r) \tag{4.88}$$

The main difference between R_{F_1} and R_{F_2} lies in the parameter Knudsen number K_r. For a small Knudsen number of radiation, R_{F_1} is proportional to K_r but for a large Knudsen number of radiation, R_{F_2} is inversely proportional to K_r. The radiative heat flux tends to zero for both a very large and a very small mean free path of radiation.

If we define the radiative flux number as a ratio between the radiative heat flux and convective heat flux, the Peclet number P_e will not occur in the radiative flux number.

7. Modeling and Similitude

In the most general case, the force coefficient, moment coefficient, heat transfer coefficient, etc., are functions of all the nondimensional parameters discussed above. But in most practical cases, there are only a few parameters which are important in a particular problem. For instance, in an incompressible viscous fluid flow without body force, the Reynolds number will be the only important parameter. In an inviscid compressible flow of an ideal gas, the Mach number will be the most important parameter. One of the main objects of fluid mechanics is to determine the relative importance among these nondimensional parameters.

After we know these important parameters, we have to find the functional relations between various coefficients C_F, C_M, N_u, etc., and those important parameters. The second main object of fluid mechanics is to determine these functional relations. We may determine these functional relations either experimentally or theoretically.

In experimental investigations, it seems to be the best and simplest way that one could make a full scale test. For instance, in aircraft design, we may make a free flight test of the aircraft. But usually we do not want to make such a full scale test because of the following reasons: one may want some data before the aircraft is built; the flight conditions of a real aircraft may not be easily controlled; and the cost of such full scale test may be very high. Because of such reasons, many experimental investigations are carried out in a laboratory with a smaller scale model than the full scale object. Now the question is: under what conditions may the results of a small model be applied to the full scale object? From our previous discussions, we see that if all the important parameters of the model are the same as those of the full size object, the nondimensional coefficients should be the same. But it is almost impossible that the important parameters of the model test will all be the same as those of the full scale body. Since the important parameters of a model may differ from those of the full scale body, we have to separate the influence of various parameters and to determine the functional relations for each of these parameters and obtain some overall effects of all the important parameters. This is very important in the application of experimental results.

Let us illustrate this modeling technique by considering the model test of an aircraft in a high speed wind tunnel. There are two types of similitude which should be fulfilled before we can apply the test results of the model to a full scale aircraft:

First is the geometrical similitude, by which we mean that there is a linear scaling between the model and the full scale aircraft. In other words, the results of this model test can be applied to an aircraft of similar shape.

Second is the dynamical similitude, by which we mean that all the important parameters in the model test should be the same as those of the full scale aircraft in flight conditions.

To obtain geometrical similitude is simple but to obtain full dynamical similitude will be difficult. For simplicity, we consider the case of a high subsonic or low supersonic flow of air. In this case, the air may be considered as an ideal gas of constant specific heat. The force coefficient may be written as follows:

$$C_F = f(\alpha, M, P_r, R_e, \gamma) \tag{4.89}$$

where α is the angle of attack of the model, M is the Mach number of the flow, R_e is the Reynolds number of the flow, P_r is the Prandtl number of the air, and γ is the ratio of specific heats of the air. Even in this simple case, it is practically impossible to find the exact functional form of Eq. (4.89). We have to separate various effects of these parameters. The parameters P_r and γ depend on the properties of the fluid and the temperature range. If we use the same fluid in the model test and in the same temperature range as those in the full scale aircraft flying conditions, both P_r and γ will be the same in the model test as those in the full scale aircraft flying conditions. If we use a different gas in the model test than air or we test the model at a different temperature range than that in the free flight conditions, we should consider the possible effects due to different P_r and γ. At present, we assume that P_r and γ are the same in model test and full scale aircraft flying conditions. In the test, we measure the force coefficients at various angles of attack α. Hence to achieve the dynamical similitude, we have to have the same Mach number and the same Reynolds number for the model test as those for the full scale aircraft flight. In general, it is difficult to obtain dynamical similitude in both the Mach number and the Reynolds number simultaneously. For instance, if we adjust the velocity so that the Mach number of the model is the same as that of the full scale aircraft, the Reynolds number of the model will be much smaller than that of the full scale aircraft. On the other hand, if the Reynolds number of the model is increased to the value of the full scale aircraft by increasing the velocity of the flow, the Mach number of the model will be larger than that for the full scale aircraft. Even though we may increase the density of the air so that the Reynolds number of the model test may increase, the enormous pressure due to the increase of density in the high speed wind tunnel may be undesirable.

In practice, we have to separate the force into various parts and find the variation of each part of the force with the Mach number and the Reynolds number. For instance, the force on a wing of an aircraft may be resolved into a lift force which is perpendicular to the direction of the flight and a drag force which is in a horizontal direction but against the flight direction. As long as there is no separation of the boundary layer, the lift coefficient C_L is practically independent of Reynolds number. Hence we have

$$C_L = C_L(\alpha, M) \tag{4.90}$$

We may measure the lift coefficient C_L vs. angle of attack α at various Mach number of the free stream in a high speed wind tunnel.

The drag force has a complicated relation with the Mach number M and the Reynolds number R_e. As a first approximation, we may divide the drag into two parts: one is the pressure drag or wave drag and the other is the skin friction. The pressure drag or wave drag is influenced more by the Mach number than by the Reynolds number and we may write approximately $C_{Dp} = C_{Dp}(\alpha, M)$. The skin friction is influenced more by the Reynolds number than by the Mach number and we may write approximately $C_{Ds} = C_{Ds}(\alpha, R_e)$. As a result, we have

$$C_D = C_{Dp}(\alpha, M) + C_{Ds}(\alpha, R_e) \tag{4.91}$$

In certain cases, such as in the hypersonic flow where there is interaction between the shock wave and the boundary layer, we may not be able to separate the drag coefficient in the simple form of Eq. (4.91). However, Eq. (4.91) gives us some simple guidance. We should check it with experimental results in order to determine whether it may be used or not. In model testing, it is very important to find out the most important parameter and we should get dynamical similitude of the model and the full size body according to this important parameter.

8. Similar Solutions of Fluid Dynamic Equations

In general, the fundamental equations of fluid dynamics are nonlinear and difficult to solve. With the application of dynamical similarity, the following useful results may be obtained:

(1) We may obtain some similarity laws which enable a family of possible solutions to be deduced from a single result;

(2) we may simplify the fundamental equations of fluid dynamics; and

(3) we may find some simple and interesting solutions of the fundamental equations.

Now we are going to give an example for each of the above three cases to illustrate the general principle of using dynamical similarity in the theoretical study of fluid dynamics.

(A) Similarity Laws of Transonic, Supersonic, and Hypersonic Flows

Let us consider the wave drag of bodies having at least one of the two body dimensions perpendicular to the main flow direction small in comparison to that in the dimension parallel to the main flow in a compressible fluid. The wave drag $C_{Dp}(\alpha, M)$ is practically independent of viscosity of the fluid, i.e., the Reynolds number. Hence we may assume that the fluid is inviscid and compressible. Let us consider a three-dimensional steady irrotational flow of a compressible fluid. The velocity potential Φ for this case satisfy the following equation for steady flow:

$$(a^2 - \Phi_x^2)\Phi_{xx} + (a^2 - \Phi_y^2)\Phi_{yy} + (a^2 - \Phi_z^2)\Phi_{zz} -$$

$$-2\Phi_x\Phi_y\Phi_{xy} - 2\Phi_y\Phi_z\Phi_{yz} - 2\Phi_z\Phi_x\Phi_{zz} = 0 \qquad (4.92)$$

where x, y, and z are the Cartesian coordinates, the subscripts denote partial differentiation, i.e., $\Phi_x = \partial\Phi/\partial x$ etc., and a is the local sound speed, determined by the equation

$$a^2 + \frac{\gamma - 1}{2}(\Phi_x^2 + \Phi_y^2 + \Phi_z^2) = a_0^2 \qquad (4.93)$$

where a_0 is the stagnation sound speed of the gas.

Now if a thin wing of finite span is placed in an otherwise uniform stream of velocity V in the x-direction, we may introduce a perturbed velocity potential ϕ such that

$$\Phi = V(x + \phi) \qquad (4.94)$$

with $\partial\phi/\partial x$, $\partial\phi/\partial y$, $\partial\phi/\partial z \ll 1$.

Eq. (4.93) becomes

$$\frac{a^2}{a_1^2} = 1 - \frac{\gamma - 1}{2}M^2(2\phi_x + \phi_x^2 + \phi_y^2 + \phi_z^2) \qquad (4.95)$$

where $M = V/a_1$ and a_1 is the sound speed of the free stream.

Substituting Eqs. (4.94) and (4.95) into Eq. (4.92) and retaining terms up to the second order only, we have, with $r^2 = (\gamma - 1)/(\gamma + 1)$,

$$\left[1 - M^2 - (\gamma + 1)M^2\phi_x - \frac{\gamma - 1}{2}M^2(\phi_y^2 + \phi_z^2)\right]\phi_{xx}$$

$$+ \left[1 - (\gamma - 1)M^2\phi_x - \frac{\gamma + 1}{2}M^2(\phi_y^2 + r^2\phi_z^2)\right]\phi_{yy}$$

$$+ \left[1 - (\gamma - 1)M^2\phi_x - \frac{\gamma + 1}{2}M^2(r^2\phi_y^2 + \phi_z^2)\right]\phi_{zz}$$

$$- 2M^2\phi_y\phi_{xy} - 2M^2\phi_z\phi_{xz} = 0 \qquad (4.96)$$

Eq. (4.96) is the fundamental differential equation for ϕ for the consideration of similarity laws. In Eq. (4.96), we have retained terms which, while not important everywhere, may be so in certain regions. For example, the hypersonic flow ϕ_x, ϕ_y^2, and ϕ_z^2 are all of the same order of magnitude. Even though we retain in Eq. (4.96) terms of second order, we seek and use those terms which give an effective first-order equation.

The first-order boundary conditions for ϕ are:

(1) At infinity:

$$\phi_x = \phi_y = \phi_z = 0 \qquad (4.97)$$

(2) On the surface of the wing which is represented by

$$z = h(x, y) \qquad (4.98)$$

the flow must follow the wing surface, i.e.,

$$\left(\frac{\partial \phi}{\partial z} \right)_{z=0} = \frac{\partial h(x, y)}{\partial x} \qquad (4.99)$$

We introduce the transformation:

$$x = c\xi; \quad y = b\eta; \quad z = l\zeta; \quad \phi = \phi' m \qquad (4.100)$$

where c is the mean chord of the wing, b, its span with $A = b/c =$ aspect ratio and l and m are conversion factors which are to be determined.

Substitution of Eq. (4.100) into Eq. (4.96) gives

$$
\begin{aligned}
&\left\{ (M^2 - 1)\left(\frac{l}{c}\right)^2 + (\gamma+1)M^2\left(\frac{m}{c}\right)\left(\frac{l}{c}\right)^2 \phi'_\xi \right. \\
&\quad + \frac{\gamma-1}{2}M^2\left(\frac{m}{c}\right)^2\left(\frac{l}{c}\right)\left[\frac{\phi'^2_\eta}{A^2} + \phi'^2_\zeta\left(\frac{c}{l}\right)^2\right] \Bigg\} \phi'_{\xi\xi} \\
&\quad + \left\{ -1 + (\gamma-1)M^2\,\frac{m}{c}\,\phi'_\xi + \frac{\gamma+1}{2}M^2\left(\frac{m}{c}\right)^2\left[\frac{\phi'^2_\eta}{A^2}\right.\right. \\
&\quad \left.\left. + r^2\left(\frac{c}{l}\right)^2\phi'^2_\zeta\right]\right\} \phi'_{\eta\eta}\left(\frac{l}{cA}\right)^2 + \left\{ -1 + (\gamma-1)M^2\,\frac{m}{c}\,\phi'_\xi \right. \\
&\quad \left. + \frac{\gamma+1}{2}M^2\left(\frac{m}{c}\right)^2\left[r^2\frac{\phi'^2_\eta}{A^2} + \left(\frac{c}{l}\right)^2\phi'^2_\zeta\right.\right] \Bigg\} \phi'_{\zeta\zeta} \\
&\quad + 2M^2\,\frac{m}{c}\left(\frac{l}{cA}\right)^2\phi'_\eta\,\phi'_{\xi\eta} + 2M^2\frac{m}{c}\phi'_\zeta\phi'_{\xi\zeta} = 0 \qquad (4.101)
\end{aligned}
$$

and the nondimensional boundary conditions are

$$\phi'_{\xi} = \phi'_{\eta} = \phi'_{\zeta} = 0 \quad \text{(at infinity)} \tag{4.102}$$

$$\left(\frac{\partial \phi'}{\partial \zeta}\right)_{\zeta=0} = \frac{l\tau}{m} f_{\xi}(\xi, \eta) \quad \text{(on the wing surface)} \tag{4.103}$$

where $Tf(\xi, \eta) = h(x, y)$ and $T = c\tau$ is the maximum thickness of the wing. We assume that $\tau \ll 1$.

To get the parameters in the similarity laws, we write

$$l = c\tau^{-n}, \quad m = c\tau^{n'} \tag{4.104}$$

where n and n' are exponents to be determined.

If we consider only the linear terms in Eq. (4.101), the necessary conditions for the existence of similarity laws from Eq. (4.101) are

$$\frac{M^2 - 1}{\tau^{2n}} = K_1^2 \tag{4.105}$$

and

$$A\tau^n = K'_2 \tag{4.106}$$

where K_1 and K'_2 are constants. The left-hand side terms of Eqs. (4.105) and (4.106) may be considered as important nondimensional parameters in our problems.

From the boundary conditions (4.103), the necessary condition for the existence of similarity laws is

$$\tau^{1-n-n'} = \text{constant}$$

or

$$n + n' = 1 \tag{4.107}$$

From Eqs. (4.105) and (4.106), we have

$$A(M^2 - 1)^{1/2} = \text{constant} = K_2 \tag{4.108}$$

The nondimensional parameter of Eq. (4.108) gives a unique similarity condition for the aspect ratio of wings of affinely related shapes.

The index n in the parameter K_1 is arbitrary because the choice of n' in Eq. (4.107) is arbitrary. Hence the similarity laws for linearized theory such as those of the Glauert-Prandtl rule for subsonic flow and that by Ackeret for supersonic flow are arbitrary. To have a unique similarity law it is necessary to study the nonlinear terms. From Eq. (4.101), we have two different groups of nonlinear terms: one group being more important for transonic-supersonic flow ($n>0$), and the other for supersonic-hypersonic flow ($n<0$).

For the transonic-supersonic flow, the important nonlinear term is $(\gamma+1) M^2 (m/c)(l/c)^2 \phi'_\xi$. The similarity condition from this term is

$$M^2 \tau^{n'-2n} = \text{constant} = K_3^2 \tag{4.109}$$

In the immediate transonic region ($M \cong 1$), we have from Eq. (4.109)

$$n' - 2n = 0 \tag{4.110}$$

From Eqs. (4.107) and (4.110), we have

$$n = \frac{1}{3}, \quad n' = \frac{2}{3} \tag{4.111}$$

This is the well known transonic similarity law of von Kármán. We have a transonic similarity parameter K:

$$K = \frac{1-M_\infty}{(\Gamma\tau)^{2/3}} = \frac{1}{2} \frac{1-M^2_\infty}{(\Gamma\tau)^{2/3}} \tag{4.112}$$

where $\Gamma = (\gamma+1)/2$. In two transonic flows over affinely related bodies, if K has the same value, we can deduce the value of the force coefficient of one case from that of the other. For instance, for the drag of a two-dimensional wing, we have

$$C_D = \frac{\tau^{5/3}}{\Gamma^{1/3}} D(K) \tag{4.113}$$

The function $D(K)$ is a function of K only.

In the case that Mach number M is not restricted to the immediate neighborhood of unity, we may eliminate τ and n from Eqs. (4.105), (4.107), and (4.109) to get a relation between n and M as follows:

$$n \left\{ 4 \log(K_3/M) + 3 \log \frac{M^2-1}{K_1^2} \right\} = \log \frac{M^2-1}{K_1^2} \tag{4.114}$$

This shows clearly that n tends to 1/3 as M tends to unity. Furthermore, by differentiating and proceeding to the limit, it may be shown that

$$\lim_{M \to 1+0} \frac{dn}{dM} = \begin{cases} 0, & K_3 = 1 \\ \infty, & K_3 > 1 \\ -\infty, & K_3 < 1 \end{cases} \qquad (4.115)$$

It is plausible to reject infinite value for dn/dM at $M = 1$, in view of the established position of the transonic similarity law which requires $n = 1/3$ to be approximately valid in the neighborhood of $M = 1$. Then $K_3 = 1$ and the law of variation of n with M is

$$n = \frac{1}{3} \frac{\log\left[(M^2 - 1)/K_1^2\right]}{\log\left[(M^2 - 1)/(K_1^2 M^{4/3})\right]} \qquad (4.116)$$

In Fig. 4.2, we show the variation of n and M for various values of K_1.

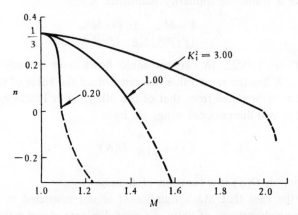

Figure 4.2   Transonic-supersonic similarity laws (solid curves)

We are now able to deduce similarity laws in the sense that a solution $\phi(\xi, \eta, \zeta)$ of the differential equation of motion may be used to give an infinity of flows in the transonic-supersonic region. Given values of the parameters K_1 and K_2, if we choose a value of M, the corresponding values of τ, A, and n are found.   For larger value of K_1, the range of velocity of the ordinary transonic similarity law (4.113) is larger.

For the supersonic-hypersonic flow region, the important non-linear terms are of the same order of magnitude as $(\gamma - 1) M^2(m/c) \phi'_\xi$ and the similarity condition is then

$$M^2 \tau^{n'} = \text{constant} = K_4^2 \qquad (4.117)$$

As M tends to be infinite, Eqs. (4.105) and (4.106) give

$$n' = -2n \qquad (4.118)$$

From Eqs. (4.107) and (4.118), we have

$$n = -1, \ n' = 2 \qquad (4.119)$$

This is the well known hypersonic similarity law due to Tsien. We have the hypersonic similarity parameter k:

$$k = M_\infty \tau \qquad (4.120)$$

In two hypersonic flows over affinely related bodies if k has the same value, we can deduce the value of the force coefficient of one case from that of the other. For instance, for the drag of a two-dimensional wing, we have

$$C_D = \tau^3 D(k) \qquad (4.121)$$

It is interesting to note that with an increasing free stream Mach number, the pressure drag coefficient is increasingly sensitive to the thickness ratio of the body τ. For subsonic flow C_{D_p} is almost independent of τ, at transonic flow C_{D_p} is proportional to $\tau^{5/3}$, at ordinary supersonic speed, C_{D_p} is proportional to τ^2, while at hypersonic speed C_{D_p} is proportional to τ^3.

Since both K_1 and K_4 are constant and as M tends to infinity, n tends to -1 and K_4 tends to K_1; hence $K_1 = K_4$. Eliminating τ and n' from Eqs. (4.105), (4.107), and (4.117) with $K_1 = K_4$, we have

$$n = -\frac{\log\left[(M^2 - 1)/K_1^2\right]}{\log\left[M^4/K_1^2(M^2 - 1)\right]} \qquad (4.122)$$

Fig. 4.3 shows the variation of n and M for various K_1 based on

Eq. (4.122). In practical cases, K_1 is usually larger than one for a transonic-supersonic region and smaller than one for a supersonic-hypersonic region.

The above two similarity laws break down when n is near to zero. In other words, in the ordinary supersonic region, the similarity law is not unique.

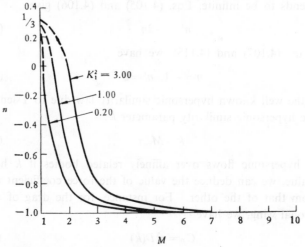

Figure 4.3   Supersonic-hypersonic similarity laws (solid curves)

(B) *Boundary Layer Equations for an Incompressible and Viscous Fluid*

Our second example is to show how we can simplify the fundamental equation by similarity consideration. We consider the case of the flow problems of an incompressible and viscous fluid at the very large Reynolds number R_e. For simplicity, we consider the two-dimensional steady flow case. Using the definition of streamfunction ψ of Eq. (3.42), Eq. (4.37a) is automatically satisfied. Eq. (4.37b) in terms of streamfunction becomes

$$\frac{\partial \psi}{\partial x_2} \frac{\partial \Delta \psi}{\partial x_1} - \frac{\partial \psi}{\partial x_1} \frac{\partial \Delta \psi}{\partial x_2} = \nu_g \Delta \Delta \psi \qquad (4.123)$$

where

$$\Delta = \frac{\partial^2}{\partial x_1^2} + \frac{\partial^2}{\partial x_2^2}$$

When we introduce the nondimensional variables:

$$x_i = x_i^* L, \quad \Delta = L^2 \Delta^*, \quad \psi = UL\psi^*, \quad R_e = \frac{UL}{\nu_g} \qquad (4.124)$$

Eq. (4.123) in nondimensional form becomes

$$\frac{\partial \psi^*}{\partial x_2^*} \frac{\partial \Delta^* \psi^*}{\partial x_1^*} - \frac{\partial \psi^*}{\partial x_1^*} \frac{\partial \Delta^* \psi^*}{\partial x_2^*} = \frac{1}{R_e} \Delta^* \Delta^* \psi^* \qquad (4.125)$$

Now we consider the flow past a semi-infinite flat plate, parallel to a uniform stream, lying on $x_2^* = 0$ and $x_1^* \geqq 0$. When the Reynolds number R_e is large, the thickness of the viscous region will be of the order of $1/(R_e)^{1/2}$, i.e., the x_2^* region is of the order of $1/(R_e)^{1/2}$. In order to investigate the boundary layer region, we must amplify the x_2^* distance in the boundary layer by the following transformation:

$$x = x_1^*, \quad y = (R_e)^{1/2} x_2^*, \quad \Phi = (R_e)^{1/2} \psi^* \qquad (4.126)$$

Substituting Eq. (4.126) into Eq. (4.125), we have

$$\frac{\partial}{\partial y} \left(\frac{\partial^3 \Phi}{\partial y^3} + \frac{\partial \Phi \partial^2 \Phi}{\partial x \partial y^2} - \frac{\partial \Phi}{\partial y} \frac{\partial^2 \Phi}{\partial x \partial y} \right)$$

$$+ \frac{1}{R_e} \left(2 \frac{\partial^4 \Phi}{\partial x^2 \partial y^2} + \frac{\partial \Phi}{\partial x} \frac{\partial^3 \Phi}{\partial x^2 \partial y} - \frac{\partial \Phi}{\partial y} \frac{\partial^3 \Phi}{\partial x^3} \right)$$

$$+ \frac{1}{R_e^2} \frac{\partial^4 \Phi}{\partial x^4} = 0 \qquad (4.127)$$

By assuming that the Reynolds number tends to infinity and that the derivatives in Eq. (4.127) remain finite, Eq. (4.127) becomes the boundary layer equation as follows:

$$\frac{\partial^3 \Phi}{\partial y^3} + \frac{\partial \Phi}{\partial x} \frac{\partial^2 \Phi}{\partial y^2} - \frac{\partial \Phi}{\partial y} \frac{\partial^2 \Phi}{\partial x \partial y} = P(x) \qquad (4.128)$$

where $P(x)$ is an arbitrary given function of x, obtained from the integration with respect to y of the first term in Eq. (4.127) because the other two terms become zero when R_e tends to infinity. In treating the boundary layer problem, the pressure gradient $P(x)$ is assumed to be known which corresponds to the inviscid solution outside the boundary layer. Hence we show that at very large Reynolds number, the Navier-Stokes equation (4.37) may be replaced by the boundary layer equation (4.128) if we are interested in the viscous layer in the neighborhood of a solid body. For very large Reynolds number, the whole flow field may be divided into two parts: one is the inviscid flow region in which we simply drop the viscous terms, i.e., Euler's equations may be used to replace the Navier-Stokes equations; and the other is the boundary layer flow region near the solid boundary in which the viscous force should be retained but the boundary layer equations may be used to replace the Navier-Stokes equations.

The same procedure may be applied for the compressible fluid case. We have both the velocity and temperature boundary layers.

Since the boundary layer equation (4.128) is only the first order approximation of the Navier-Stokes equations for very large Reynolds number, we may find other high order equations for moderately high Reynolds number flow.

(C) *Spherical and Cylindrical Shock Waves Produced by Instantaneous Energy Release*

Most of the fundamental equations of fluid mechanics are non-linear. It is difficult to find exact solutions of these nonlinear equations. One way to find simple solutions of these equations is the similar solution of these equations. By similar solution, we mean that the solutions at different stations or at different time may be reduced to a single solution. The Blasius solution of ordinary boundary layer equations for an incompressible and viscous fluid over a semi-infinite plate is a classical example of a similar solution (see problem 15 and reference[7]). In this section, we discuss the similar solution of a shock wave produced by an instantaneous energy release, such as the explosion of an atomic bomb. Sir Geoffrey Taylor[13] was the first one to analyze the case of a spherical wave and Lin[4] has extended Taylor's problem to the cylindrical case.

We assume that a finite amount of energy is suddenly released in an infinitely concentrated form in a spherical case and as energy per unit length E in the cylindrical case. If we assume the values of energy release E are the same, these two problems may be treated simultaneously.

We assume that the air is an ideal gas so that the specific heats are constant. The fundamental equations in our problem are as follows:

$$\frac{\partial u}{\partial t}+u\frac{\partial u}{\partial r} = -\frac{1}{\rho}\frac{\partial p}{\partial r} \tag{4.129}$$

$$\frac{\partial \rho}{\partial t}+u\frac{\partial \rho}{\partial r}+\rho\left(\frac{\partial u}{\partial r}+\delta\frac{u}{r}\right)=0 \tag{4.130}$$

$$\left(\frac{\partial}{\partial t}+u\frac{\partial}{\partial r}\right)\left(\frac{p}{\rho^{\gamma}}\right)=0 \tag{4.131}$$

where $\delta=1$ for cylindrical case, $\delta=2$ for spherical case and u is the radial velocity and all the variables are functions of radial distance r and time t only. The fluid is assumed to be inviscid.

For similar solutions of Eqs. (4.129) to (4.131) for an expanding blast wave of constant total energy E, we assume the expressions for the solutions of the three unknowns p, ρ, and u as follows:

$$\frac{p}{p_0}=\frac{f_1(\eta)}{R^{1+\delta}}, \quad \frac{\rho}{\rho_0}=\psi(\eta), \quad u=\frac{\phi_1(\eta)}{R^{(1+\delta)/2}} \tag{4.132}$$

where R is the radius of the shock wave which is a function of time t, and subscript 0 refers to the values of the undisturbed atmosphere. If r is the radial coordinate, $\eta=r/R$ and f_1, ϕ_1 and ψ are functions of η only. Except for the variation of R with respect to time t, the solutions of this problem depends on the variable η only. Hence we have a similar solution if we find those functions f_1, ϕ_1, and ψ.

Substituting Eq. (4.132) into Eqs. (4.129) to (4.131), we have

$$\left(\frac{1+\delta}{2}\phi_1+\eta\phi'_1\right)\frac{1}{R^{(3+\delta)/2}}\frac{dR}{dt}-\frac{\phi_1\phi'_1}{R^{2+\delta}}=\frac{p_0 f'_1}{\rho_0\psi R^{2+\delta}} \quad \Big]$$

$$\eta\psi' R^{(1+\delta)/2}\frac{dR}{dt} - \psi'\phi_1 - \psi\left(\phi'_1 + \frac{\delta\phi_1}{\eta}\right) = 0 \left.\begin{array}{c}\\\\\end{array}\right\}$$

$$\left[\gamma\eta f_1\frac{\psi'}{\psi} - \eta f'_1 - (1+\delta)f_1\right] R^{(1+\delta)/2}\frac{dR}{dt}$$

$$+\phi_1 f'_1 - \frac{\gamma\psi' f_1\phi_1}{\psi} = 0 \quad\quad\quad\quad\quad (4.133)$$

where the prime refers to derivatives with respect to η. The similarity conditions are satisfied if the following condition is fulfilled:

$$R^{(1+\delta)/2}\frac{dR}{dt} = A = \text{constant} \quad\quad (4.134)$$

Substituting Eq. (4.134) into Eq. (4.133), we have the following equations which depend on η only and independent of t explicitly

$$\phi_1\phi'_1 + \frac{p_0}{\rho_0}\frac{f'_1}{\psi} = A\left(\frac{1+\delta}{2}\phi_1 + \eta\phi'_1\right) \left.\begin{array}{c}\\\\\\\\\end{array}\right\}$$

$$\phi_1 + \left(\phi'_1 + \frac{\delta\phi_1}{\eta}\right)\frac{\psi}{\psi'} = A\eta \quad\quad\quad (4.135)$$

$$\gamma f_1\phi_1\frac{\psi'}{\psi} - \phi_1 f'_1 = A\left[\gamma\eta f'_1\frac{\psi'}{\psi} - \eta f_1 - (1+\delta)f_1\right]$$

Eqs. (4.135) may be transformed into nondimensional form by the transformation:

$$f_1(\eta) = \frac{A^2}{a^2}f(\eta); \quad \phi_1(\eta) = A\phi(\eta) \quad\quad (4.136)$$

where $a^2 = \gamma p_0/\rho_0$ is the square of the sound speed of the undisturbed air. Substituting Eq. (4.136) into Eq. (4.135), we have

$$f = \frac{(1+\delta)\eta(\eta-\phi)\psi + \frac{1}{2}(1-\delta)\gamma\eta\psi f\phi + \delta\gamma\psi f\phi^2}{\eta[f - \psi(\eta-\phi)^2]} \left.\begin{array}{c}\\\\\\\\\end{array}\right\}$$

$$\psi' = \frac{\psi[(\eta-\phi)f' + (1+\delta)f]}{\gamma f(\eta-\phi)} \quad\quad\quad (4.137)$$

$$\phi' = (\eta-\phi)\frac{\psi'}{\psi} - \frac{\delta\phi}{\eta}$$

Now we assume that the shock is very strong and then the Rankine-Hugoniot relations become

$$\frac{p_1}{p_0}=\frac{2}{\gamma+1}\frac{U^2}{a^2}, \quad \frac{\rho_1}{\rho_0}=\frac{\gamma+1}{\gamma-1}, \quad \frac{u_1}{U}=\frac{2}{\gamma+1} \tag{4.138}$$

where $U=dR/dt$ and subscript 1 refers to the values immediately behind the shock. Only in the case of very strong shock, we may have a similar solution. The boundary conditions at $\eta=1$ are

$$f(1) \cong \frac{2\gamma}{\gamma+1}, \quad \psi(1) \cong \frac{\gamma+1}{\gamma-1}, \quad \phi(1) \cong \frac{2}{\gamma+1} \tag{4.139}$$

Eqs. (4.137) may be integrated numerically from the boundary conditions (4.139). For $\gamma=1.4$, Taylor integrated the case of $\delta=2$, i.e., the spherical shock[13] and Lin[4] integrated the case of $\delta=1$, i.e., cylindrical shock.

We may express the constant A in terms of the energy release E by integrating the total energy, kinetic energy and thermal energy of the disturbance in the whole space, i.e.,

$$E=2\delta\pi \int_0^R \left(\frac{p}{\gamma-1}+\frac{1}{2} \rho u^2 \right) r^\delta \, dr=S(\gamma)\rho_0 A^2 \tag{4.140}$$

where $S(\gamma)$ is a function of γ. For $\gamma=1.4$, $S=5.36$ when $\delta=2$, and $S=3.85$ when $\delta=1$. Hence for $\gamma=1.4$ and $\delta=2$, i.e., a spherical shock in air:

$$R=1.025\left(\frac{E}{\rho_0}\right)^{1/5} t^{2/5} \tag{4.141}$$

while for $\gamma=1.4$ and $\delta=1$, i.e., a cylindrical shock in air:

$$R=1.009\left(\frac{E}{\rho_0}\right)^{1/4} t^{1/2} \tag{4.142}$$

Hence we find the speed of propagation of a very strong spherical or cylindrical shock wave in air for a given amount of energy release.

PROBLEMS

1. Find the dimensions of the following quantities:

(a) Magnetic permeability μ_e,

(b) Inductive capacity ϵ,

(c) Universal gas constant or Boltzmann constant k,

(d) Stefan-Boltzmann constant a_R.

2. From dimensional analysis, find the important parameters of the motion of a simple pendulum in which the important quantities are the period of the pendulum t, the length of the pendulum L, the mass of the weight m, and the angle of the swing θ and the gravitational acceleration g.

3. By dimensional analysis, discuss the wave motion on the surface of shallow water and find a formula for the velocity of propagation of the wave.

4. By dimensional analysis, find the important nondimensional quantities in the study of the performance of a propeller in which the important physical quantities are: the thrust of the propeller T_r, the diameter of the propeller d, the forward velocity of the airplane U, the density of the air ρ, the viscosity of the fluid μ, and the rotating speed of the propeller N.

5. Show that the skin friction of a flat plate is a function of the Reynolds number.

6. Calculate the following nondimensional parameters of air under standard atmospheric conditions at sea level and at 10,000 ft altitude with velocity $U = 200$ mph and a length $L = 10$ ft.

(a) Reynolds number R_e,

(b) Mach number M,

(c) Froude number F_r,

(d) Relativistic parameter R_r.

7. Calculate the radiation pressure number for the range of temperature from $0°C$ to $10^6 \, °K$ for the following pressures:

(a) $p = 1$ atmospheric pressure,

(b) $p = 0.1$ atmospheric pressure,

(c) $p = 0.01$ atmospheric pressure.

8. Calculate the magnetic pressure number and magnetic Reynolds number for mercury when the velocity of the flow is $U = 1$ m/sec and $L = 1$ meter and magnetic induction is 10^4 gausses.

9. The energy equation of a viscous and compressible fluid is

$$\rho c_p \frac{DT}{Dt} = \frac{Dp}{Dt} + \frac{\partial}{\partial x_i}\left(\kappa \frac{\partial T}{\partial x_i}\right) + \left[\mu\left(\frac{\partial u_i}{\partial x_j} + \frac{\partial u_j}{\partial x_i}\right)\right.$$

$$\left. - \frac{2}{3}\mu\left(\frac{\partial u_k}{\partial x_k}\right)\delta_{ij}\right]\frac{\partial u_i}{\partial x_j}$$

Find the nondimensional parameters which characterize the dynamic similarity of the equation.

10. Prove that Eq. (4.6) is generally true.

11. Show that for an inviscid and compressible ideal gas, the ratio of the variation of density to that of the variation of velocity is proportional to the square of the Mach number as a first approximation.

12. Show that for a two-dimensional steady inviscid flow of an ideal gas, the fundamental equation is of elliptical, parabolic, or hyperbolic type according to whether the Mach number of the flow is less than, equal to, or larger than unity respectively.

13. Derive the fundamental equations for the transonic flow of a steady two-dimensional isentropic flow, and find the transonic similarity laws of the lift, drag, and moment of a two-dimensional wing from the resultant equation.

14. Derive the fundamental equations for the hypersonic flow of a steady, two-dimensional isentropic flow and find the hypersonic similarity laws on the lift, drag, and moment of a two-dimensional wing from the resultant equation.

15. Find the general similar solution of the boundary layer equation (4.128) for an incompressible steady laminar flow.

16. Find the boundary layer equations for a steady laminar flow over a body of revolution in a uniform stream for the following cases:

    (a) The radius of the body is much larger than the thickness of the boundary layer,

    (b) The radius of the body is of the same order of magnitude as the thickness of the boundary layer.

REFERENCES

[1]   Alfven, H. "On the existence of electromagnetic hydrodynamic waves", *Arkiv f. Math. Astro. Ock. Fysik,* Vol. 29B, No. 2, 1943, pp. 1—7.

[2]   Birkhoff, G. Hydrodynamics. Princeton University Press, 1960.

[3]   Bridgeman, P. W. Dimensional Analysis. (rev. ed.) Yale University Press, 1931.

[4]   Lin, S. C. "Cylindrical shock waves produced by instantaneous energy release", *Jour. Appl. Phys.,* Vol. 25, No. 1, 1954, pp. 54—57.

[5]   Pack, D. C. and Pai, S. I. "Similarity laws for supersonic flows", *Quarterly Appl. Math.,* Vol. 11, No. 4, 1954, pp. 377—384.

[6]   Pai, S.-I. Magnetogasdynamics and Plasma Dynamics. Springer-Verlag, 1962.

[7]   Pai, S.-I. Viscous Flow Theory, Vol. I. Laminar Flow. D. Van Nostrand, 1956.

[8]   Pai, S.-I. Introduction to the Theory of Compressible Flow. D. Van Nostrand, 1959.

[9]   Pai, S.-I. Radiation Gasdynamics. Springer Verlag, 1966.

[10]   Pai, S.-I. "Laminar Flow", in Handbook of Fluid Dynamics, Sec. 5, V. L. Streeter (ed.) McGraw-Hill, 1961.

[11]   Reynolds, O. "An experimental investigation of the circumstances which determine whether the motion of water shall be direct or sinuous and of the law of resistance in parallel channels", *Trans. Proc. Royal Society of London,* Vol. A—174, 1883, pp. 935—982; or collected papers, Vol. II, p. 51.

[12]   Sedov, L. I. Similarity and Dimensional Methods in Mechanics. Academic Press, 1959.

[13]   Taylor, Sir Geoffrey. "The formation of a blast wave by a very intense explosion", *Proc. Royal Society of London,* Vol. A—201, No. 1065, 1958, pp. 159—186.

[14]   von Kármán, Th. From Low Speed Aerodynamics to Astronautics. Pergamon Press, 1963.

CHAPTER

V

AEROTHERMOCHEMISTRY:

FLOW WITH CHEMICAL REACTIONS

1. Introduction

When the velocity of the fluid flow is low, e.g., Mach number M is below 3, and the temperature of the fluid is also low, e.g., of the order of 1000°K, the gas or air may be considered as an ideal gas with constant specific heat. The internal molecular structure will not influence the flow phenomena except the numerical value of the ratio of the specific heats γ of the gas. Hence in classical gas-dynamics, we need not consider the molecular structure nor the chemical reaction. When the speed of the flow field increases to the hypersonic range, say M is larger than 5 or much higher, and the temperature of the gas is over 2000°K or much higher, we have to consider the molecular structure and many excitation energies other than those of translation and rotation of the molecules (see chapter I section 11). Many new phenomena occur. In this chapter, we shall limit our temperature range up to 5000°K or so. Hence we need to consider the vibration and dissociation modes of internal energy in addition to translation and rotation modes, but not ioniza-tion nor thermal radiation. The last two items will be discussed in the next two chapters.

When we include the phenomena of vibrational energy and dissociation and recombination of the gas molecules in the analysis of the flow problem, many new phenomena occur which are beyond the scope of classical gasdynamics of constant specific heat and which are mainly due to the properties of a reacting gas mixture. Some of these new phenomena are as follows:

(1) Thermodynamics of a reacting gaseous mixture. The

specific heat of each species in the mixture may be a function of temperature when the internal degrees of freedom of the molecules are included as we discussed in chapter I section 11. The total enthalpy of the mixture depends on the composition of the mixture which may be changed in the flow field. In section 2, we shall discuss the thermodynamics of a reacting gaseous mixture in some detail so that we may use them in our flow field study.

(2) Transport phenomena of a mixture of reacting gases. In a reacting gaseous mixture, we should consider the transport phenomena on viscosity, heat conductivity, and diffusion. These phenomena differ from those of classical gasdynamics in two aspects: One is that the coefficients of viscosity, heat conductivity, and diffusion depend on the composition of the mixture. We shall discuss some of the properties of the transport phenomena in section 3. The second is how to include the diffusion phenomena in the flow analysis which is the main subject of this chapter.

(3) Chemical reaction rate and heat release. The composition of a reacting gaseous mixture changes according to the chemical reaction rate which will be discussed in section 4. Here we have another two new phenomena over classical gasdynamics. One is the molecular relaxation. The times to reach thermodynamic equilibrium for various internal degrees of freedom are not the same. In general, the translational and rotational energy of molecules may reach their thermodynamic equilibrium in a very short time. Hence in classical gasdynamics, we assume that the gas is in thermodynamic equilibrium at all times in the flow field. The vibrational energy and dissociation take longer times than those of translation and rotation to reach their thermodynamic equilibrium. Hence we may assume that the translational and rotational energies are in thermodynamic equilibrium in most part of the flow field, but we have to consider the reaction time for other internal degrees of freedom such as vibration and dissociation (see sections 7, 8, and 13). This is known as molecular relaxation.

The other important phenomenon is the heat released or absorbed in the chemical reaction processes. In the analysis of flow problem, we may consider such heat release or absorption as an external heat capacity, i.e., heat addition or absorption from an external agency. Hence we have diabatic flows instead of adiabatic flows which are usually studied in classical gasdynamics. Heat addition would affect many fluid dynamic results such as Kelvin's theorem of Eq. (3.73).

Even though aerothermochemistry, which was first suggested by the late Professor Theodore von Kármán, became popular and important because of hypersonic and high temperature flow associated with the reentry problem, it has been studied for a long time with low speed combustion and detonation. We shall discuss low speed combustion aerodynamics and detonation in section 12.

We discuss the fundamental equations for a reacting gaseous mixture in section 5 by including the above three items. Even in the case without chemical reaction, it is very difficult if not impossible to study the whole flow field of a viscous and compressible fluid. The same situation remains for the case with chemical reaction. For any practical flow problem, we have to divide the flow field into inviscid flow field which covers most of the flow field away from a solid boundary or some other transition regions such as shock wave and viscous flow regions which includes the boundary layer flow and shock transition region. In sections 6 to 10, we discuss the inviscid flow problems of a reacting gaseous mixture while in sections 11 to 16, some viscous flows of a reacting gaseous mixture will be discussed.

2. Thermodynamics of a Reacting Gaseous Mixture

We consider a mixture of gases of N different species among which chemical reactions may take place. The thermodynamic state of such a mixture may be specified by $N+2$ independent variables, provided that we assume that the temperatures of all the species are the same. For the more general case that the species may have different temperature, we shall discuss it in chapter IX section 4. In the present case, we may use either the pressure p, temperature T of the mixture, and the mass concentration k_s of sth species ($s=1$, 2, $\cdots$, N) as independent variables or the pressure p and the specific entropy S of the mixture and the mass concentration.

We shall limit ourselves to the case of a mixture of N perfect gases. The total enthalpy H of the mixture is the sum of the enthalpies of all the species H_s in the mixture, i.e.,

$$H = \sum_{s=1}^{N} k_s H_s \qquad (5.1)$$

where k_s is the mass concentration of sth species [Eq. (1.67)] and H_s is the enthalpy of sth species which may be written as

$$H_s = \int_{T_0}^{T} c_{ps}\, dT + Q_s = H_{so} + Q_s \qquad (5.2)$$

where c_{ps} is the specific heat at constant pressure of sth species including translational, rotational, electron excitational, and vibrational energies which is in general a function of temperature; Q_s is dissociation or recombination energy or the heat of formation of this species s including the enthalpy at the reference temperature T_0. The specific heat of the mixture at constant pressure is then

$$c_p = \left(\frac{\partial H}{\partial T} \right)_p = \sum_{s=1}^{N} \left[k_s \left(\frac{\partial H_s}{\partial T} \right)_p + H_s \left(\frac{\partial k_s}{\partial T} \right)_p \right] \qquad (5.3)$$

Eq. (5.3) shows that the specific heat at constant pressure c_p of a reacting gaseous mixture consists of two parts: one is due to the specific heat $c_{ps} = (\partial H_s / \partial T)_p$ and the other is due to the change of mass concentration k_s. The variation of mass concentration k_s should be determined by solving the diffusion equations simultaneously with other gasdynamic equations which will be discussed later in this chapter. This is in general known as nonequilibrium flow with chemical reaction. As we shall show later, we may simplify the problem by considering two limiting cases: one is known as the frozen flow in which we assume that the change of the mass concentration is negligible and that we use the first term on the right hand side of Eq. (5.3) only for the mixture as a whole; the other is known as the equilibrium flow in which we assume that the chemical reactions take place so rapidly that they reach the thermodynamic equilibrium condition instantaneously. Hence for a given mixture of reacting gases, the concentrations of various species depend only on the pressure and temperature of the mixture. Under thermodynamic equilibrium, we may write:

$$H = H(p, S, k_s) \tag{5.4}$$

The temperature T of the mixture is then

$$T = \left(\frac{\partial H}{\partial S} \right)_{p, k_s} \tag{5.5}$$

and the density ρ of the mixture is given by

$$\frac{1}{\rho} = \left(\frac{\partial H}{\partial p} \right)_{S, k_s} \tag{5.6}$$

where the suffixes in above equations denote which of the variables are kept constant during the partial differentiation.

Thermodynamics is really thermostatics in which we assume that the flow velocity is zero. However it is usual to assume that the thermodynamic relations hold in flow problems if we consider the local values of the thermodynamic variables. It is a good approximation. In thermodynamics with chemical reactions, we may define a specific chemical potential ϕ_s:

$$\phi_s = \left(\frac{\partial H}{\partial k_s} \right)_{p, S, k_j} \tag{5.7}$$

where k_j refers to all the mass concentrations except the one considered, i.e., k_s.

The first law of thermodynamics for the reacting gaseous mixture is

$$T \, dS = dH - \frac{1}{\rho} dp - \sum_{s=1}^{N} \phi_s dk_s \tag{5.8}$$

Of course, one may derive the relations (5.5) to (5.7) from Eq. (5.8).

3. Transport Properties of a Reacting Gaseous Mixture

The transport properties of this mixture are viscosity, heat conductivity, and diffusion. The coefficients of viscosity, heat conductivity, and diffusion should be calculated from the kinetic theory. We shall briefly discuss the kinetic theory in chapter VIII. In fluid dynamics, we assume that these transport properties are given functions of the state variables, as we have discussed the case for a single gas in chapter I. In this section, we simply give the formulas for these transport coefficients which are useful in the calculation of flow problems. In general, these formulas are very complicated, and special treatises such as references [3] and [15] should be referred to.

(1) Binary diffusion coefficient D_{12}. The diffusion coefficient between two species 1 and 2 is a function of the pressure p and temperature T of the mixture as well as the molecular weights of the species m_1 and m_2 which may be written as follows:

$$D_{12} = 1.145 \times 10^{-6} T^{3/2} p^{-1} \left(\frac{m_1 + m_2}{2\, m_1 m_2} \right)^{1/2} A_{12}^{-2} \text{ ft}^2/\text{sec} \qquad (5.9)$$

where T is in degrees Rankine and p is in atmosphere. The factor $A_{12} = \frac{1}{2}(A_1 + A_2)$ depends on the properties of species and the value A_i is a constant for a given species. For instance, A_i is equal to 3.749 Å for N_2; 2.88 for N; 3.541 for O_2; 2.96 for O and 3.47 for NO. It is interesting to know that the binary diffusion coefficient between atom and molecule occurred in air is of the order of $2.4 \times 10^{-8} p^{-1} T^{3/2}$ ft²/sec.

(2) Coefficient of viscosity of a gaseous mixture. The coefficient of viscosity of a gaseous mixture depends on the coefficient of viscosity of each of the species in the mixture and their mole fraction X_s in the mixture. The following formula may be used:

$$\mu = \sum_{s=1}^{N} \frac{X_s^2}{\dfrac{X_s^2}{\mu_s} + 1.386 \displaystyle\sum_{j \neq s} \dfrac{X_s X_j RT}{p m_i D_{sj}}} \qquad (5.10)$$

where μ_s is the coefficient of viscosity of sth species, which is usually a function of temperature T. The simplest case is given by Eq. (1.53) where n may be different for different species. We may also use some polynomials of T for μ_s.

(3) Coefficient of heat conductivity of a mixture. The evaluation of the coefficient of heat conductivity for a mixture is more difficult than that of viscosity or diffusion. Usually, we may estimate the coefficient of heat conductivity from the coefficient of viscosity and the Prandtl number P_r or from the diffusion coefficient and the Lewis-Semenov number L_e. For an ordinary gaseous mixture, the value of the Prandtl number and that of the Lewis-Semenov number do not differ appreciably from unity. The estimation of the Prandtl number and the Lewis-Semenov number would be easier to obtain than that for the coefficient of heat conductivity. For a first approximation, we may even put P_r and L_e both to be unity to get some essential features of such flow with diffusion and chemical reactions. For a second approximation, we may assume that P_r and L_e are some constant values a little different from unity to get the influence of these parameters on the flow field. We shall discuss these points later in this chapter.

4. Chemical Reaction Rates and Heat Release

We have discussed briefly the chemical reaction in chapter I section 15. Here we discuss a little more but special treatises such as references [10], [11], and [24] should be referred to.

There are many types of chemical reaction. The simplest one is known as a first order chemical reaction which is proportional to the molar concentration of a single species in a mixture, e.g., the dissociation of oxygen molecule into oxygen atoms. In general, let us consider a binary mixture of gases A and B. In the first order chemical reaction process, we have

$$A \longrightarrow bB \tag{5.11}$$

where A and B are the molar concentration of the two species and b is the number of moles of B formed by decomposition of one mole of A. If the rate of transformation of gas A into gas B is K_A, we have

$$K_A = -k_f n_A \tag{5.12}$$

where k_f is the forward rate of coefficient which is a function of temperature. For all practical purpose, if the first order reaction occurs in a mixture of more than two species, the rate of transformation of ith species may be written as

$$K_i = -k_f n_i \qquad (5.13)$$

where n_i is the number density of ith species. Eq. (5.13) is the same as Eq. (1.74).

Many reactions are bimolecular and proceed as the result of reactions following binary collisions. For a second order bimolecular reaction, we have

$$b_1 A + b_2 B \longrightarrow b_3 C + b_4 D \qquad (5.14)$$

where b_1, b_2, b_3, and b_4 are the number of moles of the gases A, B, C, and D respectively in the reaction process. The rate of transformation of gas A into gases C and D is proportional to the product of molar concentration of both gas A and gas B, i.e.,

$$K_A = -k_f n_A n_B \qquad (5.15)$$

Now the rate of reaction is proportional to the product of molar concentrations of two gases. For this reason, we call it a second order reaction process. Similarly, we may have higher order reaction processes. In general, chemical reactions can proceed both in the forward direction as shown in Eqs. (5.11) and (5.15) and in the backward direction with a different rate coefficient k_b. For a mixture consisting of N_a separate atomic species $(A_1, A_2, \cdots, A_{Na})$ and N_m separate molecular species $(A_{Na+1}, \cdots, A_{Na+Nm})$, the molecules are formed from the combinations among the N_a different species of atoms. The total number of chemical species are $N = N_a + N_m$ and the total number of possible independent chemical reactions is N_m. For the rth chemical reaction, we may write the chemical reaction equations as

$$\sum_{r=1}^{N} b'_{ir} A_i \underset{k_{br}}{\overset{k_{fr}}{\rightleftharpoons}} \sum_{r=1}^{N} b''_{ir} A_i \qquad (5.16)$$

$$\text{(reactants)} \qquad \text{(products)}$$

where b'_{ir} and b''_{ir} are respectively the stoichiometric coefficients of reactants and products. Stoichiometric coefficients give the number of moles reacting. The forward and the backward specific reaction rate coefficients for the rth reaction are respectively k_{fr} and k_{br}. The net rate of chemical process in the forward direction of the rth reaction is then:

$$K_r = k_{fr} \prod_{r=1}^{N} (A_i)^{b'_{ir}} - k_{br} \prod_{r=1}^{N} (A_i)^{b''_{ir}} \qquad (5.17)$$

where

$(A_i) = m_i n_i / M_i =$ molar concentration of ith species and
$M_i =$ molecular weight of A_i

Since the rth reaction yields $(b'_{ir} - b''_{ir})$ moles of A_i, the corresponding mass rate production of A_i is [see Eq. (3.34)]

$$\sigma_{ir} = M_i (b'_{ir} - b''_{ir}) K_r \qquad (5.18)$$

The net mass rate of the production of A_i in all N_m reactions is then

$$\sigma_i = \sum_{r=1}^{N_m} \sigma_{ir} \qquad (5.19)$$

The dependence of the rate coefficient k_f and k_b on temperature was first found experimentally by Arrhenius and has been confirmed by theoretical studies. The rate coefficient may be written as

$$k_f = B \exp\left(-\frac{E_A}{kT}\right) \qquad (5.20)$$

where B is called the frequency factor and sometimes we may write $B = B_1 T^n$ where both B_1 and n are constant for special chemical reaction. E_A is the activation energy which is a measure of energy that a molecule must possess in order to react successfully. A similar expression may be written for k_b.

Substituting Eqs. (5.17) and (5.18) into Eq. (5.19) we have

$$\sigma_i = \sum_{s=1}^{N} \frac{M_i}{\tau_s} (b''_{is} - b'_{is}) \left[R_s \prod_{j=1}^{N} (k_j)^{b'_{js}} - \right.$$

$$\left. - \prod_{j=1}^{N} (k_j)^{b''_{js}} \right] \tag{5.21}$$

where

$$R_s = \frac{k_{fs}}{k_{bs}} \prod_{j=1}^{N} \left(\frac{\rho}{M_i} \right)^{b'_{js} - b''_{js}} \tag{5.22}$$

$$\tau_s = \frac{\rho}{k_{bs} \prod_{j=1}^{N} \left(\dfrac{\rho}{M_j} \right)^{b''_{js}}} \tag{5.23}$$

$$k_j = \frac{m_j n_m}{\rho} = \text{mass concentration of } j\text{th species.}$$

τ_s has the dimension of time and is the characteristic reaction time for sth reaction [see Eq. (4.69)]. If the sth reaction is at equilibrium, Eq. (5.21) gives

$$R_{se} = \prod_{j=1}^{N} (k_j)^{b''_{js} - b'_{js}} \tag{5.24}$$

or from Eqs. (5.22) and (5.24), we have

$$R_{cse} = \frac{k_{fse}}{k_{bse}} = \prod_{j=1}^{N} A_{je}^{(b''_{js} - b'_{js})} \tag{5.25}$$

Eq. (5.25) is the condition of chemical equilibrium of the sth reaction and R_{cse} is the corresponding equilibrium constant which is a known function of temperature for a given mixture obtainable from statistical thermodynamic principles.[8]

In Eq. (5.16) or (5.14), we consider the change of species only. At the same time, heat may be released or absorbed during the chemical process. The amount of heat released or absorbed during the chemical process can be obtained from the statistical mechanics. Some such chemical reactions which occur often in high temperature air are given as follows[29]:

$$O_2 + M + 5.1 \text{ e.v.} \rightleftarrows O + O + M \tag{5.26a}$$

$$N_2 + M + 9.8 \text{ e.v.} \rightleftharpoons N + N + M \tag{5.26b}$$

$$NO + M + 6.5 \text{ e.v.} \rightleftharpoons N + O + M \tag{5.26c}$$

$$N_2 + O + 3.3 \text{ e.v.} \rightleftharpoons NO + N \tag{5.26d}$$

$$NO + O + 1.4 \text{ e.v.} \rightleftharpoons O_2 + N \tag{5.26e}$$

$$N_2 + O_2 + 1.9 \text{ e.v.} \rightleftharpoons NO + NO \tag{5.26f}$$

5. Fundamental Equations of a Reacting Gaseous Mixture

In the gasdynamics of a mixture of N reacting gases, we consider $N+5$ variables, i.e., the velocity vector q with three components u^i, the temperature T, pressure p, and density ρ of the mixture and the concentration c_s (or the mass concentration k_s) where $i=1,2$, or 3 and $s=1,2, \cdots, N-1$. Hence we have to find $N+5$ relations for our problem which are as follows:

(1) Equation of state of the mixture. We consider the case that all the species in the mixture are perfect gas and then for the mixture, we have [see Eq. (1.69)]

$$p = knT = \rho RT \tag{5.27}$$

(2) The equation of continuity of the mixture as a whole is [see Eq. (3.32)]:

$$\frac{\partial \rho}{\partial t} + \nabla \cdot (\rho q) = 0 \tag{5.28}$$

(3) The diffusion equation for the sth species [see Eq. (3.40)]

$$\frac{Dk_s}{Dt} = - \sum_{q \neq s}^{N} \frac{\partial}{\partial x^j} \left(\rho D_{qs} \frac{\partial c_q}{\partial x^j} \right) - \frac{\partial}{\partial x^j} \left(D_p \frac{\partial \ln p}{\partial x^j} \right)$$

$$+ \frac{\partial}{\partial x^j} \left(D_T \frac{\partial \ln T}{\partial x^j} \right) + \sigma_s \tag{5.29}$$

(4) The equation of motion of the mixture

$$\rho \frac{Du^i}{Dt} = X^i + \frac{\partial \tau^{ij}}{\partial x^j} \tag{5.30}$$

where $\tau^{ij} = \tau_0{}^{ij} - p\delta^{ij}$ is the ijth component of the stress tensor and X^i is the ith component of the body forces per unit volume which may be written as

$$X^i = \sum_{s=1}^{N} m_s n_s X_s^i - \rho \frac{\partial P_0}{\partial x^i} \tag{5.31}$$

where X_s^i is the ith component of nonconservative force per unit mass on the molecules of the sth species and P_0 is the potential energy of the system per unit mass.

(5) Energy equation of the mixture is

$$\rho \frac{DH}{Dt} = \frac{Dp}{Dt} + \frac{\partial}{\partial x^i}\left(\frac{\kappa}{c_p} \frac{\partial H}{\partial x^i}\right) + \Phi + Q_d + Q_t \tag{5.32}$$

where the summation convention is used for the tensorial indices i or j only but not on the species indice s, q, etc. The viscous dissipation of the mixture is

$$\Phi = \tau_0{}^{ij} \frac{\partial u^i}{\partial x^j} \tag{5.33}$$

and

$$\tau_0{}^{ij} = \mu \left(\frac{\partial u^i}{\partial x^j} + \frac{\partial u^j}{\partial x^i}\right) - \frac{2}{3} \mu \left(\frac{\partial u^k}{\partial x^k}\right)\delta^{ij} \tag{5.34}$$

and Q_d is the energy due to diffusion which may be written as

$$Q_d = \sum_{s=1}^{N} w_s^i X_s^i \tag{5.35}$$

where w_s^i is the ith component of the diffusion velocity of the sth species. Q_d is the work done by the nonconservative forces of all the species. Q_t is the rate of heat produced by external agencies such as chemical reactions which is

$$Q_t = \sum_{s=1}^{N} Q_s \sigma_s \tag{5.36}$$

where Q_s is the heat release or absorption by the sth species. We may also consider Q_t due to chemical reaction as a change of its internal energy. In this manner we do not consider Q_t due to chemical reaction as a separate term in the energy equation but include it in the expression of internal energy or in terms of enthalpy such as shown in Eq. (5.2).

For the flow problems of a mixture of reacting gases, we are going to solve Eqs. (5.27) to (5.32) for the variables u^i, T, ρ, p, and c_s (or k_s) with proper initial and boundary conditions. There are $N-1$ equations of the type of Eq. (5.29) and three equations of the type of Eq. (5.30). Hence we have $N+5$ fundamental equations. Our equations are much more complicated than those of fluid dynamics without chemical reactions. We know that we cannot solve the complete Navier-Stokes equations for most flow problems of practical importance. Hence the chance to solve the fundamental equations of flow problems with chemical reactions is much smaller. We have to make further approximations. From the study of classical fluid dynamics, we know that for high Reynolds numbers (which is true for many engineering problems) we may divide the flow field into two portions, the inviscid flow field and the viscous flow field of boundary layer or other transition region style. Hence we shall also divide the problems of flows with chemical reaction into two parts: one is the inviscid flow and the other is the viscous flow. In the viscous flow problems, we shall study the boundary layer flows and the shock wave structure problems.

6. Inviscid Flow of a Reacting Gaseous Mixture

First we consider the inviscid flows of a reacting gaseous mixture in which the effects of viscosity, heat conduction, and diffusion are all negligible. This corresponds to the flow field outside the boundary layer or other transition regions. Chemical reactions are spontaneous irreversible thermodynamic processes which lead to entropy production in the flow field. From Eq. (5.8), we have the entropy production as follows:

$$T \frac{DS}{Dt} = \frac{DH}{Dt} - \frac{1}{\rho} \frac{Dp}{Dt} - \sum_{s=1}^{N} \phi_s \frac{Dk_s}{Dt} \qquad (5.37)$$

The equation of motion (5.30) without any body force nor viscous stresses becomes

$$\rho \frac{Du^i}{Dt} + \frac{\partial p}{\partial x^i} = 0 \qquad (5.38)$$

The energy equation (5.32) without any transport phenomena becomes

$$\rho \frac{DH}{Dt} = \frac{Dp}{Dt} + Q_t \qquad (5.39)$$

As we have mentioned before, we may put Q_t equal to zero if we consider the heat Q_t given by chemical reaction only and we include this heat in the expression of enthalpy. On the other hand, as we shall show later in this section, we may keep Q_t even if we treat only the chemical reaction problem without any other heat addition so that some simple analysis may be performed.

The diffusion equation (5.29) with all diffusion coefficients to be zero becomes

$$\rho \frac{Dk_s}{Dt} = \sigma_s \qquad (5.40)$$

From Eqs. (5.37) to (5.40), we have

$$T\rho \frac{DS}{Dt} = Q_t - \sum_{s=1}^{N} \phi_s \sigma_s \qquad (5.41)$$

We may express the mass source in terms of chemical reaction rate given by Eq. (5.21). Before we do this, we study first the effect of heat addition to the flow field by neglecting the detailed influence of the chemical reaction rate and the change of composition of the mixture. In this way, we may assume $\sigma_s = 0$ in the above equations (5.40) and (5.41) but assume that Q_t is different from zero. This is known as diabatic flow.

In diabatic flow, Eq. (5.41) becomes

$$T\frac{DS}{Dt} = \frac{Q_t}{\rho} = Q_t^* \qquad (5.42)$$

In terms of stagnation enthalpy H_s

$$H_s = H + \frac{1}{2} q^2 \tag{5.43}$$

We have from Eqs. (5.38) and (5.39):

$$\frac{DH_s}{Dt} = \frac{1}{\rho} \frac{\partial p}{\partial t} + Q_i^* \tag{5.44}$$

For steady flow, Eqs. (5.42) and (5.44) become respectively

$$q \cdot \nabla S = \frac{Q_i^*}{T} \tag{5.45a}$$

$$q \cdot \nabla H_s = Q_i^* \tag{5.45b}$$

Eqs. (5.45) show that in diabatic flow, both the entropy and stagnation enthalpy are no longer constant along a streamline but the variations of entropy S and stagnation enthalpy H_s along a streamline are given by Eqs. (5.45) which depend on the heat addition Q_i^*. In adiabatic flows, both entropy S and stagnation enthalpy H_s are constant along a streamline. The vorticity ω in a diabatic steady flow is governed by the equation:

$$q \times \omega = T \nabla S - \nabla H_s \tag{5.46}$$

In general, the flow will be rotational for diabatic flows.

For diabatic flows, it is convenient to use the Crocco vector W defined by[14]

$$W = q/q_m \tag{5.47}$$

where q_m is the maximum possible velocity for a given stagnation temperature T_0, i.e.,

$$q_m = (2c_p T_0)^{\frac{1}{2}} \tag{5.48}$$

We have then

$$p = p_0 (1 - W^2)^{\frac{r}{\gamma - 1}} \tag{5.49}$$

and

$$T = T_0 (1 - W^2) \tag{5.50}$$

In terms of W, the equations of motion, continuity, and energy for steady diabatic flow become, respectively:

$$\ln p_0 = \frac{2\gamma}{\gamma-1}[(1-W^2)^{-1}W \times (\nabla \times W) - q_w W] \quad (5.51a)$$

$$\nabla \cdot \left[(1-W^2)^{-\frac{1}{\gamma-1}} W \right] = q_w(1-W^2)^{-\frac{1}{\gamma-1}} \left(1 + \frac{\gamma+1}{\gamma-1} W^2 \right) (5.51b)$$

$$W \cdot \nabla \ln q_m = (1-W^2)q_w \quad (5.51c)$$

where

$$q_w = \frac{Q_t^*}{q_m^3(1-W^2)} \quad (5.52)$$

In diabatic flow, we solve Eqs. (5.51) for the unknowns W, p_0, and q_m. The interesting point is that the stagnation pressure p_0 and the stagnation temperature T_0 are no longer constant as those in the adiabatic case. Some new phenomena occur. For instance, in adiabatic flow, there is a limiting circle or sphere at Mach number $M=1$ for the source flow, while in diabatic flow, there is no limiting circle or sphere for the source flow (see problem 1).

Even though the simple approach of diabatic flow gives us some new results, much more new information may be obtained if we include the chemical reaction rate in our analysis. From now on, we shall put $Q_t=0$ in Eq. (5.41), i.e., there is no addition of heat except those due to chemical reaction which are included in the definition of enthalpy.[17]   Then from Eqs. (5.41) and (5.21), we have

$$\frac{DS}{Dt} = -k \sum_{s=1}^{N} \frac{1}{\tau_s} \left[R_s \prod_{j=1}^{N} (k_j)^{b'_{js}} - \prod_{j=1}^{N}(k_j)^{b''_{js}} \right]$$

$$\times \ln \prod_{j=1}^{N} \left(\frac{x_j}{x_{je}} \right)^{b''_{js} - b'_{js}} \quad (5.53)$$

where

$$x_j = k_j(M/M_j) \text{ mole fraction of } j\text{th species}$$

$M=$ average molecular weight of the mixture

$x_{j_e}=$ equilibrium mole fraction of jth species.

There are two special cases for the reacting processes which are of interest: one is known as equilibrium flow in which the characteristic reaction time τ_s tends to be zero, and the other is known as frozen flow in which the characteristic reaction time τ_s tends to be infinite. For these two limiting cases, the variation of entropy along a fluid element is zero. When τ_s is finite, the flow is known as nonequilibrium flow and the variation of entropy along a fluid element is not zero. It is interesting to compare Eqs. (5.42) and (5.53). In Eq. (5.42), we can not distinguish the three cases of equilibrium, nonequilibrium, and frozen flows as we can in Eq. (5.53). Furthermore, in Eq. (5.42), we do not have the detailed information about the heat release as we have in Eq. (5.53).

In the nonequilibrium flows, it is convenient to represent the total variation of a quantity by the sum of the equilibrium quantity and the nonequilibrium part of the quantity, i.e.,

$$A=A_e+A' \qquad (5.54)$$

where A may be any thermodynamic quantity such as specific entropy or internal energy, etc., the subscript e refers to the equilibrium value and prime represent the part deviated from the equilibrium value. For instance, we consider the case in which the translational, rotational, and vibrational energies are assumed to be their equilibrium values according to the local temperature T while the dissociation energy does not reach its equilibrium value yet. Hence the total specific internal energy is then

$$U_m=U_{me}+U'_m=U_{n_e}-\sum_{s=1}^{N} k_s\,e_s^* \qquad (5.55)$$

where e_s^* is the heat evolved in the formation of the sth species per unit mass.

The entropy change dS may be written as

$$dS=dS_e+dS' \qquad (5.56)$$

where

$$dS_e = \frac{1}{T} dH_e - \frac{1}{\rho T} dp \qquad (5.57)$$

$$dS' = \frac{1}{T'} dH' - \frac{de}{T'} \qquad (5.58)$$

where subscript e refers to the values for translational, rotational, and vibrational energies corresponding to the local temperature T, while prime refers to the dissociation effect. Because dissociation is not in thermodynamic equilibrium with respect to local temperature T, we introduce another temperature T' which characterizes the dissociation state such that the local dissociation state corresponds to the equilibrium state at the temperature T'. Substituting Eqs. (5.57) and (5.58) into Eq. (5.56), we have

$$T \, dS = dH - \frac{1}{\rho} dp - \left(\frac{T}{T'} - 1 \right) \sum_{s=1}^{N} e_s^* \, dk_s \qquad (5.59)$$

Comparing Eqs. (5.37) and (5.59), we have

$$\phi_s = \left(\frac{T}{T'} - 1 \right) e_s^* \qquad (5.60)$$

and then Eq. (5.41) becomes

$$T\rho \frac{DS}{Dt} = \left(1 - \frac{T}{T'} \right) \sum_{s=1}^{N} e_s^* \sigma_s \qquad (5.61)$$

where we put $Q_t = 0$. Eq. (5.61) shows the variation of entropy in nonequilibrium dissociating flow. For equilibrium flow $T = T'$, then $(DS/Dt)_e = 0$; and for frozen flow $\sigma_s = 0$, $(DS/Dt)_f = 0$.

After we know the variation of entropy in the nonequilibrium flow, we may calculate the variation of vorticity and circulation in a reacting gas flow in the usual manner (see problem 2).

In a general nonequilibrium flow, a steady homoenergic irrotational flow is not homoentropic. Only in the equilibrium flow or the frozen flow, a steady homoenergic flow is homoentropic. Furthermore, in a general nonequilibrium flow, the variation of circulation is not zero in a region of constant entropy or constant temperature because of the chemical reaction terms.

7. Frozen and Equilibrium Flows of a Simple Dissociating Gas

Since the two limiting cases of chemical reaction (frozen and equilibrium flows) give upper and lower bounds of the flow with chemical reactions, they have been extensively studied. In this section, we give some detailed results for a simple dissociating gas which is a reacting gaseous mixture resulting from the dissociation of a pure diatomic gas: $A_2 \rightleftharpoons 2A$. We also assume that the translational, rotational, and vibrational energies are in equilibrium with respect to the local temperature T. Furthermore, we assume that the temperature is not high enough so that the electronic excitation and ionization are negligible. The enthalpy for this simple dissociating gas is then

$$H = \left[\frac{7}{2} + \frac{3}{2}\alpha + (1-\alpha)f(\theta) \right] \frac{p/\rho}{1+\alpha} + \alpha d \qquad (5.62)$$

where

$f(\theta) = \theta/(e^\theta - 1)$
$\theta = h\nu/(kT)$
α = degree of dissociation
d = dissociation energy per unit mass
h = Planck constant
k = Boltzmann constant
ν = vibrational frequency.

The equation of state for this simple dissociating gas is

$$p = (1+\alpha)\rho R_2 T \qquad (5.63)$$

where R_2 is the gas constant for A_2.

The equation of continuity for the atom obtained from the diffusion equation (5.40) with the help of Eq. (5.21) may be written in terms of the degree of dissociation α as follows:

$$\frac{D\alpha}{Dt} = \frac{1}{\tau}[K(1-\alpha) - \alpha^2] \qquad (5.64)$$

where

$$K = [M_2/(4\rho)](k_f/k_b) \qquad (5.65)$$

and

$$\tau = M_2^2/[4\rho^2 k_b(1+\alpha)] \qquad (5.66)$$

For the case of steady inviscid flows, the equations of continuity, that of motion and that of energy give:

$$(u^2-a_f^2)\,\frac{\partial u}{\partial x} + (v^2-a_f^2)\,\frac{\partial v}{\partial y} + (w^2-a_f^2)\,\frac{\partial w}{\partial z}$$

$$+uv\left(\frac{\partial u}{\partial y}+\frac{\partial v}{\partial x}\right) + uw\left(\frac{\partial u}{\partial z}+\frac{\partial w}{\partial x}\right)$$

$$+vw\left(\frac{\partial v}{\partial z}+\frac{\partial w}{\partial y}\right) - a_f^2\,G\left(u\,\frac{\partial\alpha}{\partial x} + v\,\frac{\partial\alpha}{\partial y}\right.$$

$$\left. + w\,\frac{\partial\alpha}{\partial z}\right) = 0 \qquad (5.67)$$

where a_f is known as the frozen speed of sound of the mixture and is given by the following formula:

$$a_f^2 = \left(\frac{\partial p}{\partial \rho}\right)_{s,\,\alpha} = -\frac{\left(\dfrac{\partial H}{\partial \rho}\right)_{p,\,\alpha}}{\left(\dfrac{\partial H}{\partial p}\right)_{\rho,\,\alpha} - \dfrac{1}{\rho}} \qquad (5.68)$$

and G is a factor of thermodynamic variables defined by the equation:

$$-G = \frac{\left(\dfrac{\partial H}{\partial \alpha}\right)_{p,\,\rho}}{\left(\dfrac{\partial H}{\partial \rho}\right)_{p,\,\alpha}} = -\frac{1}{\rho}\left(\frac{\partial \rho}{\partial \alpha}\right)_{p,\,T} - \rho\beta_f\,\frac{1}{c_{pf}}\left(\frac{\partial H}{\partial \rho}\right)_{p,\,T}$$

$$(5.69)$$

where

$$c_{pf} = \left(\frac{\partial H}{\partial T}\right)_{p,\,\alpha} = \text{specific heat at constant pressure and}$$

$$\text{frozen composition}$$

$$\beta_f = -\frac{1}{\rho^2}\left(\frac{\partial\rho}{\partial T}\right)_{p,\,\alpha} = \text{volume expansion coefficient at constant pressure and frozen composition.}$$

Eq. (5.67) should be solved simultaneously with Eq. (5.64). Eqs. (5.64) and (5.67) are much more complicated than the corresponding equations of ordinary gasdynamics, because of the interaction between the flow field and the chemical reaction. However, this interaction disappears in the two limiting cases: frozen flow and equilibrium flow.

In the frozen flow, the degree of dissociation is constant and then

$$\alpha = \alpha_0 = \text{constant} \tag{5.70}$$

In the equilibrium flow, the degree of dissociation is a given function of the local thermodynamic state and we have

$$K_e = \frac{\alpha_e^2}{1-\alpha_e^2} \tag{5.71}$$

The equilibrium constant K_e is a function of temperature and density. For an ideal dissociating gas, Lighthill suggested the following simple formula for K_e:

$$K_e = \frac{\rho_d}{\rho}\exp\left(-\frac{d}{R_2 T}\right) \tag{5.72}$$

where ρ_d is a constant reference density of dissociation.

Eq. (5.67) reduces to the simple form for the frozen flow and the equilibrium flow as follows:

$$(u^2-a_i^2)\frac{\partial u}{\partial x} + (v^2-a_i^2)\frac{\partial v}{\partial y} + (w^2-a_i^2)\frac{\partial w}{\partial z}$$
$$+uv\left(\frac{\partial u}{\partial y}+\frac{\partial v}{\partial x}\right)+vw\left(\frac{\partial v}{\partial z}+\frac{\partial w}{\partial y}\right)$$
$$+uw\left(\frac{\partial u}{\partial z}+\frac{\partial w}{\partial x}\right) = 0 \tag{5.73}$$

where a_i is equal to a_f for the frozen flow and a_i is equal to a_e for equilibrium flow. The sound speed for the equilibrium flow a_e is defined by the following formula:

$$\frac{1}{a_e^2} = \frac{1}{a_f^2} + \rho G \left(\frac{\partial \alpha_e}{\partial p}\right)_s \tag{5.74}$$

We use the equilibrium degree of dissociation α_e to calculate the sound speed. Eq. (5.73) is of the same form as that for classical gasdynamics whose methods of solution are well known[22] (see problem 2). The most interesting thing is to extend the method of classical gasdynamics for the nonequilibrium flow as we shall discuss below.

8.   Sound Waves in a Dissociating Gas[4, 25]

In order to bring out the essential features of chemical reaction, particularly in the nonequilibrium flow, we consider first the propagation of waves of small amplitude in the x-direction in a simple dissociating gas. Since the amplitude of the wave is small, we may linearize the fundamental equations; and since we consider waves propagated in the x-direction only, all the unknowns are functions of time t and the x-coordinate only. The fundamental equation for this one dimensional acoustic wave is[4]

$$\left(\frac{1}{a_{e0}^2}\frac{\partial^2 u'}{\partial t^2} - \frac{\partial^2 u'}{\partial x^2}\right) + \tau_0 \frac{\partial}{\partial t}\left(\frac{1}{a_{f0}^2}\frac{\partial^2 u'}{\partial t^2} - \frac{\partial^2 u'}{\partial x^2}\right) = 0 \tag{5.75}$$

where

$$\tau'_0 = \tau_0 \frac{1 - \alpha_0^2}{2\alpha_0^2} = \text{characteristic reaction time}$$

and subscript 0 refers to the values in the undisturbed state.

It is evident that $\tau'_0 = 0$, the sound wave propagates at the equilibrium sound speed a_{e0} and if $\tau'_0 = \infty$, the sound wave propagates at the frozen sound speed a_{f0}. For the nonequilibrium flow with finite τ'_0, the small disturbance will be propagated at a speed different from both a_{e0} and a_{f0}. Furthermore, even in an inviscid fluid, dispersion and absorption phenomena occur in the nonequilibrium flow case.

Let us investigate how a disturbance of a given frequency ω will be propagated in a simple dissociating gas. Let us assume that the disturbance is of the following type:

$$t>0, \ x=0: \ u'=U \exp(i\omega t)$$
$$x\to\infty: \ u'=0 \tag{5.76}$$

where U and ω are positive real constants and ω is the frequency of the initial disturbance. The solution of Eq. (5.75) which satisfies the boundary conditions (5.76) is

$$u' = U \exp\left[-\frac{\omega m}{a_{f_0}}(\sin A)\,x\right]\cdot\exp\left\{i\omega\left[t-\frac{m}{a_{f_0}}(\cos A)\,x\right]\right\} \tag{5.77}$$

where

$$m = \frac{(a_{f_0}/a_{e_0})^4+(\omega\tau'_0)^2}{1+(\omega\tau'_0)^4} \tag{5.78}$$

$$\tan 2A = \frac{[(a_{f_0}/a_{e_0})^2-1]\,\omega\tau'_0}{(a_{f_0}/a_{e_0})^2+\omega^2\tau'^2_0} \tag{5.79}$$

Eq. (5.77) shows that the disturbances are propagated as a speed

$$a = \frac{a_f}{m\cdot\cos A} \tag{5.80}$$

The interesting points are that the sound speed a in a nonequilibrium flow depends on not only the characteristic reaction time but also the frequency ω of the disturbance. Thus we have the dispersion phenomenon; the amplitude of the disturbance is decaying with increase x according to the first exponential term in Eq. (5.77) even though there is no transport phenomenon; all the transport coefficients are zero and the gas is considered as an inviscid fluid. The damping factor λ_i may be defined by the following formula:

$$\exp(-\lambda_i x)=\exp\left(-\frac{\omega\cdot m\cdot\sin A}{a_{f_0}}\,x\right)$$

or

$$\lambda_i = \frac{\omega\cdot m\cdot\sin A}{a_{f_0}} \tag{5.81}$$

We see that the damping factor depends on both τ'_0 and ω.

The damping factor will be zero in both frozen and equilibrium flows and the sound speed reduces to the corresponding sound speed in these two limiting cases.

9.   Linear Theory of a Steady Flow in a Simple Dissociating Gas[5, 32]

It is possible to reexamine all the flow problems in classical gasdynamics including the effects of chemical reaction. We shall only discuss a few typical flow problems so that the essential features of chemical reaction may be brought out.

First we consider the steady flow of a uniform velocity U passing over a thin body.   Under this condition, the fundamental equations may be linearized and we have the following equation for the perturbed velocity components u', v', and w':

$$\tau \frac{\partial}{\partial \xi} \left[(1 - M_{f0}^2) \frac{\partial u'}{\partial \xi} + \frac{\partial v'}{\partial \eta} + \frac{\partial w'}{\partial \zeta} \right]$$
$$+ (1 - M_{e0}^2) \frac{\partial u'}{\partial \xi} + \frac{\partial v'}{\partial \eta} + \frac{\partial w'}{\partial \zeta} = 0 \qquad (5.82)$$

where $M_{f0} = U/a_{f0}$ and $M_{e0} = U/a_{e0}$ are respectively the frozen and equilibrium Mach number of the freestream: ξ, η, and ζ are respectively the nondimensional $x, y,$ and z coordinates based on a reference length L and $\tau = \tau'_0 U/L$ is the nondimensional reaction time for the freestream.   Eq. (5.82) reduces to the same form as that of ordinary gasdynamics for the frozen or the equilibrium flow when the nondimensional reaction time τ is zero or infinite.   For the nonequilibrium flow, we have to use Eq. (5.82).   For linearized theory, we may neglect the variation of entropy in the flow field and introduce a velocity potential ϕ such that

$$q = \nabla \phi \qquad (5.83)$$

Eq. (5.82) in terms of ϕ becomes

$$\tau \frac{\partial}{\partial \xi} \left[(1 - M_{f0}^2) \frac{\partial^2 \phi}{\partial \xi^2} + \frac{\partial^2 \phi}{\partial \eta^2} + \frac{\partial^2 \phi}{\partial \zeta^2} \right]$$
$$+ (1 - M_{e0}^2) \frac{\partial^2 \phi}{\partial \xi^2} + \frac{\partial^2 \phi}{\partial \eta^2} + \frac{\partial^2 \phi}{\partial \zeta^2} = 0 \qquad (5.84)$$

Eq. (5.84) has been studied extensively for the two-dimensional cases such as wavy wall in subsonic and supersonic flow field (see problem section). It was found that wave drag may occur at subsonic speeds which is in contrast to the results of classical gasdynamics.

Let us consider the flow past a sharp corner in a supersonic flow of an ideal dissociating gas based on the linearized equation (5.84). This is the well known Prandtl-Meyer flow in ordinary gasdynamics. In ordinary gasdynamics, it is known that linearized equations similar to Eq. (5.84) are not good for studying the Prandtl-Meyer flow and one should use the perturbation theory of hyperbolic characteristics to study such a flow. However, in the present case, the flow Mach number changes during the expansion zone, some approximate results may be obtained from Eq. (5.84). Fig. 5.1 shows a sketch of this flow field which is bounded upstream by the frozen Mach line with a Mach angle $\alpha_f = \sin^{-1}(1/M_{f_0})$. Without chemical reaction or in the limiting cases of frozen and equilibrium flows, the pressure on the wall downstream of the corner is a constant depending on the freestream Mach number and the turning angle θ. However, for nonequilibrium flow, the pressure on the wall

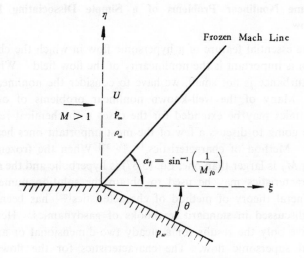

Figure 5.1 Supersonic flow around a sharp corner

downstream of the corner is no longer constant but varies continuously in the relaxation zone from the value of the frozen flow immediately near the corner to the equilibrium flow value far away from the corner. In general, we may write

$$p_w = p_\infty - \frac{U^2\theta}{(M_{f0}^2-1)^{\frac{1}{2}}} K(\xi, \tau) \tag{5.85}$$

where $K(\xi, \tau)$ is a function of the distance from the corner ξ and the chemical reaction time τ. If either ξ or τ is zero, $K(\xi,\tau)$ will be unity and we have the frozen flow result. If either ξ or τ is infinite,

$$K(\xi, \tau) = \frac{(M_{f0}^2-1)^{\frac{1}{2}}}{(M_{e0}^2-1)^{\frac{1}{2}}}$$

and we have the equilibrium flow result. The formula (5.85) is applicable for both the expansion $(+\theta)$ corresponding to Prandtl-Meyer flow and compression $(-\theta)$ to the wedge flow case, i.e., weak shock case. With the help of such a formula, we may easily calculate the lift and drag force on the airfoil in a similar manner as those in ordinary gasdynamics (see problems).

10. Some Nonlinear Problems of a Simple Dissociating Inviscid Flow

The essential feature of a hypersonic flow in which the chemical reaction is important is the nonlinearity of the flow field. Whenever the disturbance is not small, we have to consider the nonlinear problems. Many of the well-known nonlinear problems of ordinary gasdynamics may be extended for the cases with chemical reaction We are going to discuss a few of the most important ones here.

(1) Method of characteristics.[4, 25, 37] When the frozen Mach number M_f is larger than one, Eq. (5.67) is hyperbolic and the method of characteristics may be used to obtain the solution numerically. The general theory of method of characteristics[22] has been extensively discussed in standard textbooks of gasdynamics. Hence we shall give only the results for a steady two-dimensional or axi-symmetrical supersonic flow. The characteristics for the flows of a simple dissociating gas are as follows:

(a) Hyperbolic characteristics

$$\left(\frac{dy}{dx}\right)_{\pm} = \tan(\theta \pm \alpha_f) \tag{5.86}$$

where y is the radial coordinate in the axi-symmetrical case and θ is the flow angle with respect to the x-axis, i.e., $\tan\theta = v/u$, and α_f is the frozen Mach angle, i.e., $\sin\alpha_f = 1/M_f$ and $M_f = q/a_f$ and $q^2 = u^2 + v^2$.

(b) Streamlines

$$\frac{dy}{dx} = \tan\theta = \frac{v}{u} \tag{5.87}$$

Along the above characteristics, we should have some compatibility relations. There are two frozen Mach number corresponding to the plus and minus sign of Eq. (5.86) respectively. The compatibility conditions along these two families of Mach line are as follows:

$$\frac{(M_f^2 - 1)^{\frac{1}{2}}}{q^2} \, dp \pm d\theta + \delta \frac{\sin\theta}{y} \frac{1}{M_f} dL_{\pm} + \frac{1}{M_f} \frac{Z\sigma_s}{q} dL_{\pm} = 0 \tag{5.88}$$

where $dL_{\pm}$ is the elementary arc along the plus and minus frozen characteristics given by the plus and minus sign of Eq. (5.86), $\delta = 0$ for two-dimensional flow and $\delta = 1$ for axi-symmetrical flow and

$$Z = -\frac{1}{\rho T}\left(\frac{\partial \rho}{\partial S}\right)_{p,\,\alpha}\left[\phi_2 - \phi_1 + T\left(\frac{\partial S}{\partial \alpha}\right)_{p,\,\rho}\right] \tag{5.89}$$

The last term of Eq. (5.88) is due to the entropy variation by non-equilibrium chemical reaction.

The streamlines in the present problem are three-fold characteristics. There are three compatibility conditions along the streamlines which are as follows:

$$dS = -[\sigma_s/(\rho Tq)](\phi_1 - \phi_2)\, dL \tag{5.90}$$

$$d\alpha = [\sigma_s/(\rho q)]\, dL \tag{5.91}$$

$$q\, dq + (1/\rho)\, dp = 0 \tag{5.92}$$

where σ_s is the chemical source function for the atomic species, and ϕ_1 and ϕ_2 are respectively the specific chemical potential for atomic and molecular species, and dL is the elementary arc length along a streamline.

From Eqs. (5.86) to (5.92), we may develop the finite difference method of characteristics in the same manner as that for ordinary gasdynamics.[22]

(2) Hypersonic flow in a nozzle.[35] The hypersonic flow in a nozzle has been studied both experimentally and theoretically. In the theoretical analysis, we may use the quasi-one-dimensional analysis, such that all the flow variables are assumed to be a function of the distance along the axis of the nozzle only in the case of steady flow. Let us consider the case of a simple dissociating gas in a nozzle of cross-sectional area $A(x)$ where x is the distance along the axis of the nozzle. The fundamental equations are:

$$\rho Au = \text{constant} \tag{5.93a}$$

$$\rho u\, du + dp = 0 \tag{5.93b}$$

$$\frac{d\alpha}{dx} = \frac{1}{\tau u}\left[K(1-\alpha)-\alpha^2\right] = \frac{\sigma_s}{\rho u} \tag{5.93c}$$

$$\frac{dS}{dx} = -\frac{\sigma_s}{\rho Tu}(\phi_1-\phi_2) \tag{5.93d}$$

$$p = (1+\alpha)\,\rho R_2 T \tag{5.93e}$$

Eqs. (5.93) are similar to those equations of nozzle flow in ordinary gasdynamics except that a new physical quantity τ is introduced. We may form a new nondimensional parameter X from the chemical reaction time τ to characterize the present problem as follows:

$$X = \frac{1}{\tau}\left(\frac{u}{A}\frac{dA}{dx}\right)^{-1} \tag{5.94}$$

For adiabatic expansion through the nozzle, in regions where X approaches zero, frozen flow will prevail and in regions where X approaches infinity, equilibrium flow will prevail. In an accelerating flow region with falling pressure and thus falling density, X tends to decrease along the streamtube, and deviation from equilibrium flow would tend to appear. Once the flow in a nozzle deviates appreciably from its equilibrium condition, a return to the equilibrium is unlikely to take place within the nozzle. The shock wave shape in the nozzle is sensitive to the departures from the equilibrium, and the flow behind a normal shock is changed only slightly by freezing in the nozzle. It is possible to design a nozzle shape such that the departure from equilibrium flow is minimized.

(3) The dissociating flow past a blunt body. Another interesting dissociating flow is the hypersonic flow past a blunt body in which the temperature is very high and dissociation effects are important. The simple case of a blunt body is a sphere which has been extensively investigated. Successive approximation method may be used in such an investigation. The zeroth order approximation is the case where the free stream Mach number M approaches infinity and the ratio of the free stream density to the density immediately behind the shock ϵ tends to be zero. In this zeroth order approximation, the Newtowian plus centrifugal pressure law may be used. The first order approximation is for the case of small but finite density ratio ϵ. The small perturbation theory may be used. The effects of dissociation are an increase of entropy and a decrease of temperature of the resultant flow in comparison with that without dissociation.

11. Viscous Flows of a Reacting Gaseous Mixture[7]

The analysis of inviscid flow problems discussed in the last five sections is good only in the region far away from the solid surface or other region of large variation of the physical quantities of the fluid, such as the shock wave region and flame region. Whenever there is a large variation of physical quantities of the fluid, we should consider the diffusion phenomena in the fluid. In aerothermochemistry, there are three kinds of diffusivity: the diffusivity of momentum which is characterized by the kinematic viscosity $\nu_g = \mu/\rho$; the diffusivity of heat energy which is characterized by the thermal diffusivity $\nu_T = \kappa/(c_p\rho)$; and the diffusivity of mass which is characterized by the coefficient of diffusion D. All these diffusivity coefficients have the same dimensions $UL = L^2/t$. The ratio of any two of these diffusivity coefficients gives us one of the important nondimensional parameters in aerothermochemistry as we have discussed in chapter IV section 6. Here we have three important nondimensional parameters as follows:

$$\text{Prandtl number } P_r = \frac{\text{kinematic viscosity}}{\text{thermal diffusivity}} = \frac{\mu c_p}{\kappa} \quad (5.95a)$$

$$\text{Lewis-Semenov number } L_e = \frac{\text{thermal diffusivity}}{\text{mass diffusivity}} = \frac{\kappa}{c_p \rho D} \quad (5.95b)$$

$$\text{Schmidt number } S_c = \frac{\text{mass diffusivity}}{\text{kinematic viscosity}} = \frac{D}{\nu_g} \quad (5.95c)$$

It is well known that when the diffusivity is small, the diffusion phenomena will be limited in a narrow region known as boundary layer region or other transition region. Let us consider the boundary layer flow over a solid body. For flow with chemical reactions, we have three types of boundary layers: one is the velocity boundary layer which is determined by the diffusivity of momentum; another is the thermal boundary layer which is determined by the thermal diffusivity; and the third is the mass or concentration boundary layer which is determined by the coefficient of diffusion. The three nondimensional parameters of Eqs. (5.95) show the relative magnitude of these three boundary layer thicknesses. For instance, the Prandtl number P_r represents the ratio of the square of the thermal boundary layer thickness to that of the velocity boundary layer thickness. The thickness of velocity boundary layer is inversely proportional to the square root of the Reynolds number $R_e = UL/\nu_g$, the thickness of thermal boundary layer is inversely proportional to the square root of the Peclet number $P_e = P_r R_e$, and the thickness of the concentration boundary layer is inversely proportional to the diffusion number $R_D = R_e/S_c$. If the values of the Prandtl number P_r, Lewis-Semenov number L_e, and Schmidt number S_c are of the order of unity, these three boundary layer thicknesses will be of the same order of magnitude. Then for high Reynolds number flows, we may apply the boundary layer approximations to all of these three boundary layers as we shall discuss later in this chapter.

From our discussion in section 4, the chemical reactions are characterized by two physical quantities: one is the chemical reaction time τ_i and the other is the amount of heat release Q due to chemical reaction. Hence we have two more nondimensional parameters for aerothermochemistry in addition to the above three parameters P_r, L_e, and S_c. These two new parameters are the Damkoehler numbers[6, 20] of Eqs. (4.68) and (4.70). Of course the reaction frequency U_i is equal to $1/\tau_i$. We have

$$D_{am_1} = \frac{L}{\tau_i U} \tag{5.95d}$$

$$D_{am_2} = \frac{Q}{c_p T} \tag{5.95e}$$

If the Damkoehler number D_{am_1} is very small, we have the frozen flow condition, i.e., the chemical reaction will practically not take place in the flow field; while if D_{am_1} is very large, we have the equilibrium flow condition, i.e., the chemical reaction takes place in such a short time that we may assume that it reaches the equilibrium condition everywhere in the flow field instantaneously.

The second Damkoehler number D_{am_2} represents how large will be the effect of the chemical reaction in the flow field. If D_{am_2} is very small, the effect of chemical reaction is negligible; while if D_{am_2} is very large, the effect of chemical reaction in the flow field is predominant.

In general, we should use the complete set of equations of aerothermochemistry discussed in section 5 to study the viscous flow of a reacting gaseous mixture. However, in most practical problems, we consider the cases where the variation of the physical quantities is large on one direction only and hence some simplifications may be made. Ordinarily two types of problems of viscous flow have been extensively studied: one is the case where a large variation in the direction of the flow velocity occurs, such as the case of a shock wave or a flame front and the other is the case where a large variation in the direction perpendicular to the main flow direction occurs such as the boundary layer flows. We are going to discuss some typical cases for these two types of flow problems in the following sections.

12. Flame Front Structure and Detonation Wave[34]

We consider one-dimensional flow shown in Fig. 5.2. The unburned mixture and the burned product are separated by a chemical reaction zone which moves towards the unburned mixture at a speed u_s. In this chemical reaction zone, chemical reaction or combustion takes place. If the velocity u_s is very small, we have a flame front and if the velocity u_s is very large, we have a detonation wave. If we know the value of u_s, the relation between the variables in front of and behind the reaction zone may be found in the exact same

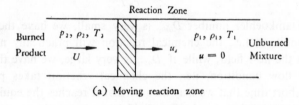

(a) Moving reaction zone

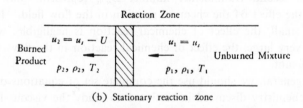

(b) Stationary reaction zone

Figure 5.2   Detonation wave or flame front

manner as those for a normal shock with heat addition.[22]   However, in general, the velocity of the reaction zone u_s is not known from purely theoretical analysis. It may be obtained either by experiment or in the following manner:

For a given amount of total heat release in the reaction zone, we may plot the pressure p against the specific volume $1/\rho$ curve such as curve *BNGFK* in Fig. 5.3. The initial condition A is in general not on this Hugoniot curve *BNGFK*. Without heat release, the initial condition *A* is always on the Hugoniot curve. With heat release, the Hugoniot curve is in fact one of the one-parameter family of curves depending on the heat release. In other words, we have a series of curves similar to *BNGFK* and each of these curves corresponds to a given amount of heat release. The final state of the

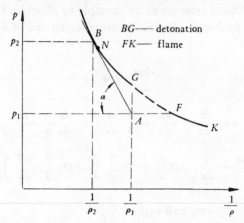

Figure 5.3 Hugoniot curve of detonation and flame front

flame front or detonation wave must be on the Hugoniot curve corresponding to the total amount of heat release. The point *G* represents the final state of combustion at constant volume, i.e., $\rho_1 = \rho_2$, where the point *F* represents the final state of combustion at constant pressure $p_1 = p_2$. When the final state lies on the portion *FK*, we have a flame front. The dotted portion *GF* does not correspond to any possible physical process. For detonation wave, the Chapman-Jouquet condition is usually assumed for determination of the detonation velocity. Under this condition, the detonation velocity is given by the point *N* which is the tangent point on the Hugoniot curve (Fig. 5.3) drawn from the initial state *A*. The point *N* is the most stable condition for detonation and the velocity of detonation is then

$$u_s = \frac{1}{\rho_1} (\tan \alpha)^{\frac{1}{2}} \tag{5.96}$$

After the velocity of detonation u_s is known, we may calculate the velocity, the temperature, and the concentration distributions in the reaction zone. It is usually convenient to choose the coordinate system such that the reaction zone is stationary [Fig. 5.2(b)]. Under this coordinate system, the flow is steady and all variables are function of x only. We shall consider the case of binary mixture with first order chemical reaction as an example to illustrate the essential points of this problem. The fundamental equations of our problem are then:

$$\frac{d\rho u}{dx} = 0 \qquad (5.97a)$$

$$\rho u \, \frac{du}{dx} + \frac{dp}{dx} - \frac{4}{3} \frac{d}{dx} \left(\mu \, \frac{du}{dx} \right) = 0 \qquad (5.97b)$$

$$\frac{dm_1 cnu}{dx} = \frac{d}{dx} \left(\rho D \, \frac{dc}{dx} + D_p \frac{1}{p} \frac{dp}{dx} - D_T \frac{1}{T} \frac{dT}{dx} \right)$$
$$- m_1 \, cnB \exp\left(-\frac{E_A}{kT} \right) \qquad (5.97c)$$

$$\rho u \frac{dH_s}{dx} - \frac{4}{3} \frac{d}{dx}\left(\mu u \, \frac{du}{dx} \right) - \frac{d}{dx}\left(\kappa \, \frac{dT}{dx} \right)$$
$$- m_1 cnBQ_1 \exp\left(-\frac{E_A}{kT} \right) = 0 \qquad (5.97d)$$

where n is the number density of the mixture and m_1 is the mean mass of a particle in the unburned mixture. Subscript 1 refers to the value in the unburned mixture and subscript 2 refers to the value in the burned product. The symbols without subscript refer to the value in the reaction zone.

Eqs. (5.97) may be integrated with respect to x and we have

$$\rho u = \text{constant} = m \qquad (5.98a)$$

$$mu + p - \frac{4}{3} \mu \, \frac{du}{dx} = mK = \text{constant} \qquad (5.98b)$$

$$m_1 cnu - \rho D \frac{dc}{dx} - D_p \frac{1}{p} \frac{dp}{dx} + D_T \frac{1}{T} \frac{dT}{dx}$$

$$+ \int_{-\infty}^{x} m_1 \, cnB \exp \left(-\frac{E_A}{kT} \right) dx$$

$$= \text{constant} = m_1 n_1 u_1 = \rho_1 u_1 \tag{5.98c}$$

$$mH_s - \frac{4}{3} \mu u \frac{du}{dx} - \kappa \frac{dT}{dx} - \int_{-\infty}^{x} m_1 \, cnBQ_1 \exp \left(-\frac{E_A}{kT} \right) dx$$

$$= \text{constant} = mK^* = mH_{01} \tag{5.98d}$$

where $H_s = H + \frac{1}{2} u^2 = $ stagnation enthalpy.

Eqs. (5.98) may be considered as a system of three first order differential-integral equations. We may eliminate du/dx in Eq. (5.98d) by the help of Eq. (5.98b) and obtain an equation of dT/dx only. Similarly, we may eliminate dp/dx and dT/dx in Eq. (5.98c) and obtain an equation with the differential term dc/dx only.

These three first order differential-integral equations may be solved with the following conditions:

At $x = -\infty$: in front of the reaction zone, we have a uniform state

$$u = u_1, \quad \rho = \rho_1, \quad T = T_1, \quad c = c_1 = 1$$

where c is the concentration of the unburned mixture. Without chemical reaction, the shock speed u_1 may be considered as the given value. With chemical reaction, we have to determine the speed u_1 from the amount of heat release and the Hugoniot curve of Fig. 5.3. Since the total amount of heat release depends on the solution of the problem, the successive approximation method has to be used to solve this problem.

At $x = +\infty$, far behind the reaction zone, we have another uniform state

$$u = u_2, \quad \rho = \rho_2, \quad T = T_2, \quad c = c_2 = 0$$

and

$$\int_{-\infty}^{\infty} cm_1\, nB \exp\left(-\frac{E_A}{kT}\right) dx = \rho_1 u_1 \tag{5.99a}$$

$$\int_{-\infty}^{\infty} m_1\, cnBQ_1 \exp\left(-\frac{E_A}{kT}\right) dx = mQ^* \tag{5.99b}$$

In the uniform state 2, Eqs. (5.98) give

$$\rho_1 u_1 = \rho_2 u_2 \tag{5.100a}$$

$$\rho_1 u_1^2 + p_1 = \rho_2 u_2^2 + p_2 \tag{5.100b}$$

$$H_1 + \frac{1}{2} u_1^2 + Q^* = H_2 + \frac{1}{2} u_2^2 \tag{5.100c}$$

Since Q^* of Eq. (5.99b) depends on the solution of Eq. (5.98), we have to use the method of successive approximation. First we may assume a value of $Q^* = Q_1^*$ with the known value of Q_1^*, Eqs. (5.100) are exactly the well known equations to determine the condition across a normal shock with a given amount of heat release. In general there are two solutions: one corresponds to a shock wave with heat release, i.e., a detonation wave and the other, an ordinary wave with heat release, i.e., a flame front. For a given initial state 1, we may have either a flame front or a detonation wave depending on the Mach number of the state 1, as we have seen in Fig. 5.3. If the Mach number M_1 is very small, we have a flame front across which the pressure is almost constant. If the Mach number M_1 is large, we have a detonation wave across which a large pressure jump occurs. After we obtain the final state 2, we may integrate the three first order differential equations for u, T, and c numerically. After we obtain the first approximate solution of u, c, and T, we may calculate the value of $Q^* = Q_2^*$ and with this new Q_2^*, we may repeat the computation process and so on, until the desired accuracy is obtained.

It is interesting to point out the difference of the present problem from that of the shock structure of ordinary gasdynamics without chemical reaction. In the first place, we have three equations for u, T, and c in the present problem while there are only two equations for u and T in the corresponding problem of ordinary gasdynamics. In ordinary gasdynamics, it is possible to eliminate the spatial coordinate x and to obtain a single equation for $T(u)$. After we obtain the solution of $T(u)$, we may calculate the spatial coordinate x. In the present problem, we do not have the simple relation $T(u)$ but we have $T(u, x, c)$ because of the integral terms in Eqs. (5.98). Hence we have to solve these three equations simultaneously. The calculation is rather lengthy. Since the Prandtl number and Lewis-Semenov number are of the order of unity, the thickness for the velocity, temperature, and concentration distributions in the reaction zone are of the same order of magnitude, no further simplification can be made in this calculation. The thickness of the detonation wave is still small while that of the flame front will be large in comparison with the typical dimension of the flow field.

13. Shock Wave Structure with Relaxation[30]

In the last section, we consider the chemical reaction zone according to a first order reaction in a combustion problem. Similar analysis may be carried out for other types of chemical reaction. For instance, in the dissociation of a diatomic gas, the second order chemical reaction should be used. The shock wave structure as well as the Rankine-Hugoniot relations are affected by the modes of various internal energies of a gas. At high temperature over $2000°K$, the specific heat of a gas is no longer constant because of the excitation of vibrational energy, dissociation, and ionization as we have discussed in chapter I section 11. Furthermore the time to reach the equilibrium condition for various modes is not the same. Usually, the translational and rotational energies of a gas molecule will reach their equilibrium conditions in a very short time, such as a few mean free paths. The vibrational energy will reach its equilibrium condition in a longer time and the dissociation will need an even longer time than that for vibrational energy to reach its equilibrium condition. In order to determine these relaxation time effects on the shock wave structure, we may divide the transition region of a shock wave into various subdivisions (Fig. 5.4) as a first approximation. In region 1, there is no change of vibrational energy nor dissociation and that the translational and rotational energies are at their equilibrium values at all points in the flow field. The flow field in this region may be obtained by integration of Eqs. (5.98b) and (5.98d) without those integral terms and

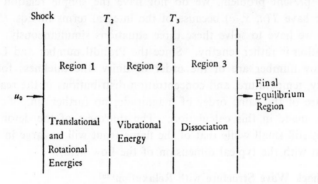

Figure 5.4   Shock transition regions

with a constant specific heat. At the end of the region 1, the translational temperature of the gas increases from the temperature in front of the shock T_1 to a value of T_2, but the vibrational temperature T_v is still equal to T_1. Hence the local vibrational temperature T_v is not equal to the translational temperature T_2. In region 2, the vibrational temperature T_v of the gas gradually increases from the value T_1 to its final value of equilibrium and at the same time the translational temperature T_t of the gas decreases from its highest value T_2 to its final equilibrium temperature T_e. In the equilibrium condition, we have $T_t = T_r = T_v = T_e = T_3$, where T_3 is the equilibrium temperature at the end of region 2. In region 3, we may assume that translational, rotational, and vibrational temperatures of the gas are equal and that dissociation of the gas gradually takes place. At first, the composition of the gas will not be its equilibrium value corresponding to the local translational temperature, but as we go down the stream, the composition of the gas will eventually reach its equilibrium value corresponding to the local translational temperature, i.e., finally all the internal energies have their equilibrium values corresponding to the local temperature. Of course, in exact theoretical analysis, we do not have to divide the regions behind the initial shock front arbitrarily, as shown in Fig. 5.4, and sometimes these regions are overlapped. However, such a simple picture gives qualitative results of a shock wave structure with relaxation, i.e., the temperature in the shock wave first increase from T_1 to a high value T_2 and then gradually decreases to its final equilibrium value (Fig. 5.5). Ordinarily, the region 1 is controlled by viscosity and heat conductivity, which are proportional to the mean free path of the gas. Hence the thickness of region 1 is of the order of a few mean free paths of the gas. The region 2 or 3 is governed by the reaction time τ or a relaxation length $L_r = \tau U$. The reaction time τ in the present case may be introduced in the following manner:

For simplicity, we consider the case of shock wave structure where the vibrational relaxation is important and where both the

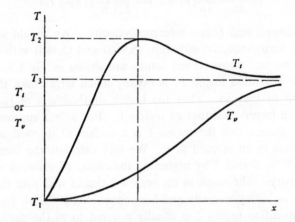

Figure 5.5 Shock wave structure with vibrational relaxation

translational and rotational energies are in equilibrium and their temperature is denoted by a single translational temperature T_t. Since the vibrational energies is not in equilibrium, we may assume that there is a vibrational temperature T_v which is in general different from T_t. Now we assume that there is no dissociation of the gas molecules. Hence the change of internal energy of the gas is

$$dU_m = c_{v0}dT_t + c_{vv}dT_v \qquad (5.101)$$

The expression of Eq. (5.101) should be used for the total internal

energy of the gas in the energy equation. Since we introduce a
new variable: the vibrational temperature T_v, we have to add a
reaction equation in our fundamental equations:

$$\frac{DT_v}{Dt} = \frac{1}{\tau}(T_t - T_v) \qquad (5.102)$$

where c_{v0} is the specific heat at constant volume of both the translational and rotational energies, which is a constant, and c_{vv} is the
specific heat at constant volume of the vibrational energy which is
a function of temperature T_v. The factor τ is the vibrational relaxation time. For the one-dimensional flow Eq. (5.102) becomes

$$\frac{dT_v}{dx} = \frac{1}{\tau u}(T_t - T_v) = \frac{1}{L_r}\left(\frac{U}{u}\right)(T_t - T_v) \qquad (5.102a)$$

where $L_r = \tau U$ and U is a reference velocity. We should solve Eq.
(5.102a) simultaneously with Eqs. (5.98b) and (5.98d) without these
integral terms. Some typical results are shown in Fig 5.5.

The relaxation length L_r is usually much larger than the mean
free path of the gas. Thus the length of thickness of the region
2 is much larger than that of region 1. For a first approximation,
we may assume that the region 1 is a surface of discontinuity, i.e.,
shock wave in an inviscid flow. We may calculate the temperature
T_t and T_v in region 2 by neglecting the effects of viscosity and heat
conductivity. The result in the region 2 checks well with that exact
solution including viscosity and heat conductivity. Such an inviscid
shock transition region 2 is usually referred to as the partially dispersed shock. If the relaxation length is long enough, it is possible
to obtain the solution of both region 1 and region 2 without viscosity
and heat conductivity. This case is known as fully dispersed shock.
In the fully dispersed shock, the transition region is so wide that
the gradients of velocity and temperature are small and the viscous
and heat conductive effects are thus negligible.

14. Boundary Layer Flow with Chemical Reaction[23]

It is well known that the viscous effect of a fluid with small coefficient of viscosity is usually limited in the narrow region known as

the boundary layer.   This is still true for the viscous flow with chemical reaction.   For instance, when we study the heat transfer of the body at hypersonic speed when dissociation is important, the viscous effects are important only in the thin layer near the body. Similar situation occurs in the combustion problem near a surface or in the mixing zone of two streams.   In these cases, we may apply the boundary layer approximations to our fundamental equations of section 5 so that much more simplified boundary layer equations with chemical reactions may be obtained.

For simplicity, we first consider a two-dimensional steady flow in which the y-direction is the direction of boundary layer thickness and x-direction is that of the main flow velocity.   We further consider the case of binary mixture with first order chemical reaction and neglect the pressure and thermal diffusion terms.   The boundary layer equations for this simple case are as follows:

$$p = knT \tag{5.103a}$$

$$\frac{\partial \rho u}{\partial x} + \frac{\partial \rho v}{\partial y} = 0 \tag{5.103b}$$

$$\rho u \frac{\partial k_1}{\partial x} + \rho v \frac{\partial k_1}{\partial y} = \frac{\partial}{\partial y}\left(\rho D \frac{\partial k_1}{\partial y}\right) + \sigma_1 \tag{5.103c}$$

$$\rho u \frac{\partial u}{\partial x} + \rho v \frac{\partial u}{\partial y} = -\frac{\partial p}{\partial x} + \frac{\partial}{\partial y}\left(\mu \frac{\partial u}{\partial y}\right) \tag{5.103d}$$

$$\rho u \frac{\partial H}{\partial x} + \rho v \frac{\partial H}{\partial y} = u \frac{\partial p}{\partial x} + \frac{\partial}{\partial y}\left(\frac{\kappa}{c_p} \frac{\partial H}{\partial y}\right)$$

$$+ \mu \left(\frac{\partial u}{\partial y}\right)^2 + Q^* \sigma_1 \tag{5.103e}$$

where σ_1 is the source term of the unburned mixture with concentration k_1.

Eq. (5.103c) gives the boundary layer of concentration k_1, Eq. (5.103d) gives the boundary layer of velocity u, and Eq. (5.103e) gives the boundary layer of temperature T.   If we transform these three equations in nondimensional form, we will obtain Reynolds number,

Schmidt number, and Prandtl number as the important nondimensional parameters in the present problem.

Now we solve Eqs. (5.103) for the case of two uniform streams of two different gases: one is the combustible gas 1 and the other is the combustion product 2 (Fig. 5.6). When the temperature of the combustion product is high enough, combustion may occur in the mixing zone which may be regarded as a laminar jet flame in which sharp flame front occurs. For such problems, we may assume that the chemical reaction is a first order reaction such as shown in Eq. (5.97), i.e., $\sigma_1 = -m_1 cnB \exp[-E_A/(kT)]$. In this problem, it is convenient to use the von Mises transformation in which the variables x and streamfunction ψ are used instead of x and y. The streamfunction ψ is defined as follows:

$$\frac{\partial \psi}{\partial y} = \rho u, \qquad \frac{\partial \psi}{\partial x} = -\rho v \qquad (5.104)$$

In terms of x and ψ, Eq. (5.103b) is automatically satisfied. Eqs. (5.103c) to (5.103e) become respectively:

$$\frac{\partial c}{\partial x} = \frac{\rho^2}{n^2 m_1 m_2} \frac{\partial}{\partial \psi} \left(\rho^2 u D \frac{\partial c}{\partial \psi} \right)$$

$$- \frac{c(\beta c + 1)B}{u} \exp\left(-\frac{E_A}{kT} \right) \qquad (5.105a)$$

$$\frac{\partial u}{\partial x} = \frac{\partial}{\partial \psi} \left(\rho \mu u \frac{\partial u}{\partial \psi} \right) \qquad (5.105b)$$

$$\frac{\partial H}{\partial x} = \frac{\partial}{\partial \psi} \left(\frac{\kappa \rho u}{c_p} \frac{\partial H}{\partial \psi} \right) + \mu \rho u \left(\frac{\partial u}{\partial \psi} \right)^2$$

$$+ Q^* \frac{m_1 cnB}{u} \exp\left(-\frac{E_A}{kT} \right) \qquad (5.105c)$$

where $\beta = (m_1/m_2) - 1$.

Eqs. (5.105) are a set of three generalized heat conduction equations which may be easily integrated by stepwise numerical procedure from the following initial conditions:

$$x=0: \quad c= \frac{n_1}{n} = c_0(\psi), \quad u=u_0(\psi), \quad T=T_0(\psi) \qquad (5.106)$$

After $c(x, \psi)$, $u(x, \psi)$, and $T(x, \psi)$ are obtained, the y-coordinate for any given x may be obtained by the following formula:

$$y-y_0=\int_0^\psi \frac{d\psi}{\rho u} \qquad (5.107)$$

where y_0 is a constant corresponding to $\psi=0$ at a given x.

In general we have to solve the three partial differential equations (5.105) simultaneously. As an example, we consider a simple case of isovel jet mixing in which the velocities of the two streams shown in Fig. 5.6 are the same and the whole flow field has a constant velocity U. We need to solve Eqs. (5.105a) and (5.105c) only for c and T. We consider a two-dimensional jet with the initial conditions at $x=0$:

$$\psi<1 \text{ and } \psi>-1: \quad c=1, \quad T^*= \frac{T}{T_1} = 1$$

$$\psi>1 \text{ and } \psi>-1: \quad c=0, \quad T^*=3.5$$

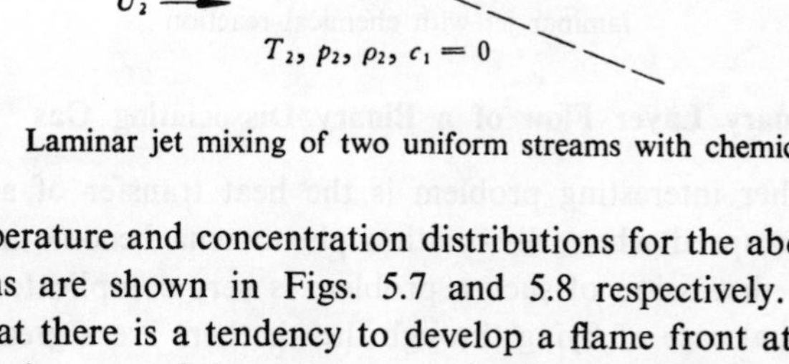

Figure 5.6 Laminar jet mixing of two uniform streams with chemical reaction

The temperature and concentration distributions for the above initial conditions are shown in Figs. 5.7 and 5.8 respectively. Fig 5.7 shows that there is a tendency to develop a flame front at the edge of the jet, i.e., a peak in temperature distribution occurs.

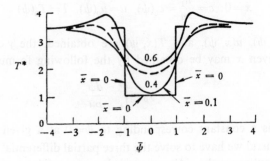

Figure 5.7   Temperature distributions in isovel mixing of a two-dimensional laminar jet with heat release

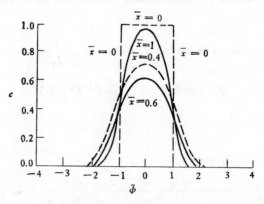

Figure 5.8   Concentration distributions in isovel mixing of a two-dimensional laminar jet with chemical reaction

15. Boundary Layer Flow of a Binary Dissociating Gas

Another interesting problem is the heat transfer of a body at hypersonic speed where dissociation phenomena become important. A general discussion of such a problem is very complicated because in the actual case of flying through the air there are a great number of chemical reactions which should be considered and these chemical reaction rates are still not fully known. Since the heat transfer at hypersonic speeds is a very important problem, various simplified analyses have been made in order to bring out the essential features of heat transfer in a dissociating gas. One of the simple assumptions which has been extensively used is that the local composition of the air may be determined by the thermodynamical equilibrium condition.[21] In this case, we assume that the chemical reaction rates are so large that local thermodynamical equilibrium is maintained all the time. Under this condition, we may replace the diffusion equation (5.103c) by the condition of equilibrium flow (5.25), where the temperature is always the local temperature. Such analysis has been carried out for both the boundary layer flow over a flat plate and that near a stagnation point of a blunt body. The equilibrium flow corresponds to infinite chemical reaction rate. In actual condition, the chemical reaction rate is finite and we have to solve those equations similar to Eqs. (5.103).

An improvement of the equilibrium flow analysis is to consider the nonequilibrium flow with the assumption that the air may be considered as a two-species or binary mixture: one of them is the molecule species and the other is the atom species.[9] Such an analysis has been extensively studied.[16, 27] The equilibrium flow is a special case of the nonequilibrium flow in which the chemical reaction rate is infinite. The other extreme case is the frozen flow case in which the chemical reaction rate is zero. The results of the nonequilibrium flow lie between the results of equilibrium and frozen flows. When we solve Eqs. (5.103), besides the ordinary boundary conditions of velocity and temperature on the body surface, we have to have boundary conditions for the concentration which depend on the type of surface condition of the body. There are two types of wall which give the limiting conditions of the surface: one is the catalytic wall on which the concentration of the atom is zero, i.e.,

$$k_A(y=0)=0 \qquad (5.108)$$

In other words, all the atoms are recombined into molecules on the surface of a catalytic wall. The other is the noncatalytic wall on which the gradient of the concentration of atom is zero, i.e.,

$$\frac{\partial k_A}{\partial y}(y=0)=0 \tag{5.109}$$

where k_A is the mass concentration of the atom. On the surface of a noncatalytic wall, there is no recombination of atoms into molecules and then there is no change of mass concentration of the atom. In general, the property of a surface may lie between the above two limiting cases.

One of the most interesting practical problems for the boundary layer flow of a dissociating gas is the stagnation point boundary layer flow near a blunt body. We may choose the coordinate system shown in Fig. 5.9. The boundary layer equations for a body of revolution with a radius $r_0(x)$ which is much larger than the thickness of the boundary layer are as follows:

$$\frac{\partial \rho r_0 u}{\partial x}+\frac{\partial \rho r_0 v}{\partial y}=0 \tag{5.110a}$$

$$\rho u\frac{\partial k_s}{\partial x}+\rho v\frac{\partial k_s}{\partial y}+\sum_{r \neq s}\frac{\partial}{\partial y}\left(\rho D_{rs}\frac{\partial c_r}{\partial y}\right)=\sigma_s \tag{5.110b}$$

$$\rho u\frac{\partial u}{\partial x}+\rho v\frac{\partial u}{\partial y}=-\frac{\partial p}{\partial x}+\frac{\partial}{\partial y}\left(\mu\frac{\partial u}{\partial y}\right) \tag{5.110c}$$

$$\rho u\frac{\partial H}{\partial x}+\rho v\frac{\partial H}{\partial y}=u\frac{\partial p}{\partial x}+\mu\left(\frac{\partial u}{\partial y}\right)^2+\frac{\partial}{\partial y}\left(\frac{\kappa}{c_p}\frac{\partial H}{\partial y}\right)$$
$$-\sum_s\frac{\partial H_s}{\partial y}\sum_{r \neq s}\rho D_{rs}\frac{\partial c_r}{\partial y}-\sum_s\sigma_s H_s \tag{5.110d}$$

If we consider the flow near a stagnation point, the free stream velocity $u_e=\text{constant}\cdot x=(du_e/dx)x$, Eqs. (5.110) may be transformed into a similar solution by the well-known Mangler-Dorodnitsyn transformation:

$$\xi=\int_0^x \mu_w\rho_w u_e r^2 dx; \quad \eta=\frac{\rho_e u_e}{\sqrt{2\xi}}\int_0^y r\frac{\rho}{\rho_e}dy \tag{5.111}$$

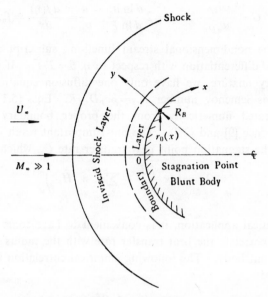

Figure 5.9 Boundary layer near a stagnation point of a blunt body in a hypersonic flow

Eqs. (5.110) becomes

$$\left(\frac{G}{P_r} \sum_{r \neq s} L_{sr} c_{rn} \right)_n - f k_{sn} = \frac{\beta}{\frac{du_e}{dx}} \frac{\sigma_s}{\rho} \qquad (5.112a)$$

$$(G f_{nn})_n + f f_{nn} + \beta \left(\frac{\rho_e}{\rho} - f_n^2 \right) = 0 \qquad (5.112b)$$

$$\left(\frac{c_p G \theta_n}{p_r} \right)_n + c_p f \theta_n - \sum_s \frac{c_{ps} G}{p_r} \theta_n \sum_{r \neq s} L_{sr} c_{rn}$$
$$+ \frac{u_e^2}{T_e} \left[G(f_{nn})^2 + \beta f_n \left(\theta - \frac{\rho_e}{\rho} \right) \right] = \frac{\beta}{\frac{du_e}{dx}} \frac{\sigma_s H_s}{\rho T_e} \qquad (5.112c)$$

where subscript e refers to the free stream value,

$$G = \frac{\mu\rho}{\mu_w\rho_w}, \quad \beta = 2\frac{d\ln u_e}{d\ln \xi}, \quad \frac{u}{u_e} = \frac{df(\eta)}{dx} = f_\eta,$$

$f(\eta)$ is the nondimensional stream function, subscript η refers to the partial differentiation with respect to η, $\theta = T/T_e$. If we consider the binary mixture, we have only one diffusion equation (5.112a). The Lewis-Semenov number $L_{rs} = \rho c_p D_{sr}/K$. Eqs. (5.112) have to be integrated numerically from the proper boundary conditions (see reference [9] and [27]). The most important result of this problem is the stagnation point heat transfer rate Q_w which is

$$Q_w = \left(\kappa\,\frac{\partial T}{\partial y} - \sum_s \sigma_s\, w_s\, H_s \right)_w \qquad (5.113)$$

For practical application, it is convenient to have some simple formula to correlate the heat transfer rate with the radius of the nose of the blunt body. The following empirical correlation formula has been used:

$$Q_w(R_B)^{\frac{1}{2}} = C\rho_\infty^{\frac{1}{2}} U_\infty^3 \qquad (5.114)$$

where Q_w is the stagnation point heat transfer rate and R_B is the nose radius of the blunt body. Subscript ∞ refers to the values in front of the shock. The coefficient C depends on the contribution of the actual chemical species and which is not known *a priori*. Scala and Gilbert[27] suggested that the following formula may be used:

$$\frac{Q_w R_B^{\frac{1}{2}}}{(p_0)^{\frac{1}{2}}(H_0 - H_w)} = (12.0 + 0.866 M_\infty) \times 10^{-3}\ \text{lb}/(\ \text{ft}^{3/2}\cdot\text{sec}\cdot\text{atm}^{\frac{1}{2}})$$

$$(5.115)$$

where Q_w is in B.t.u./ft$^2\cdot$sec; R_B is in feet; p_0, stagnation pressure is in atmospheres; M_∞ is the molecular weight of the ambient gas; H_0, stagnation enthalpy is in B.t.u./lb, and H_w, static enthalpy of the gas evaluated at surface temperature B.t.u./lb. For a first approximation, Eq. (5.115) is applicable to the earth's atmosphere as well as some other planetary atmosphere.

From the results of nonequilibrium flow near a stagnation

point, the main deviation in the heat transfer parameter from that of low temperature perfect gas value is due to the variation of the product of the density and coefficient of viscosity $\mu\rho$ across the boundary layer. If the wall catalyzes atomic recombination, the total heat transfer is not much affected by a nonequilibrium state of the boundary layer if the Lewis-Semenov number is near one. If the wall is noncatalytic, the heat transfer may be appreciably reduced when the boundary layer is frozen throughout, i.e., when the recombination reaction time become much longer than the time for a particle to diffuse through the boundary layer.

16. Turbulent Boundary Layer with Chemical Reactions[16]

In some engineering problems, the flow is turbulent rather than laminar as we discussed in previous sections. Our knowledge of turbulent flow without chemical reaction is still very meager and hence our knowledge of turbulent flow with chemical reactions is even more meager. However, we may extend the semi-empirical treatment of turbulent flow without chemical reaction to the case with chemical reaction.

The effects of turbulence are twofold: one is to increase the transport coefficients values and the other is to increase the chemical reaction rate. Since we have three different transport coefficients in flow with chemical reactions, we have three turbulent transport coefficients. The only successful semi-empirical theory of turbulent shear flow is that for the two-dimensional or axi-symmetrical case. We can only expect that a similar situation exists in a turbulent flow with chemical reaction. For a turbulent boundary layer flow, we have the following transport phenomena:

(1) Momentum transport. Shearing stress consists of eddy shearing stress and viscous shearing stress as follows:

$$\tau_v = -\rho\overline{u'v'} + \mu\frac{\partial u}{\partial y} = \rho\left(-\frac{\overline{u'v'}}{\frac{\partial u}{\partial y}} + \nu_g\right)\frac{\partial u}{\partial y} = (\rho\epsilon_t + \mu)\frac{\partial u}{\partial y}$$

$$(5.116a)$$

where prime refers to turbulent fluctuating value; without prime refers to the mean value of a variable and ϵ_t is the well known turbulent exchange coefficient. Strictly speaking, the turbulent stress is much more complicated than the simple form $-\rho \overline{u'v'}$ which is true for incompressible fluid only. For compressible fluids, we have contribution due to fluctuations of density too. However, since the turbulent exchange coefficient is essential a symbolic form, we include implicitly the effects of density fluctuation in the turbulent exchange coefficient when we give the empirical formula for ϵ_t for a turbulent flow with chemical reaction.

(2) Heat transport. Heat conductive flux consists of one due to turbulent fluctuations and the other due to molecular transport:

$$-q_c = \rho \left(-\frac{\overline{vH'}}{\dfrac{\partial H}{\partial y}} + \frac{\nu_q}{P_r} \right) \frac{\partial H}{\partial y} = \rho \left(\epsilon_T + \frac{\nu_q}{P_r} \right) \frac{\partial H}{\partial y}$$

$$= \rho \left(\frac{\epsilon_t}{P_{rt}} + \frac{\nu_g}{P_r} \right) \frac{\partial H}{\partial y} \tag{5.116b}$$

where ϵ_T is known as the thermal turbulent exchange coefficient which may be different from the turbulent exchange coefficient ϵ_t. For a first approximation, we may assume that $\epsilon_t = \epsilon_T$ but in a more accurate investigation, ϵ_T is usually a little larger than ϵ_t. P_{rt} is the turbulent Prandtl number which is unity if we assume $\epsilon_t = \epsilon_T$.

(3) Mass transport. The flux of mass diffusion consists also of two parts: one is the turbulent part and the other is the laminar part as follows:

$$-q_m = \rho \left(-\frac{\overline{v'k_s'}}{\dfrac{\partial k_s}{\partial y}} + \frac{\nu_g}{S_c} \right) \frac{\partial k_s}{\partial y} = \rho \left(\epsilon_c + \frac{\nu_g}{S_c} \right) \frac{\partial k_s}{\partial y}$$

$$= \rho \left(\frac{\epsilon_t}{S_{ct}} + \frac{\nu_g}{S_c} \right) \frac{\partial k_s}{\partial y} \tag{5.116c}$$

where ϵ_c is the concentration turbulent exchange coefficient, which may be different from the turbulent exchange coefficient ϵ_t and or the thermal turbulent exchange coefficient ϵ_T and S_{ct} is the turbulent Schmidt number.

If we apply the Reynolds analogy to our present problem, we have

$$\epsilon_t = \epsilon_T = \epsilon_c; \quad P_{rt} = S_{ct} = 1 \qquad (5.116d)$$

For a first approximation, if we replace the laminar transport coefficients by the corresponding sum of the turbulent and laminar transport coefficients, we may use the solution of the laminar boundary layer flow for the case of turbulent flow to estimate the heat transfer rate, etc., particularly when we assume that ordinary Prandtl number and Schmidt number are one too.

The chemical reaction rate will be influenced by turbulent mixing. A special model may be assumed to determine this influence but it is not possible to use one model for various kinds of turbulent flow. As a first approximation, we may assume that Eq. (5.20) may be used for a turbulent flow where T is the mean temperature of the turbulent flow and the constants B and E_A should be determined experimentally or empirically so that the influence of turbulent fluctuations is included. In conclusion, our knowledge of turbulent flow in general is still in a very unsatisfactory manner and a great effort is needed to understand such a complicated and facilitating phenomenon whether there is chemical reaction or not.

PROBLEMS

1. Discuss the simple radial flow, both two-dimensional and three-dimensional, in a diabatic flow such that the amount of heat addition is a constant Q_t.
2. Discuss the variation of vorticity and circulation in an inviscid reacting gas flow.
3. Derive Eq. (5.64) and discuss the equilibrium constant K in terms of partition function and heat release for a simple dissociating gas.
4. By the help of reference [8], discuss the equilibrium constant R_{cse} of Eq. (5.25).
5. Calculate the velocity and pressure distribution of a uniform flow of a simple dissociating gas over an infinitely wavy wall.

6. Calculate the lift and drag force on a flat plate airfoil at a small angle of attack α in a uniform supersonic inviscid flow of a simple dissociating gas under the condition of (a) equilibrium flow and (b) frozen flow.

7. Consider the propagation of weak waves generated by a piston moving in an infinitely long tube so that Eq. (5.75) holds. Find the solution of the wave equation (5.75) with the initial conditions:

$$\text{at } t=0: \quad u=C=\text{constant} \qquad \text{for } |x|<x_0$$
$$u=0 \qquad \text{for } |x|>x_0$$
$$\frac{\partial u}{\partial t}=0$$

8. Derive the one-dimensional wave equation similar to Eq. (5.75) in cylindrical coordinates where the only velocity component different from zero is the radial velocity component. All the variables depend only on the radial coordinate r and the time t. Find the general solution of the resultant equation.

9. Derive the one dimensional wave equation similar to Eq. (5.75) in spherical coordinates where the only velocity component different from zero is the radial velocity component. All the variables are functions of radial coordinate r and time t only. Find the general solution of this wave equation.

10. Derive the second order approximation of the velocity potential equation for a steady flow in a simple inviscid dissociating gas.

11. Compute the lift, drag, and moment coefficients for an airfoil of rhomboidal cross section in a supersonic two-dimensional steady flow of an inviscid simple dissociating gas at a small angle of attack α.

12. A jet of uniform velocity U_0 is discharged into a medium at rest. If the density of the jet at the exit of the nozzle is slightly different from that of the surrounding stream at rest, derive the linearized equation for the density in the jet where the gas is a simple inviscid dissociating gas. Assume that the flow is two-dimensional and steady. Discuss the density distribution in the jet if (a) the flow is subsonic and (b) the

flow is supersonic. Also in both (a) and (b) cases, consider the three cases of reaction time (i) $\tau=0$, (ii) $\tau=\infty$, and (iii) τ is finite.

13. Discuss the general Prandtl-Meyer flow, the expansion of a supersonic flow around a corner, for a simple dissociating gas.

14. Derive the characteristic relations in supersonic flow of a simple dissociating gas [Eq. (5.88)].

15. Write down the fundamental equations of the one-dimensional steady flow in a nozzle of variable cross section for a simple dissociating gas (5.93) and find an equation for the velocity u. Show that the parameter X of Eq. (5.94) is one of the important parameters in this problem and discuss the solution of this nozzle equation for a de Laval nozzle.

16. Calculate the flow field in a stagnation point region of a two-dimensional blunt body in a hypersonic flow of a simple dissociating gas.

17. Write down the fundamental equations of a viscous and chemical reacting gaseous mixture in nondimensional form and determine the important parameters of these equations.

18. Calculate the pressure ratio, velocity ratio, and temperature ratio across a flame front for a given amount of heat release and a small Mach number M_1 in terms of the power series of M_1^2.

19. Calculate a typical flame structure where $M_1=0.1$ and the temperature across the flame is $T_2/T_1=4.0$.

20. Calculate the shock wave structure for $M_1=4.0$ where the vibrational relaxation phenomenon is taken into account.

21. Write down the fundamental equations of one-dimensional steady flow of a viscous, heat conducting, and simple dissociating gas and discuss the singularities of this system of equations.

22. By von Mises transformation, i.e., using variables x and stream-function ψ instead of x and y, write down the two-dimensional steady flow equations of the boundary layer flow of a binary mixture with first order chemical reaction.

23. Solve the equations obtained in problem 22 for the case of

an isovel mixing between two uniform streams of same velocities but different temperature and concentration, using the method of power series in x. Find the solution of the zeroth order and the first order approximations.

24. Find the similar solution at the stagnation point of a blunt body in a hypersonic flow in which the binary gas mixture is in nonequilibrium dissociating condition.

REFERENCES

[1] Adamson, T. C. Jr. "Ignition and combustion in a laminar mixing zone". Rept. 20—79, JPL, Calif. Inst. of Tech., 1954.

[2] Cambel, A. B., Duclos, D. P., and Anderson, T. P. Real Gases. Academic Press, 1963.

[3] Chapman, S. and Cowling, T. O. The Mathematical Theory of Non-uniform Gases (2nd ed.) Cambridge University Press, 1952.

[4] Chu, B. T. "Wave propagation and the method of characteristics in reacting gas mixtures with applications to hypersonic flow". Brown University report WADC TN 57—213 AD 118, 350, May, 1957.

[5] Clarke, J. F. "The linearized flow of a dissociating gas", *Jour. Fluid Mech.*, Vol. 7, Pt. 4, 1960, pp. 577—595.

[6] Damkoehler, G. *Elektrochem.*, Vol. 46, 1940, p. 601.

[7] Fay, J. A. and Riddell, F. D. "Theory of stagnation point heat transfer in dissociated air", *Jour. Aero. Sci.*, Vol. 25, No. 2, 1958, pp. 73—85.

[8] Fowler, R. H. and Guggenheim, E. A. Statistical Thermodynamics. Cambridge University Press, 1956.

[9] Freeman, N. C. "Non-equilibrium flow of an ideal dissociating gas", *Jour. Fluid Mech.*, Vol. 4, Pt. 4, 1958, pp. 407—425.

[10] Frost, A. A. and Pearson, R. G. Kinetics and Mechanism. John Wiley, 1953.

[11] Glasstone, S., Laidler, K. J., and Eyring, H. The Theory of Rate Processes. McGraw-Hill, 1941.

[12] Gross, R. A. and Esch, R. "Low speed combustion aerodynamics", *Jet Propulsion*, Vol. 24, No. 2, 1954, pp. 95—101.

[13] Hall, J. G., Eschenroeder, A. Q., and Marrone, P. V. "Inviscid hypersonic air flows with coupled nonequilibrium processes", IAS paper, No. 62—67, Inst. of Aero. Sci., 1962.

[14] Hicks, B. L. "Diabatic flow of a compressible fluid". *Quar. Appl. Math.* Vol. 6, No. 3, 1948, pp. 221—237.

[15] Hirschfelder, J. D., Curtiss, C. F., and Bird, R. B., Molecular Theory of Gases and Liquids. John Wiley, 1954.

[16] Lees, L., "Convective heat transfer with mass addition and chemical reac-

tions", in Combustion and Propulsion, Third Agard Coll. Pergamon Press, 1958, pp. 451—499.

[17] Li, T. Y. "Recent advances in nonequilibrium dissociating gasdynamics", *ARS Jour.* Vol. 31, No. 2, 1961, pp. 170—178.

[18] Lighthill, M. J. "Dynamics of a dissociating gas. Part I. Equilibrium flow", *Jour. Fluid Mech.*, Vol. 2, Pt. 1, 1957, pp. 1—32.

[19] Lin, S. C. "A bimodal approximation for reacting turbulent flows. I. Description of the model, II. Example of quasi-one dimensional wake flow", *AIAA Jour.* Vol. 4, No. 2, 1966, pp. 202—216.

[20] Logan, J. G. "Relaxation phenomena in hypersonic aerodynamics", IAS preprint No. 778, Inst. Aero. Sci., 1957.

[21] Moore, J. J. "A solution of the laminar boundary layer equations for a compressible fluid with variable properties including dissociation", *Jour. Aero. Sci.*, Vol. 19, No. 8, 1952, pp. 508—518.

[22] Pai, S.-I. Introduction to the Theory of Compressible Flow. D. Van Nostrand, 1959.

[23] Pai, S.-I. "Laminar jet mixing of two compressible fluids with heat release", *Jour. Aero. Sci.*, Vol. 23, No. 11, 1956, pp. 1012—1018.

[24] Penner, S. S. "Chemical reactions in flow systems", AGARD NATO Agardograph, No. 7, Butterworths Sci. Pub. 1955.

[25] Resler, E. L., Jr. "Characteristics and sound speed in nonisentropic gas with nonequilibrium thermodynamic states", *Jour. Aero. Sci.*, Vol. 24, No. 11, 1957, pp. 785—791.

[26] Scala, S. M. and Baulknight, C. W. "Transport and thermodynamic properties in a hypersonic laminar boundary layer, Part I. Properties of the pure species", *ARS Jour.*, Vol. 29, No. 1, 1959, pp. 39—45.

[27] Scala, S. M. and Gilbert, L. M. "Theory of hypersonic laminar stagnation region heat transfer in dissociating gases. Developments in Mech.", Proc. 8th Midwestern Conf. of Mech., Pergamon Press, 1965.

[28] Shchelkin, K. I. and Troshin, Ya. K. Gasdynamics of Combustion. Mono Book Corp., 1965.

[29] Stein, A. M. "High temperature thermodynamics and chemistry for entry into planetary atmospheres — the state of the art", Report SID 64—1549, North American Aviation, Inc. Space & Information System Div., 1964.

[30] Talbot, L. and Scala, S. M. "Shock wave structure in a relaxing diatomic gas". Tech. Inf. Ser. R608D850, Space Sci. Lab. General Electric Co., 1960.

[31] Tsien, H. S. "Influence of flame front on the flow field", *Jour. Appl. Mech.*, Vol. 18, No. 2, 1951, pp. 188—194.

[32] Vincenti, W. G. "Nonequilibrium flow over a wavy wall", *Jour. Fluid Mech.*, Vol. 6, Pt. 4, 1959, pp. 481—496.

[33] von Kármán, Th. From Low Speed Aerodynamics to Astronautics. Pergamon Press, 1963.

[34] von Kármán, Th., Emmons, H. W., Tankin, R. S., and Taylor, G. I. "Gasdynamics of combustion and detonation". Sec. G of Fundamentals of Gas

Dynamics, H. W. Emmons (ed.), vol. III of High Speed Aerodynamics and Jet Propulsion. Princeton University Press, 1958.

[35]   Wegener, P. P. "Supersonic nozzle flow of a reacting gas mixture", *Phys. Fluids*, Vol. 2, No. 3, 1959, pp. 264—275.

[36]   Witteman, W. J. "Vibrational relaxation in carbon dioxide", *Jour. Chem. Phys.*, Vol. 35, No. 1, 1961, pp. 1—9.

[37]   Wood, W. W. and Kirkwood, J. K. "Characteristic equations for reacting flow", *Jour. Chem. Phys.*, Vol. 27, No. 2, 1957, p. 596.

CHAPTER

MAGNETOFLUID DYNAMICS
AND PLASMA DYNAMICS

1. Plasma Dynamics and Magnetofluid Dynamics

At high temperature, above 10,000 °K, the gas will be ionized. We may use the term Plasma for ionized gas. The properties of an ionized gas or a plasma differ considerably from neutral gas. Hence we may consider the plasma as the fourth state of matter. The most important difference between the properties of a plasma and those of ordinary gas is that in a plasma the electromagnetic force plays a major role. We should consider simultaneously electromagnetic forces and gasdynamical forces in our analysis of the flow problems of a plasma. Many new phenomena occur due to the interaction of the gasdynamical and electromagnetic forces.

The science which deals with the flow problems of a plasma is called Plasma Dynamics. The scope of plasma dynamics is very broad. It contains problems from electrical discharge in rarefied gas, propagation of electromagnetic waves in ionized media, to the so-called Magnetofluid Dynamics. In this chapter, we consider only those problems of plasma dynamics in which the plasma may be considered as a continuum and the electromagnetic forces are of the same order of magnitude as gasdynamic forces and their interaction is important, particularly for those cases known as magnetofluid dynamics.

The plasma may be considered as a mixture of N species which consist of ions, electrons, and neutral particles. In the flow field, ionization and recombination of ions and electrons may occur. Hence in plasma dynamics, we have also the effect of chemical reactions if we consider the ionization process as a chemical reaction. However, since the mass of electron is much smaller than that of ion or neutral particle, the diffusion velocity of electrons is very large and the treatment of diffusion coefficient approximation will not be satisfactory. A better treatment is known as multifluid theory of magnetofluid dynamics which will be discussed in chapter IX and in which we study the temperature, pressure, density, and velocity vector of every species in a plasma. But for a first approximation, if the variation of the composition of the plasma in a flow field is small, it is sufficient to consider the plasma as a single fluid and we use the temperature, pressure, density, and velocity vector of the plasma as a whole to describe the flow field. We shall use this point of view in this chapter. We may call it the classical theory of magnetofluid dynamics. In chapter IX, we shall show under certain condition, we may reduce the multifluid theory results to that of the classical theory.

The most popular name for the branch of fluid dynamics dealing with the flow problems of electrically conducting fluids is known as magnetohydrodynamics (MHD). In the early days of investigations, we could not produce in the laboratory the flow phenomena of gases where the electromagnetic forces are of the same order of magnitude as the gasdynamical forces. Most of the investigations are theoretical. We could check these theoretical results with astronautical observations. However, it was found that these interaction phenomena could be obtained in the laboratory by investigating the electromagnetic phenomena of the flow of electrically conducting liquid such as mercury. Hence the term Magnetohydrodynamics was found and extensively investigated. In magnetohydrodynamics, the compressibility effects of the medium is negligible. Magnetohydrodynamics is derived from the term hydrodynamics which is concerned mainly with water or liquids.

After 1950, engineers and applied physicists began in earnest the investigation of the flow of an electrically conducting fluid because the importance of controlled fusion research and that of space technology became evident. It is possible to produce the flow of a plasma in laboratory in which the electromagnetic forces are of the same order of magnitude as those gasdynamical forces by means of shock tube, electrical discharge, or a combination of them. In such cases, we should consider the effects of compressibility as well as the effects due to electromagnetic force. Hence we should call it Magnetogasdynamics which is derived from the term gasdynamics. However, the term Magnetohydrodynamics has been used in such a very loose manner in current practice that it covers all the branches of plasma dynamics. It is not correct. The best name should be called magnetofluid dynamics which is derived from the general term of fluid dynamics including the effects of magnetic forces. It is evident that both the magnetohydrodynamics and the magnetogasdynamics are branches of magnetofluid dynamics.

When we study the interaction between the fluid dynamical forces and the electromagnetic forces in a plasma, the influence of electric field and magnetic field are, in general, of the same importance. Hence the name of this branch of fluid dynamics should be called electromagnetofluid dynamics. However as we shall show in section 3, there are many problems in which the energy in the electric field is much smaller than that in the magnetic field. As a result, all the electromagnetic variables may be expressed in terms of magnetic field. Hence we consider only the interaction between the magnetic field and the fluid dynamical variables. Hence the name of magnetofluid dynamics is used. Most of the problems discussed in this chapter belong to magnetofluid dynamics. But there are important cases in which the electric field may be the dominant term rather than the magnetic field such as the problems of electrogasdynamics treated in section 12. In such a case, we should treat the electric field as an independent variable from the magnetic field. As we shall show later, even in the frame of ordinary magnetofluid dynamics, the effects of electric field still play an important role in the flow field which will be discussed in detail in following sections.

2. Fundamental Equations of Electromagnetofluid Dynamics

We consider the plasma as a single fluid of definite composition. In order to describe the complete flow field for this case, we have to know the following 16 variables:

(1) the velocity vector of the plasma q which has three components in general, i.e., u^i, $i = 1, 2,$ or 3,

(2) the temperature of the plasma T,

(3) the pressure of the plasma p,

(4) the density of the plasma ρ,

(5) the electrical field strength E which has three components E^i,

(6) the magnetic field strength H which has three components H^i,

(7) the excess electrical charge ρ_e, and

(8) the electrical current density J which has three components J^i.

In order to find these 16 variables, we have to use 16 relations between them as our fundamental equations which are as follows:

(1) The equation of state of the plasma.   Plasma may be considered as an ideal gas; hence the simple equation of a perfect gas may be used, i.e.,

$$p = \rho RT \tag{6.1}$$

where R is the gas constant of the plasma which depends on the composition of the plasma.

(2) Equation of continuity.   The conservation of mass of the plasma gives

$$\frac{\partial \rho}{\partial t} + \nabla \cdot (\rho q) = 0 \tag{6.2}$$

This is the same as that for ordinary gas [see Eq. (3.32)].

(3) Equation of motion.   The equation of motion of a plasma is in vector form as follows:

$$\rho \, \frac{Dq}{Dt} = -\nabla p + \nabla \cdot \tau' + F_e \tag{6.3}$$

The viscous stress tensor τ' is given in Eq. (3.52) and the only body force F which will be considered here is the electromagnetic force F_e which is

$$F_e = \rho_e E + J \times B \tag{6.4}$$

where $B = \mu_e H$ is the magnetic induction.   We neglect the gravitational force F_g and radiation stresses force F_R here [see Eq. (3.56)].

(4) Energy equation. The conservation of energy gives

$$\rho \frac{DU_m}{Dt} = -p(\nabla \cdot q) + \Phi + \nabla \cdot (\kappa \nabla T) + \frac{I^2}{\sigma_e} \tag{6.5}$$

Here we have the Joule heat term I^2/σ_e where I is the electrical conduction current and σ_e is the electrical conductivity of the gas [see Eq. (3.104)].

(5) Maxwell's equations of the electromagnetic fields are

$$\nabla \times H = J + \frac{\partial D}{\partial t} = J + \frac{\partial \epsilon E}{\partial t} \tag{6.6}$$

$$\nabla \times E = -\frac{\partial B}{\partial t} = -\frac{\partial \mu_e H}{\partial t} \tag{6.7}$$

where $D = \epsilon E$ is the dielectric displacement and $B = \mu_e H$ is the magnetic flux density or magnetic induction, ϵ is inductive capacity and μ_e is the magnetic permeability. To Eqs. (6.6) and (6.7), the MKS unit system is used so that in free space, we have

$$\mu_e = 4\pi \cdot 10^{-7} \frac{\text{kg} \cdot \text{m}}{(\text{coulomb})^2} \tag{6.8}$$

and

$$\epsilon = 8.854 \times 10^{-12} \frac{(\text{coulomb})^2 \text{sec}^2}{\text{kg} \cdot \text{m}^3} \tag{6.9}$$

The relation between the electric current density J and the electric conduction current I is

$$J = I + \rho_e q \tag{6.10}$$

where $\rho_e q$ is known as electrical convection current and the electrical conduction current is due to the relative motion of the charged particles which is a very complicated molecular phenomenon.

(6) Equation of electric current density. The exact equation of the electrical current J is very complicated because the electric current is due to the complicated motion of all the charged particles, and thus the electric current must depend on the electromagnetic

fields as well as all the gasdynamic variables. The exact differential equation for the electric current density will be discussed in the multifluid theory section of chapter IX. However, if the strength of the magnetic field is not very large and the density of the plasma is not too low, the major influence on the electric current density is due to the electromagnetic field and the so-called generalized Ohm's law may be used as the electric current equation as a first approximation. The generalized Ohm's law is

$$I=\sigma_e(E+q\times B)=\sigma_e E_u \tag{6.11}$$

where σ_e is the electrical conductivity of the plasma. In the first approximation, we may assume that σ_e is a scalar quantity. However, if the magnetic field strength is large and the density of the plasma is low, we should consider the electrical conductivity as a tensor quantity. We shall discuss the tensor electric conductivity in section 13. When the electric conductivity is a tensor, the direction of the conduction current I may be different from that of the electric field E_u in the moving coordinates. The best definition of electric conductivity will be that the appearance of conduction current leads to a transformation of electric energy into heat to the amount:

$$\text{Joule heat} = \frac{I^2}{\sigma_e} \tag{6.12}$$

(7) Equation of conservation of electric charge:

$$\frac{\partial \rho_e}{\partial t}+\nabla\cdot J=0 \tag{6.13}$$

The fundamental equations (6.1), (6.2), (6.3), (6.5), (6.6), (6,7), (6.11), and (6.13) are known as the classical equations for electromagneto-gasdynamics. From these equations, we may obtain the magnetogasdynamical equations which will be studied in sections 3 to 10 and the electrogasdynamical equations which will be discussed in section 12. We shall replace Eq. (6.11) by the corresponding equation with tensor electrical conductivity in section 13. We derive the equations of turbulent flow of magnetohydrodynamics in section 14. We shall also discuss how to modify and extend these fundamental equations to include phenomena which have been neglected in these equations in section 15 as well as in chapter IX.

Before we discuss the fundamental equations as a whole, let us first consider those electromagnetic equations (6.6), (6.7), (6.11), and (6.13) which are new as far as mechanics of fluid is concerned.

In the electromagnetic theory, besides the three fundamental dimensions of mechanics, e.g., length, mass, and time, we have another fundamental dimension, an electromagnetic unit. There are a number of different systems of units in electromagnetic theory in common use. This fact complicates the electromagnetic theory. It is often found that the same equation, e.g., Maxwell's equations (6.6) and (6.7), has different forms in different papers or books. Recently the most common unit system is the MKS unit system which is used in this book and in which the meter is used for unit length, kilogram for unit mass, second for unit time, and coulomb for unit electrical charge. For other unit systems, the readers should refer to standard books of electromagnetic theory[22, 33] or problems 2 and 3.

In electromagnetic theory, sometimes the vector and scalar potentials are used instead of electromagnetic fields. These potentials may be defined as follows:

The divergence of Eq. (6.7) gives

$$\frac{\partial}{\partial t}(\nabla \cdot B) = 0 \tag{6.14}$$

or $\nabla \cdot B =$ constant at every point in the field. This constant must be zero if ever in its past or future history, the magnetic induction B may vanish. Hence we have

$$\nabla \cdot B = \nabla \cdot (\mu_e H) = 0 \tag{6.15}$$

Because of Eq. (6.15), we may introduce a vector potential A such that

$$B = \nabla \times A \tag{6.16}$$

or

$$H = \frac{1}{\mu_e} \nabla \times A \tag{6.16a}$$

with the condition

$$\nabla \cdot A = 0 \tag{6.17}$$

Similarly the divergence of Eq. (6.6) with the help of Eq. (6.13) gives

$$\frac{\partial}{\partial t}(\nabla \cdot D - \rho_e) = 0 \tag{6.18}$$

If in the past or future history, the electric field E and the excess electric charge ρ_e may vanish simultaneously, we have

$$\nabla \cdot D = \nabla \cdot (\epsilon E) = \rho_e \tag{6.19}$$

Eq. (6.19) is known as Poisson's equation.

Substituting Eq. (6.16) into Eq. (6.7), we have

$$\nabla \times \left(E + \frac{\partial A}{\partial t} \right) = 0 \tag{6.20}$$

Because of Eq. (6.20), we may introduce a scalar potential ϕ such that

$$-\nabla \phi = E + \frac{\partial A}{\partial t} \tag{6.21}$$

Substituting the above relations of these vector and scalar potentials into Eq. (6.6), we have

$$\frac{1}{c^2} \frac{\partial^2 A}{\partial t^2} - \nabla^2 A + \frac{1}{\nu_H}\left[\frac{\partial A}{\partial t} - q \times (\nabla \times A) \right]$$

$$= -\frac{1}{c^2}\left[\frac{\partial \nabla \phi}{\partial t} + q \nabla^2 \phi \right] - \frac{1}{\nu_H} \nabla \phi \tag{6.22}$$

where the generalized Ohm's law (6.11) has been used and

$$\nu_H = \frac{1}{\mu_e \sigma_e} = \text{magnetic diffusivity} \tag{6.23}$$

The Maxwell's equations (6.6) and (6.7) and their equivalent equations are valid only for those points in whose neighborhood the physical properties of the medium vary continuously. On the boundary of the flow field, the physical properties of the medium may exhibit discontinuities. For instance, at a solid boundary, the electromagnetic properties of the plasma will change to those of a solid. Across such a surface of discontinuity of electromagnetic properties, the following four boundary conditions of electromagnetic fields hold:

(1) The transition of the normal component of the magnetic induction $B = \mu_e H$ is continuous, i.e.,

$$(B_2 - B_1) \cdot n = 0 \qquad (6.24)$$

where n is the unit normal of the surface of discontinuity. Subscript 1 and 2 refer to the values immediately on each side of the surface of discontinuity.

(2) The behavior of the magnetic field H at the boundary is

$$n \times (H_2 - H_1) = J_s \qquad (6.25)$$

where J_s is the surface current density. For finite electrical conductivity $\sigma_e \neq \infty$, J_s is zero; while for infinite electrical conductivity $\sigma_e = \infty$, J_s may be different from zero.

(3) The transition of the tangential component of the electric field E is continuous, i.e.,

$$n \times (E_2 - E_1) = 0 \qquad (6.26)$$

(4) The behavior of the dielectric displacement $D = \epsilon E$ at this boundary is

$$n \cdot (D_2 - D_1) = \rho_{eS} \qquad (6.27)$$

where ρ_{eS} is the surface free charge density.

For most of our problems of magnetogasdynamics, we may neglect the surface current density J_s and the surface free charge density ρ_{eS}. Hence our boundary conditions of the electromagnetic fields are that the tangential components of H and E and the normal component of B and D are all continuous across a surface of discontinuity which separates a solid body and a fluid or two different fluids. The distinctions between H and B and between D and E should be noticed because the values of μ_e and ϵ may be different on both sides of the boundary.

Because of the electromagnetic boundary conditions, the flow field for a given configuration depends on the electromagnetic properties of the body, e.g., whether it is a conducting body or an insulated body. The gasdynamical boundary conditions remain the same as those discussed in chapter III section 12.

The electrical conductivity σ_e of the fluid is one of the most important physical quantities in magnetofluid dynamics, which determines the value of the magnetic Reynolds number. The value of the electrical conductivity depends on the degree of ionization of the plasma as well as the state variables of the plasma which has been discussed in chapter I section 16. We shall further discuss the tensor electrical conductivity in section 13.

3.   Magnetogasdynamic Approximations

For many important flow problems, the following conditions are satisfied:

(1)   The time scale t_0 of our problem is of the same order of magnitude as L/U where L is the characteristic length and U is the characteristic velocity of the flow field. In other words, the time or frequency parameter R_t of Eq. (4.63) is of the order of unity. It means that we shall not consider the phenomena of very high frequency in which the time scale t_0 is very small.

(2)   The electrical field which may be characterized by a value E_0 is of the same order of magnitude as the induced electric field $q \times B$. In other words, the electric field parameter:

$$R_E = \frac{E_0}{UB_0} \tag{6.28}$$

is of the order of unity or smaller where B_0 is the characteristic magnetic induction. This is a good assumption for very large electric conductivity σ_e, because as $\sigma_e \to \infty$, we would expect that

$$E \cong -q \times B \tag{6.29}$$

otherwise the electrical conduction current I will become very large for a slight motion of the plasma.

(3) The velocity of the flow of the plasma is much smaller than the velocity of light c, i.e., the relativistic parameter R_r of Eq. (4.74) is much smaller than unity.

Under the above three conditions, the displacement current $\partial D/\partial t$ and the terms with excess electric charge ρ_e in the fundamental equations of electromagnetofluid dynamics are negligible in comparison with those terms with magnetic field. For instance

$$\rho_e E \cong \frac{U^2}{c^2} J \times B \tag{6.30}$$

Hence we may neglect all the terms with ρ_e and the displacement current and then express all the electrical current density and the electric field in terms of magnetic field and velocity field as follows:

$$\left.\begin{array}{l} \rho_e = 0 \\[4pt] J \cong I = \nabla \times H \\[4pt] E = \dfrac{J}{\sigma_e} - q \times B = \dfrac{1}{\sigma_e}(\nabla \times H) - q \times B \end{array}\right\} \tag{6.31}$$

As a result, we need to consider the interaction between the magnetic field strength H and the gasdynamic variables only. It is this reason that we call this subject magnetofluid dynamics or magnetohydrodynamics or magnetogasdynamics. Substituting Eqs. (6.31) into Eq. (6.7), we have a single vector equation for the magnetic field H which replaces the ten electromagnetic equations (6.6), (6.7), (6.11), and (6.13) as follows:

$$\frac{\partial H}{\partial t} = \nabla \times (q \times H) - \nabla \times [\nu_H(\nabla \times H)] \tag{6.32}$$

With the help of Eqs. (6.31), the equation of motion (6.3) becomes

$$\frac{\partial q}{\partial t} + (q \cdot \nabla) q - \frac{1}{\rho}(B \cdot \nabla) H = -\frac{1}{\rho}\nabla(p + \frac{1}{2} B \cdot H) + \frac{1}{\rho}(\nabla \cdot \tau') \tag{6.33}$$

and the energy equation (6.5) becomes

$$\rho \frac{DH_s}{Dt} = \frac{\partial p}{\partial t} + \nabla \cdot (q \cdot \tau') + \nabla \cdot (\kappa \nabla T)$$

$$+ (\nabla \times H) \cdot (\nu_H \nabla \times B - q \times B) \qquad (6.34)$$

where $H_s = U_m + RT + \frac{1}{2}q^2$ is the stagnation enthalpy of the plasma. Eqs. (6.32) to (6.34) together with Eqs. (6.1) and (6.2) are the fundamental equations of magnetogasdynamics in which there are nine unknowns: H, p, ρ, T, and q.

4. Some Properties of Magnetofluid Dynamic Equations

First we introduce the following nondimensional parameters:

$$p^* = p/(\rho_0 U^2), \quad \rho^* = \rho/\rho_0, \quad T^* = T/T_0, \quad \tau^* = \tau'[L/(U\mu_0)],$$

$$\mu^* = \mu/\mu_0, \quad \kappa^* = \kappa/\kappa_0, \quad \nu_H^* = \nu_H/\nu_{H_0}, \quad c_p^* = c_p/c_{p_0},$$

$$q^* = q/U, \quad x_i^* = x_i/L, \quad t^* = t/t_0, \quad \nabla^* = L\nabla \qquad (6.35)$$

where star refers to the nondimensional quantities and subscript 0 refers to some reference value.

The Fundamental equations of magnetofluid dynamics in terms of the nondimensional quantities of Eq. (6.35) are:

(1) Equation of state (6.1):

$$\gamma M_0^2 p^* = \rho^* T^* \qquad (6.36a)$$

(2) Equation of continuity (6.2):

$$\frac{1}{R_t} \frac{\partial \rho^*}{\partial t^*} + \nabla^* \cdot (\rho^* q^*) = 0 \qquad (6.36b)$$

(3) Equation of magnetic field (6.32):

$$\frac{1}{R_t} \frac{\partial H^*}{\partial t^*} = \nabla^* \times (q^* \times H^*) - \frac{1}{R_\sigma} \{\nabla^* \times [\nu_H^*(\nabla^* \times H^*)]\} \qquad (6.36c)$$

where $H^* = H/H_0$.

(4) Equation of motion (6.33):

$$\frac{1}{R_t} \frac{\partial q^*}{\partial t^*} + (q^* \cdot \nabla^*)q^* - R_H(H^* \cdot \nabla^*)H^* =$$

$$= -\nabla^*(p^* + \frac{1}{2} R_H H^{*2}) + \frac{1}{R_e} (\nabla^* \cdot \tau^*) \qquad (6.36\text{d})$$

(5) Energy equation (6.34):

$$\left(\frac{1}{R_t} \frac{\partial}{\partial t^*} + q^* \cdot \nabla^*\right)[c_p^* T^* + (\gamma - 1) M_0^2 \frac{1}{2} q^{*2}]$$

$$= (\gamma - 1) M_0^2 \frac{\partial p^*}{\partial t^*} + \frac{(\gamma - 1) M_0^2}{R_e} \nabla^* \cdot (q^* \cdot \tau^*)$$

$$+ \frac{1}{P_r R_e} \nabla^* \cdot (\kappa^* \nabla^* T^*) + (\gamma - 1) M_0^2 R_H (\nabla^* \times H^*)$$

$$\cdot \left\{\frac{\nu_H^*}{R_\sigma} (\nabla^* \times H^*) - (q^* \times H^*)\right\} \qquad (6.36\text{e})$$

In the nondimensional equations of magnetofluid dynamics (6.36), we have the following important nondimensional parameters:

$$R_t = \frac{t_0}{L/U} = \text{time parameter} \qquad (6.37\text{a})$$

$$\gamma = \frac{c_{p0}}{c_{v0}} = \text{ratio of specific heats} \qquad (6.37\text{b})$$

$$M_0 = \frac{U}{a_0} = \text{Mach number} \qquad (6.37\text{c})$$

$$R_e = \frac{\rho_0 UL}{\mu_0} = \text{Reynolds number} \qquad (6.37\text{d})$$

$$P_r = \frac{c_{pe} \mu_0}{\kappa_0} = \text{Prandtl number} \qquad (6.37\text{e})$$

$$R_H = \frac{\mu_e H_0^2}{\rho_0 U^2} = \text{Magnetic pressure number} \qquad (6.37\text{f})$$

$$R_\sigma = \frac{UL}{\nu_{H_0}} = \text{Magnetic Reynolds number} \qquad (6.37\text{g})$$

There are seven important nondimensional parameters in magnetogasdynamics of which the first five in Eqs. (6.37) are those of ordinary gasdynamics while the last two are the new parameters for magnetofluid dynamics. We have discussed the significance of these parameters in chapter IV section 6. Now we shall discuss in this section as well as in the next few sections the influence of these parameters on the properties and solutions of the fundamental equations (6.36).

(a) Case of very small magnetic Reynolds number R_σ. If R_σ tends to zero, we will have no influence of the electromagnetic field on the gasdynamic equations (6.36a), (6.36b), (6.36d), and (6.36e). We have the ordinary gasdynamics. However, it is interesting to consider the case that R_σ is small but not negligible. In this case, we see that the first term on the right-hand side of Eq. (6.36c) will be negligibly small in comparison to the second term. We have then

$$\frac{1}{R_r} \frac{\partial H^*}{\partial t^*} = -\frac{1}{R_\sigma} \{\nabla^* \times [\nu_H^*(\nabla^* \times H^*)]\} \tag{6.38}$$

Eq. (6.38) shows that the magnetic field will not be influenced by the velocity field. Hence we may neglect the induced magnetic field in such problems and use the given external magnetic field in solving the magnetofluid dynamic problems by considering only those gasdynamic equations in Eq. (6.36). Many engineering magnetofluid dynamic problems belong to this class. For small R_σ, the electric current J or I is mainly controlled by the electrical conductivity. Hence it is convenient to use the relation $I = \sigma_e (E + q \times B)$ instead of $I = \nabla \times H$. If we use $I = \sigma_e (E + q \times B)$ instead of $I = \nabla \times H$ in equations of motion and of energy, we will find the following two nondimensional parameters R_E [Eq. (6.28)] and $R_m = (R_H R_\sigma)^{\frac{1}{2}}$ [Eq. (4.82)] in the electromagnetic force terms instead of R_H. For small magnetic Reynolds number R_σ, the parameter R_M is more important than the parameter R_H.

(b) Case of small magnetic pressure number R_H. If the magnetic pressure number R_H is very small, we will have no influence on the gasdynamic field by the electromagnetic forces. Hence in the limit, we may solve the flow field without considering the electromagnetic field. After the velocity field is known, we may then solve Eq. (6.36c) for the magnetic field from the known velocity distribution q.

(c) Analogy of magnetic field of magnetofluid dynamics and vorticity of ordinary gasdynamics. If we assume that the electrical conductivity σ_e is a constant, the equations which govern the magnetic field H in magnetofluid dynamics are Eqs. (6.15) and (6.32):

$$\nabla \cdot H = 0 \qquad (6.39a)$$

$$\frac{\partial H}{\partial t} + (q \cdot \nabla) H - (H \cdot \nabla) q + H (\nabla \cdot q) = \nu_H \nabla^2 H \qquad (6.39b)$$

Eqs. (6.39) are formally identical to the equations which govern the vorticity $\omega = \nabla \times q$ in ordinary gasdynamics if we assume that the coefficient of viscosity μ is constant and that the fluid is barotropic, i.e., the pressure is a function of density only. The equations which govern the vorticity ω under these conditions are

$$\nabla \cdot \omega = 0 \qquad (6.40a)$$

$$\frac{\partial \omega}{\partial t} + (q \cdot \nabla)\omega - (\omega \cdot \nabla)q + \omega (\nabla \cdot q) = \nu_g \nabla^2 \omega \qquad (6.40b)$$

where $\nu_g = \mu/\rho$ is the coefficient of kinematic viscosity of the gas. Since Eqs. (6.39) and (6.40) are identical in form, we may apply all the known theorems of vorticity in a barotropic fluid of ordinary gasdynamics to the magnetic field of magnetofluid dynamics. Particularly, by Helmholtz's theorem, we know that the vortex lines move with the fluid if $\nu_g = 0$; hence we have also that the lines of magnetic force move with the fluid if $\nu_H = 0$, i.e., $\sigma_e = \infty$, the frozen-in field in an infinitely large electrically conducting fluid.

(d) Ideal plasma. When the parameters R_e, P_e, and R_σ are very large, outside the boundary layer region, the fluid may be considered as inviscid, nonheat-conducting, and infinitely electrical conducting. Such a fluid may be referred to as an ideal plasma. The system of the fundamental equations of an ideal plasma is hyperbolic and it is similar to that of ordinary gasdynamics for an inviscid fluid. For hyperbolic equations, disturbances are propagated with finite speeds. For an inviscid fluid in ordinary gasdynamics, disturbances travel with the speed of sound relative to the motion of the fluid. In magnetogasdynamics of an ideal plasma, there is more than one mode of disturbance at each point. We shall discuss these waves in section 8. Actually, the flow field of an ideal plasma represents the conditions outside the boundary layer regions which will be discussed in section 10.

5.   One Dimensional Flows of Magnetofluid Dynamics

We are going to discuss a few typical flow problems in magnetofluid dynamics which illustrate the new features of magnetofluid dynamics different from those of ordinary gasdynamics.

The first problem is the one-dimensional flow problem in which only one component of velocity u is different from zero and all the variables are functions of one spatial coordinate x which is in the direction of u and time t. Such a problem has many engineering applications such as magnetofluid dynamic (MFD) power generation and MFD propulsion, etc. The new features of this problem of MFD are that the components of some vectors such as that of the magnetic field strength H in the direction perpendicular to u should be considered. In other words, we consider the one-dimensional flow under transverse electromagnetic fields. If we apply the one-dimensional flow to the flow in a nozzle, we should use the average value of MFD variables over each cross section of the nozzle. In the resultant equations of such a quasi-one-dimensional flow, many correlation coefficients occur[23]. For simplicity, we consider here two limiting cases only, as follows:

(1) Approximately one dimensional flow under transverse electromagnetic fields. We consider the steady inviscid flow of an electrically conducting fluid in a nozzle of a slowly varying cross section $A(x)$. By approximately one dimensional flow, we mean that we may consider the externally applied magnetic field $B_y = B(x)$ and electric field $E_z = E(x)$, where the subscripts y and z refer to the y- and z-components respectively, as given functions of spatial coordinate x. Our fundamental equations are:

$$A\rho u = \text{constant} \tag{6.41a}$$

$$\rho u \frac{du}{dx} = -\frac{dp}{dx} - \sigma_e B(E + uB) \tag{6.41b}$$

$$\rho u \frac{dU_m}{dx} + \frac{p}{A} \frac{du A}{dx} = \sigma_e (E + uB)^2 \tag{6.41c}$$

$$p = \rho RT \tag{6.41d}$$

Eqs. (6.41) may be considered as a system of two first order total differential equations, as follows:

$$\frac{du}{dx} = \frac{1}{M^2 - 1} \left\{ \frac{u}{A} \frac{dA}{dx} - \frac{\sigma_e B^2}{p} (u - u_1)(u - u_3) \right\}$$

$$= \frac{F_1 (u, M, x)}{M^2 - 1} \tag{6.42a}$$

$$\frac{dM}{dx} = \frac{1}{M^2 - 1} \left\{ \left(1 + \frac{\gamma - 1}{2} M^2 \right) \frac{M}{A} \frac{dA}{dx} \right.$$

$$\left. - \left(1 + \frac{\gamma - 1}{2} M^2 \right) \frac{\sigma_e B^2 M}{up} (u - u_2)(u - u_3) \right\} = \frac{F_2 (u, M, x)}{M^2 - 1} \tag{6.42b}$$

where

$$u_1 = -\frac{\gamma - 1}{\gamma} \frac{E}{B}; \quad u_2 = \frac{(1 + M^2)u_1}{2 + (\gamma - 1)M^2}; \quad u_3 = \frac{\gamma u_1}{\gamma - 1} \tag{6.42c}$$

The velocities u_1, u_2, and u_3 are characteristic velocities of our present problems. For instance, if we consider a nozzle of constant cross-sectional area $A = $ constant, Eqs. (6.42) become

$$\frac{du}{dx} = -\frac{\sigma_e B^2}{(M^2 - 1)p} (u - u_1)(u - u_3) \tag{6.43a}$$

$$\frac{dM}{dx} = -\left(1 + \frac{\gamma - 1}{2} M^2 \right) \frac{\sigma_e B^2 M}{up} (u - u_2)(u - u_3) \tag{6.43b}$$

It is easy to show that the characteristics of the flow field depend on the initial value of velocity u_0 relative to the characteristic velocities u_1, u_2, and u_3. For instance, if the initial Mach number $M = M_0$ is less than one, and $u_0 < u_2 < u_1 < u_3$, both du/dx and dM/dx are positive initially. This is the case of magnetofluid dynamic acceleration. If the initial Mach number M_0 is sufficiently close to unity, the acceleration will soon become infinite and the nozzle is choked.

For other initial conditions, the velocity u may increase monotonically toward u_1, while the Mach number M first increases and then decreases ($u > u_2$) and finally approaches an asympotic value $M_a < 1$. The third possibility is the case of a particular set of initial conditions so that M reaches unity and u reaches u_1 simultaneously. We have a smooth acceleration from a subsonic to a supersonic flow in a nozzle of constant cross section area which is not possible in ordinary gasdynamics.

For a nozzle of variable cross sectional area $A(x)$, we have to solve Eqs. (6.42a) and (6.42b) simultaneously. The well known method of nonlinear mechanics may be used in solving Eqs. (6.42). For instance, in the phase plane M vs. u, Eqs. (6.42) give

$$\frac{dM}{du} = \frac{F_2(u, M, x)}{F_1(u, M, x)} \tag{6.44}$$

Since both F_1 and F_2 are functions of x as well as u and M, the trajectory through any point (u, M) is not unique. However, for any given shape of nozzle $A(x)$ and given electromagnetic fields $B(x)$ and $E(x)$, we can draw a definite set of trajectories for each station x in the nozzle. Furthermore, the most interesting features of Eq. (6.44) are the behaviors near the singular points of this equation. If we restrict our attention to the neighborhood of these singular points, we may consider the equation at the singular point $x = x_c$ and neglect the x-dependence of F_1 and F_2. To illustrate this point, let us consider the case of zero electric field $E = 0$. In this case Eq. (6.44) becomes

$$\frac{dM}{du} = \frac{M}{u}\left(1 + \frac{\gamma-1}{2} M^2\right) \tag{6.45}$$

Integration of Eq. (6.45) gives

$$1 + \frac{\gamma-1}{2}\frac{u^2}{a^2} = \frac{a_0^2}{a^2} \tag{6.46}$$

where a_0 is the stagnation sound speed which is a constant. The relations (6.45) and (6.46) are the same as those in ordinary gasdynamics. We need to consider Eq. (6.42a) alone which for $E = 0$ becomes

$$\frac{du}{dx}=\frac{u}{M^2-1}\left(\frac{1}{A}\ \frac{dA}{dx}-\frac{\sigma_e B^2 u}{p}\right) \tag{6.47}$$

Eq. (6.47) shows that the ponderomotive force in the present case retards the velocity of the flow in the nozzle if the flow is supersonic $M>1$ and accelerates the flow if it is subsonic $M<1$. If either $\sigma_e=0$ or $B=0$, Eq. (6.47) reduces to the form of ordinary gasdynamics and at the singular point $M=1$ and $dA/dx=0$, it is a saddle point. When both σ_e and B are different from zero, the flow behaves quite different from that of ordinary gasdynamics. For pure subsonic flow, the velocity in the nozzle for the same pressure drop in the MFD case is always higher than the corresponding velocity in ordinary gasdynamics, and the maximum velocity occurs at a station downstream of the minimum section in MFD case instead of at the minimum section in the ordinary gasdynamic case. The location of the critical station where maximum velocity occurs is given by the condition:

$$\frac{L}{A}\ \frac{dA}{dx}=\gamma R_\sigma\ \frac{M^2}{M_m^2} \tag{6.48}$$

where L is a reference length, $R_\sigma=\sigma_e\mu_e uL$ is the local magnetic Reynolds number, $M=u/a$ is the local Mach number and $M_m=u/V_H$ is the local magnetic Mach number and $V_H=H_y(\mu_e/\rho)^{1/2}$ is the local Alfven's wave speed.

If the condition (6.48) and $M=1$ occur simultaneously, we have the critical case $x=x_c$ where the singular point of Eq. (6.47) occurs. For ordinary gasdynamics, this singular point is a saddle point. For the present case, depending on the variation of $A(x)$ and $B(x)$, we may have a saddle point, a nodal point or a spiral point. Hence the MFD flow has much more variations than those in ordinary gasdynamics.

(2) Strictly one-dimensional flow under transverse electromagnetic fields. In this type of analysis, we assume that all variables are strictly independent of the transverse coordinates y and z and Maxwell's equations must be obeyed. In quasi-one-dimensional flow or approximately one-dimensional flow, we consider the average value over a section and for local values of variables, they may be

functions of x, y, and z.

The fundamental equations for one-dimensional steady flow in a nozzle of cross sectional area $A(x)$ are

$$A\rho u = \text{constant} \tag{6.49a}$$

$$\rho u \frac{du}{dx} = -\frac{dp}{dx} - B\frac{dH}{dx} \tag{6.49b}$$

$$\rho u c_p \frac{dT}{dx} + \rho u^2 \frac{du}{dx} = \frac{dB}{dx}\left(\nu_H \frac{dH}{dx} - uH\right) \tag{6.49c}$$

$$p = \rho RT \tag{6.49d}$$

$$\nu_H \frac{dH}{dx} = uH + E_0 \tag{6.49e}$$

The main difference between Eqs. (6.49) and Eqs. (6.41) is that the magnetic field H in Eqs. (6.49) is not a given function of x but should be determined by solving this set of equations. For the case of infinitely electrical conductivity $\sigma_e = \infty$ or $\nu_H = 0$, we obtain from Eqs. (6.49):

$$\frac{du}{dx} = \frac{ua^2}{u^2 - a_e^2}\frac{1}{A}\frac{dA}{dx} \tag{6.50}$$

where a_e is the effective sound speed given by the relation:

$$a_e^2 = a^2 + V_H^2 \tag{6.51}$$

In this case, Eq. (6.50) is identical to that of ordinary gasdynamics except that the effective sound speed a_e is used instead of ordinary sound speed a in the denominator. Hence at the throat of the nozzle, the choke speed will be a_e instead of a. The general conclusion of ordinary gasdynamics may be applied here with a_e for a as the critical speed, or with effective Mach number $M_e = u/a_e$ instead of ordinary Mach number $M = u/a$ as the important nondimensional parameter here.

6.   MFD Power Generation[33]

The basic principle of the magnetofluid dynamic (MFD) power

generation, whose popular name is magnetohydrodynamic generator or MHD generator, is the use of the motion of a fluid conductor in the presence of a magnetic field to generate the electric power. The idea was known to Faraday[7] in 1831 who experimented with mercury flowing in a glass tube between the poles of a magnet and proposed use of tidal currents in the terrestrial magnetic field for power generation. Only recently experiments with ionized gases as the conducting fluid have been extensively carried out.

The gaseous MHD generator has some outstanding advantages:

(1) The working fluid acts as the conductor in the generator. It is conceptually simple.

(2) The lack of highly stressed, hot moving parts makes it possible to envisage power cycles with very high peak cycle temperature. Thus, it increases the conversion efficiency.

(3) The absence of heat exchange surfaces interposed between the heat source and the working fluid may reduce the overall size and expense of the generating equipment.

(4) The strong scaling laws inherent in any MHD device make large equipment attractive. For instance, for high power application, such as the power over one megawatt, the MHD power generator seems to be more compact and lighter than other devices.

A simple form of a MHD generator, known as the Faraday MHD generator, is sketched in Fig. 6.1. It consists of a duct through which the gaseous working fluid flows, coils which produce a magnetic field across the duct, and electrodes at the top and bottom of the duct. These electrodes serve much the same purposes as the brushes in a conventional generator. The ionized gas, by virtue of its motion through the magnetic field, has an electromotive force generated in it which drives a current through it, the electrodes, and the external load. Since the magnetic Reynolds number in the duct is usually small we may apply the approximately one-dimensional flow in a transverse magnetic field analysis discussed in the last section to this case. Let us consider the case of constant velocity generator in which the flow velocity is more or less constant and the effect being a drop in pressure as the gas forces itself through the magnetic

field. The fundamental equations for constant velocity generator
are

$$\rho u \frac{dh}{dx} = JE \tag{6.52a}$$

$$\frac{dp}{dx} = -JB \tag{6.52b}$$

$$\rho A = \text{constant} \tag{6.52c}$$

where $h = c_p T$ is the enthalpy of the gas.

The electrical efficiency of this constant velocity generator is

$$\eta_e = \frac{\rho u (dh/dx)}{u(dp/dx)} = \frac{E}{uB} \tag{6.53}$$

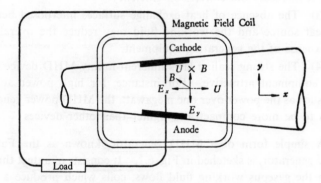

Figure 6.1   A sketch of a Faraday MHD power generator

The turbine efficiency of the generator due to the pressure loss
in the generator is

$$\eta_t = \frac{(h_i - h_f)_{\text{actual}}}{(h_i - h_f)_{\text{isentropic}}} = \frac{1 - (p_f/p_i)^{\eta_e(\gamma-1)/\gamma}}{1 - (p_f/p_i)^{(\gamma-1)/\gamma}} \tag{6.54}$$

where subscript i refers to the initial value and f refers to the final
value in the process of the generator.

The length L of the generator under the assumption of constant
σ_e, u, η_e, and B is

$$L = \frac{p_i - p_f}{(1 - \eta_e)uB^2} = \frac{p_i}{(1 - \eta_e)uB^2} \left[1 - \left(\frac{h_f}{h_i}\right)^{\gamma/(\gamma-1)\eta_e} \right] \quad (6.55)$$

Eq. (6.55) gives a simple relation between the length of the constant velocity MHD generator and several working parameters. For instance, if we take $p_i = 3$ atmospheres, $B = 20$ kilogausses, the maximum temperature of $3000\,°K$, $\sigma_e = 20$ to 40 mhos/meter, $u = 1000$ meters/sec, $\eta_e = 0.7$ to 0.8, we have $L = 6$ to 20 meters. Such a length is acceptable. However, if a large power extraction is desired, a large initial pressure p_i is required and the length may become excessive.

In the formula (6.54), we consider the ohmic heating loss of the gas only. For actual operating conditions, the following losses will also occur and they will decrease the turbine efficiency accordingly.

(1) Eddy current losses. Eddy current may occur whenever there is a rapid space variation in electric or magnetic field strength or in gas velocity. The places where eddy currents usually occur are the entrance and exit sections of the generator. Under typical conditions, they will be roughly 10% of the entrance stagnation pressure.

(2) Aerodynamic losses. Viscous drag in the boundary layer of the duct and diffuser losses will occur in a MHD generator. Such losses may be relatively small.

(3) Electrode losses. The electrodes will have a space charge sheath and potential drop associated with them much as in the conventional arc-type gas discharge. Such a loss is usually not more than 1%.

(4) Heat transfer losses. Generator wall heat transfer is apt to be one of the largest single source of loss in MHD generator. Thus, the necessity for minimizing it will play an important role in generator design. For a simple straight duct generator, one should avoid a duct length to diameter greater than about 20 and a high aspect ratio cross section.

(5) Magnetic field coil losses. Another important loss is the Joule dissipation in the coils producing the magnetic field. The

prospect of reducing the Joule loss on the field coil is greatly improved by the recent development of superconductors. The successful development of a MHD power generator depends greatly on the success of the development of superconductor.

One of the most important factors in the development of the MHD generator is how to increase the electrical conductivity of the gas without increasing the temperature of the gas to a very high temperature. Combustion gases may be seeded with additives of low ionization potential (e.g., cesium or potassium) to obtain high electrical conductivity. At the present time, it is possible to obtain the gas electrical conductivity of the order of 20 mhos/meter at the temperature of 2400°K, 1% potassium and one atmosphere pressure. It is desirable to increase the electrical conductivity to 100 mhos/meter at less than 2500°K temperature.

Another possibility in the future of MHD power generation is to produce high temperature plasma from fusion reaction. The design of such a MFD power generator will be entirely different from the current concept shown in Fig. 6.1 because of the confinement and stability problems.

7. Channel Flow of Magnetofluid Dynamics

In the previous section, we assumed that the fluid is inviscid. At a low Reynolds number, the viscous effect may not be neglected. In this section, we consider the viscous flow in a two-dimensional channel of an electrical conducting liquid. Such a flow had been first studied by Hartmann and Lazarus[13] in the 1930's. We are going to discuss a generalized Hartmann flow as follows:

We consider the fully developed flow between two parallel plates under transverse electromagnetic fields of a steady two-dimensional laminar flow of an incompressible and electrically conducting fluid of constant viscosity and constant electrical conductivity. We assume that the two walls situated at $y = \pm L$. One of the walls, at $y = -L$, is at rest and an insulated wall; while the other wall is moving with a speed u_w in the x-direction or at rest (Fig. 6.2). There is a constant magnetic field in the y-direction $H_{y0} = H_0$ and a constant electric field in the z-direction $E_{z0} = E_0$. All the variables are functions of y only except the pressure p.

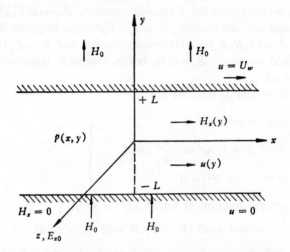

Figure 6.2 Generalized Hartmann flow of magnetohydrodynamics

We use the following nondimensional variables with star*:

$$u = U_0 u^*(y^*); \quad v = w = 0; \quad E_x = E_y = 0; \quad E_z = E_0 \, (E_z^* = 1);$$

$$y = Ly^*; \quad H_x = H_0 H_x^*(y^*); \quad H_y = H_0 (H_y^* = 1); \quad H_z = 0;$$

$$p_x = \frac{\partial p}{\partial x}; \quad x = Lx^*; \quad p = \rho U_0^2 p^*(x^*, \, y^*) = \rho U_0^2 [x^* p_x^* + p_1^*(y^*)]$$

$$\text{(6.56)}$$

where U_0 and L are reference velocity and length respectively. Our fundamental equations of magnetohydrodynamics in the present problem in nondimensional form are:[25]

$$\frac{1}{R_h^2} \frac{d^2 u^*}{dy^{*2}} - u^* = \frac{R_e}{R_h^2} p_x^* + R_E = C = \text{constant} \qquad \text{(6.57)}$$

$$\frac{dp_1^*}{dy^*} = R_\sigma R_H H_x^* (u^* + R_E) \qquad \text{(6.58)}$$

$$\frac{du^*}{dy^*} + \frac{1}{R_\sigma} \frac{d^2 H_x^*}{dy^{*2}} = 0 \tag{6.59}$$

where $R_e = (\rho U_0 L)/\mu =$ the Reynolds number, $R_H = (B_0 H_0)/(\rho U_0^2) =$ the magnetic pressure number, $R_\sigma = \sigma_e \mu_e U_0 L =$ the magnetic Reynolds number, $R_h = (R_e R_H R_\sigma)^{1/2} =$ Hartmann number and $R_E = E_0/(U_0 B_0) =$ electric field number. $B_0 = \mu_e H_0$ is the magnetic induction of the external field.

The boundary conditions are:

(a)   at $y^* = -1$, $u^* = 0$

(b)   at $y^* = +1$, $u^* = \dfrac{u_w}{U_0} = u_w^*$ $\left.\begin{array}{c} \\ \\ \\ \\ \\ \end{array}\right\}$ (6.60)

(c)   at $y^* = -1$, $H_x^* = 0$

The general solution of Eq. (6.57) for u^* is

$$u^* = A \cosh(R_h y^*) + B \sinh(R_h y^*) + C \tag{6.61}$$

where A and B are arbitrary constants to be determined by the boundary conditions. The constant C is the effective electric field parameter or the effective x-wise pressure gradient which is given in our problem and defined in Eq. (6.57). From boundary conditions (6.60), we have

$$u^* = \frac{1}{2} u_w^* \left[\frac{\cosh(R_h y^*)}{\cosh R_h} + \frac{\sinh(R_h y^*)}{\sinh R_h} \right] + C \left[1 - \frac{\cosh(R_h y^*)}{\cosh R_h} \right] \tag{6.62}$$

Eq. (6.62) shows that the nondimensional velocity distribution u^* depends on the Hartmann number R_h, effective x-wise pressure gradient $R_e p_x^*$, electric field number R_E, and the velocity of the upper plate u_w^*. When we plot u^* vs y^*, we have to specify these parameters. Furthermore, the shape of the curves u^* vs y^* depends on the choice of the reference velocity U_0.

If we use the velocity of the upper wall as the reference velocity, we have $u_w^* = 1$ and $u^*(y^*)$ is a function of R_h and C as shown in Fig. 6.3.

If we take the spatial mean velocity V flowing through the

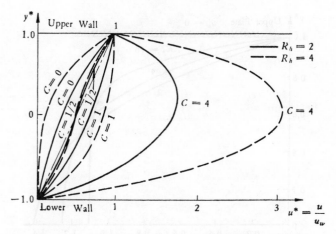

Figure 6.3 Velocity distributions of MHD Couette flow

channel as a reference velocity, we have

$$\int_{-1}^{1} u^* dy^* = 2 \tag{6.63}$$

Substituting Eq. (6.62) into (6.63), we have

$$C = \frac{R_h - \frac{1}{2} u^* \tanh R_h}{R_h - \tanh R_h} \tag{6.64}$$

Then the velocity distribution u^* depends on R_h and u_w^*. Fig. 6.4 shows the velocity distributions for the case $u_w^* = 0$. Even though the velocity distribution u^* of Fig. 6.4 is dependent of C explicitly, the actual mass flow rate depends on both R_h and C because the characteristic velocity is

$$U_0 = \left(\frac{E_0}{B_0} - \frac{p_x}{\sigma_e B_0^2}\right) \Big/ \left(\frac{R_h - \frac{1}{2} u_w^* \tanh R_h}{R_h - \tanh R_h}\right) \tag{6.65}$$

For given values of p_x and E_{z0}, the mass flow Q decreases as the strength of magnetic field H_0 increases.

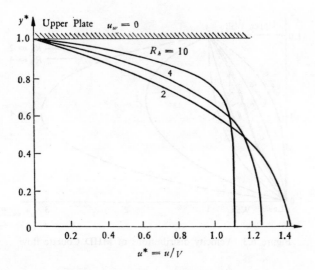

Figure 6.4   Velocity distributions of MHD Hartmann flow $(u_w = 0)$

The z-wise current density is

$$J_z = \sigma_e(E_z + uB_y) = \sigma_e U_0 B_0 (R_E + u^*) \qquad (6.66)$$

The x-wise magnetic field component may be calculated from the equation

$$\frac{dH_x}{dy} = -J_z = -\sigma_e U_0 B_0 (R_E + u^*) \qquad (6.67)$$

With the boundary condition Eq. (6.60c), we have

$$H_x^* = -R_\sigma \left\{ \frac{1}{2} u_w^* \left[\frac{\sinh (R_h y^*) + \sinh R_h}{R_h \cosh R_h} + \frac{\cosh (R_h y^*) - \cosh R_h}{R_h \sinh R_h} \right] \right.$$
$$\left. + C \left[(y^* + 1) - \frac{\sinh (R_h y^*) + \sinh R_h}{R_h \cosh R_h} \right] + R_E (y^* + 1) \right\} \qquad (6.68)$$

It is interesting to notice that the electric field E_0 may be considered as a free parameter which may be chosen by the investigator. Some values of R_E have special significance:

(a) Without electric field, we have $R_E = 0$. In this case, the total current through the channel is not zero and the x-wise magnetic field on the upper wall is also not zero.

(b) If the total current flowing through the channel is zero, we have

$$I_t = \int_{-1}^{1} J_z dy^* = 0 \tag{6.69}$$

or

$$(u_w^* - 2C) \frac{\sinh R_h}{R_h \cosh R_h} + (C + R_E) = 0 \tag{6.69a}$$

If we consider the case of $u_w^* = 0$ and take the mean flow velocity as the reference velocity $U_0 = V$, Eq. (6.69) gives $R_E = -1$. It should be noticed that this critical value of R_E depends on the choice of U_0. For MHD Couette flow of Fig. 6.3, the critical value of $R_E = \frac{1}{2}$ when there is no total current, i.e., when both walls are insulated walls.

After u^* and H_x^* are obtained, we may calculate p_1^* from Eq. (6.58) by simple quadrature. In ordinary hydrodynamics, p is independent of y but in magnetohydrodynamics, p is a function of both x and y.

Also after we obtain u^* and H_x^*, we may calculate the temperature distribution T from the energy equation:[29]

$$\rho u c_p \frac{\partial T}{\partial x} = \frac{\mu c_p}{P_r} \frac{\partial^2 T}{\partial y^2} + \frac{J^2}{\sigma} + \mu \left(\frac{\partial u}{\partial y} \right)^2 \tag{6.70}$$

where $J = \nabla \times H = i_z(-dH_x/dy)$. Because of the pressure p is a function of x and y, the temperature T will also be a function of x and y in general. Only in the case of zero pressure gradient and when the walls have constant temperature, may we neglect the term $\partial T/\partial x$ and assume that the temperature T is a function of y only. For this case, Eq. (6.70) can be easily integrated to obtain $T(y)$. In this case, the temperature distributions depend on both R_h and R_E.

Since the problems of magnetohydrodynamics for electrically conducting liquid, the temperature gradient in the x-direction will not be too large. We may solve Eq. (6.70) for temperature approximately by the following method:

Let Q be the rate of heat transfer at the walls, i.e.,

$$\left(\frac{\partial T}{\partial y}\right)_{y=\pm L} = \frac{Q}{\kappa} \qquad (6.71)$$

By simple heat balance in the channel, we have

$$\rho V L c_p \frac{\partial T}{\partial x} = Q = \text{constant} \qquad (6.72)$$

where V is the average velocity across a cross section of the channel. Then we may write

$$T - T_0 = \frac{Qx}{LV\rho c_p} + f(y) \qquad (6.73)$$

Substituting Eq. (6.73) into (6.70), we have a total differential equation of $f(y)$ which can be easily integrated. This approach has been used in ordinary hydrodynamic Poiseuille flow and is here extended into the MHD case.

8   Waves and Shocks in Magnetofluid Dynamics

The study of wave propagation in an electrically conducting fluid has both academic interest and practical application. The wave motion will bring out many special features in magnetofluid dynamics which may differ significantly from those in ordinary fluid dynamics. The practical applications of wave propagations are numerous. Some important applications are: (i) space communication systems, (ii) wave propagation in ionosphere, (iii) MHD power generation for ac current, (iv) many astrophysical and geophysical problems, etc. Because of these wide range of interests, wave motion in a plasma has been extensively studied.

The properties of a wave in an electrically conducting fluid depend on the amplitude of the wave. The simplest type of wave is the wave of infinitesimal amplitude. Ordinary sound waves and radio waves belong to such a group and they may be considered as special case of magnetogasdynamic waves. Magnetogasdynamic wave is a resultant wave due to the interaction of gasdynamic waves and electromagnetic waves by means of an externally applied magnetic field. Such an interaction will give us many new phenomena which are not in either ordinary gasdynamics nor ordinary electrodynamics. Mathematically speaking, we may linearize the equations which govern the wave of infinitesimal amplitude and the superposition principle is applicable to this wave. We shall study this type of wave first.

For waves of finite amplitude, the shape of the wave will distort as the wave propagates, while the wave of infintesimal amplitude will maintain its shape when it propagates. When the distortion is large, ordinary waves will develop into a shock wave in which a large change of physical variables occurs in a very thin region.

We assume that originally the plasma is at rest with a pressure p_0, a temperature T_0, and a density ρ_0, and that it is subjected to an externally applied uniform magnetic field $H_0 = iH_x + jH_y + k0$, where we choose the coordinate system such that $H_z = 0$ and H_x and H_y are constants, but may be zero. There is no electric field, nor electrical current, nor excess electric charge. The plasma is perturbed by a small disturbance so that in the resultant disturbed motion we have:

$$\left.\begin{array}{l} u = u(x,\ t),\ \ v = v(x,\ t),\ \ w = w(x,\ t),\ \ p = p_0 + p'(x,\ t), \\ T = T_0 + T'(x,\ t),\ \ \rho = \rho_0 + \rho'(x,\ t),\ \ E = E(x,\ t), \\ J = J(x,\ t),\ \ \rho_e = \rho_e(x,\ t),\ \ H = H_0 + h(x,\ t) \end{array}\right\} \quad (6.74)$$

For simplicity, we assume that all the perturbed quantities are functions of one spatial coordinate x and time t only. Thus we consider the wave propagation in the x-direction. We have 16 perturbed quantities. If we substitute Eqs. (6.74) into the fundamental equations of electromagnetogasdynamic equations of section 2, we have 16 linear equations for these variables, if we neglect the higher order terms of the perturbed quantities. These linearized equations may be divided into three independent groups:

(a) $h_x = 0$. The x-component of h is independent of all the other variables and may be set equal to zero without loss of generality.

(b) The equations governing the variables w, h_z, J_x, J_y, E_x,

E_y, and ρ_e are coupled where subscripts x, y, and z refer respectively to the corresponding component of a vector quantity. These equations may be called the equations of transverse waves, because they deal with the velocity and magnetic field component perpendicular to the applied magnetic field H_0.

(c) The rest of the equations which govern the variables u, v, p', ρ', T', h_y, J_z, and E_z are known as the equations of longitudinal waves, in which the sound wave is a special case.

We are looking for a periodic solution, in which all the perturbed quantities are proportional to

$$\exp\left[i(\omega t - \lambda x)\right] = \exp\left[-i\lambda_R(x - Vt)\right]\exp\left(\lambda_i x\right) \qquad (6.75)$$

where ω is a real given angular frequency, $\lambda = \lambda_R + i\lambda_i$ is the complex wave number, $i = (-1)^{\frac{1}{2}}$ and

$$V = \frac{\omega}{\lambda_R} = \text{speed of wave propagation} \qquad (6.76)$$

Substituting the expression in the form of Eq. (6.76) into the linearized equations of electromagnetogasdynamics, we obtain the dispersion relation $\lambda(\omega)$ for both the transverse and the longitudinal waves.

(A) Transverse Waves

The dispersion relation for the transverse wave is

$$\left(i\omega - \nu_H\frac{\omega^2}{c^2}\right)\left[\left(i\omega + \nu_g\lambda^2\right)\left(i\omega + \nu_H\lambda^2 - \nu_H\frac{\omega^2}{c^2}\right) + V_x^2\left(\lambda^2 - \frac{\omega^2}{c^2}\right)\right]$$
$$-\frac{\omega^2}{c^2}V_y^2\left(i\omega + \nu_H\lambda^2 - \nu_H\frac{\omega^2}{c^2}\right) = 0 \qquad (6.77)$$

where ν_g is the kinematic viscosity, ν_H is the magnetic diffusivity, $V_x = H_x(\mu_e/\rho_0)^{\frac{1}{2}}$ is the x-component of the Alfven's wave velocity, $V_y = H_y(\mu_e/\rho_0)^{\frac{1}{2}}$ is the y-component of the Alfven's wave velocity. Eq. (6.77) is a quadratic equation of λ^2 and hence we have two transverse waves. Under MFD approximation, the terms with speed of light c may be neglected and Eq. (6.77) becomes

$$\nu_g\nu_H\lambda^4 + [V_x^2 + i(\nu_g + \nu_H)\omega]\lambda^2 - \omega^2 = 0 \qquad (6.78)$$

If there is no external magnetic field $H_x = 0$, Eq. (6.78) gives two simple modes: one is the viscous wave depending on ν_g, and the other is the magnetic wave depending on ν_H. These are the two basic transverse waves. If H_x is different from zero, there are couplings between these two basic waves and we have two new magnetofluid dynamic transverse waves. For an ideal plasma with $\nu_g = \nu_H = 0$, Eq. (6.78) gives

$$V = \frac{\omega}{\lambda_R} = V_x \qquad (6.79)$$

This is known as Alfven's wave[1] which has a speed of propagation V_x. It was Alfven who first showed that if there is a homogeneous magnetic field H_x in an incompressible and inviscid fluid of density ρ_0 and of infinite electrical conductivity $\sigma_e = \infty$ or $\nu_H = 0$, the disturbance in this fluid will propagate as a wave in the direction of H_x with a speed of V_x. The Alfven's wave and the corresponding velocity of propagation play an important role in magnetofluid dynamics.

(B) *Longitudinal Waves*

The dispersion relation for the longitudinal waves is

$$\left[\kappa \left(\frac{1}{\rho_0} + \frac{i\omega 4\nu_g}{3p_0} \right) \lambda^4 - \left\{ \frac{\omega^2 \kappa}{p_0} + \frac{4\nu_g \omega^2}{3T_0(\gamma-1)} - i c_p \omega \right\} \lambda^2 \right.$$
$$\left. - \frac{i\omega^3}{T_0(\gamma-1)} \right] \left[(\nu_H \lambda^2 + i\omega)(\nu_g \lambda^2 + i\omega) + V_x^2 \lambda^2 \right]$$
$$- \lambda^2 V_y^2 (i\omega + \nu_g \lambda^2) \left[\frac{\omega^2}{T_0(\gamma-1)} - \frac{i\omega \kappa \lambda^2}{p_0} \right] = 0 \qquad (6.80)$$

The first square bracket of Eq. (6.80) represents the sound waves in a viscous and heat-conducting medium and the second square bracket of Eq. (6.80) represents the magnetofluid dynamic transverse wave. The terms with $(1/c^2)$ have been neglected in Eq. (6.80). If there is no transverse magnetic field $H_y = 0$, there is no coupling between the sound waves and the magnetic waves. If H_y is different from zero, i.e., $V_y \neq 0$, there are couplings between the gasdynamic and magnetic waves, and Eq. (6.80) gives in general four coupled waves from the four basic waves: ordinary sound wave, heat wave, viscous wave, and magnetic wave.

The effect of H_y on the sound waves may be seen clearly by considering an ideal plasma in which $\kappa = \nu_g = \nu_H = 0$, Eq. (6.80) gives

$$(a_0^2 - V^2)(V_x^2 - V^2) - V^2 V_y^2 = 0 \qquad (6.81)$$

In this case there are two longitudinal waves because Eq. (6.81) gives two solutions for the velocity of wave propagation $V = \omega/\lambda_R$. We may call one as a fast wave with $V = V_f$ and the other as a slow wave with $V = V_s$. Eq. (6.81) gives the inequality:

$$V_s \leq a_0 = (\gamma R T_0)^{\frac{1}{2}} \leq V_f \qquad (6.82a)$$

and

$$V_s \leq V_x = H_x(\mu_e/\rho_0)^{\frac{1}{2}} \leq V_f \qquad (6.82b)$$

For an ideal plasma, we have three characteristic velocities of wave propagation, i.e., V_x, V_f, and V_s, where V_x is the velocity of propagation for the transverse wave, and the other two, V_f and V_s, are the speeds of wave propagation for the longitudinal waves. The wave patterns in magnetofluid dynamics are considerably different from those in ordinary gasdynamics.

The wave patterns in magnetogasdynamics of an ideal plasma may be shown in Friedrichs' diagram[9,31] of Fig. 6.5 in which the shape of a disturbance that propagates from a point source at the origin is shown. In Fig. 6.5, the abscissa is in the direction of the magnetic field and the flow direction may be in any arbitrary direction. Basically we have two characteristic speeds: one is the ordinary sound speed $a = (\gamma R T)^{\frac{1}{2}}$, and the other is the Alfven's wave speed $V_H = H(\mu_e/\rho)^{\frac{1}{2}}$, where H is the magnitude of the magnetic field strength. The speed of wave propagation in magnetogasdynamics V may be expressed in terms of a and V_H as follows [see Eq. (6.81)]:

$$V = \left[\frac{(a^2 + V_H^2) \pm \sqrt{(a^2 + V_H^2)^2 - 4a^2 V_H^2 \cos^2 \theta}}{2} \right]^{\frac{1}{2}} \qquad (6.83)$$

where θ is the angle between the magnetic field H and the velocity of the point source $OP = U$. The plus sign is for the fast wave, while the minus sign is for the slow wave. Fig. 6.6 shows some typical

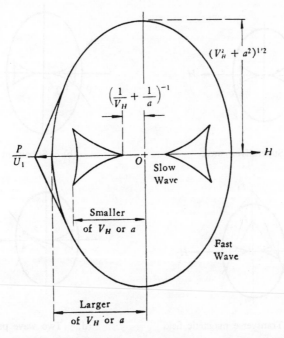

Figure 6.5 Friedrichs' diagram of magnetogasdynamic wave in an ideal plasma

wave patterns for various velocities of the point source. When the velocity of the point source $U = OP_1$ which is larger than both a and V_H and which is parallel to H, we have the wave pattern shown in Fig. 6.6a which is known as super-Alfven flow and which is a set of backward inclined wave same as that in ordinary supersonic flow. If $U = OP_2$, there will be no standing wave which is similar to the subsonic case of ordinary gasdynamics. If $U = OP_3$ in the curved triangle of the slow wave, we have a set of forward inclined wave pattern which is not possible in ordinary gasdynamics (Fig. 6.6b). When the velocity U is perpendicular to H, we have one set of wave if $U = OP_4 < (a^2 + V_H^2)^{1/2}$ and two set of waves if $U = OP_5 > (a^2 + V_H^2)^{1/2}$. The case of two sets of waves is not possible in ordinary gasdynamics (Fig. 6.6c). Finally, if $U = OP_6$, we have two sets of waves: one set is inclined forward and the other, backward.

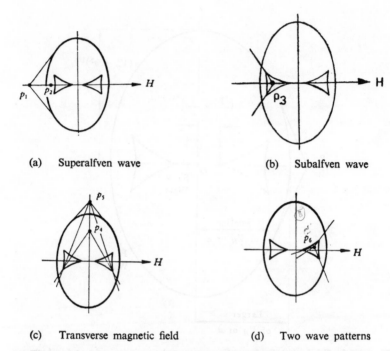

(a)   Superalfven wave   (b)   Subalfven wave

(c)   Transverse magnetic field   (d)   Two wave patterns

Figure 6.6   Wave patterns in magnetogasdynamics of an ideal plasma

This is also a case which cannot occur in ordinary gasdynamics.

It should be noticed that not only we have slow and fast waves of infinitesimal amplitude, we have also slow and fast waves of finite amplitude and the corresponding shock waves in magnetogasdynamics.

If we study the three dimensional unsteady flow of an ideal plasma, we will easily find that the characteristic velocities are V_x, V_f, and V_s which are given by Eq. (6.81) but the local values of a, V_x, and V_y should be used. Similarly, we have three types of magnetogasdynamic shock.

For a magnetogasdynamic oblique shock, we have to consider both the direction of the magnetic field and that of the flow velocity with respect to the shock front. In ordinary gasdynamics, by proper choice of the coordinate system, it is always possible to reduce the oblique shock to a corresponding normal shock case. However.

it is in general not possible to do so in magnetogasdynamics. However, it is possible to choose the coordinate system such that the velocity and the magnetic field are parallel both in front of and behind the shock front. In this way we may classify the shock waves in magnetogasdynamics.

Let us consider the case that the velocity and magnetic field are parallel on both sides of a shock wave front (Fig. 6.7). Let subscript 1 refer to the value in front of the shock and subscript 2, those behind the shock. Let θ be the turning angle of the magnetic field across the shock and ω be the shock angle. Fig. 6.7 shows the orientation of the fast and slow waves:[2, 3]

(1) If the shock angle is less than 90°, we have the fast shock which is similar to that in ordinary gasdynamics.

(2) If the shock angle is between 90° and 180°, we have the slow shock which is different entirely from those of ordinary gasdynamics. We have upstream inclined shock wave here.

(3) When the shock angle is 90°, the tangential component of the magnetic field is zero in front of the shock but different from zero behind the shock. It is known as Switch-on shock Sn.

(4) When the shock angle is equal to $90° + \theta$, the tangential component of the magnetic field is different from zero in front of the shock but vanishes behind the shock. It is known as Switch-off shock Sf.

(5) When the shock angle is equal to $90° + \tfrac{1}{2}\theta$, it is Alfven's shock or the shock corresponds to the transverse wave.

The Rankine-Hugoniot relations across a magnetogasdynamic shock are different from those of an ordinary shock in gasdynamics. To illustrate this difference, let us consider a normal shock under a transverse magnetic field which corresponds to a special case of fast shock with flow velocity u normal to the shock front with the magnetic field parallel to the shock front. In this case, the ratio of the velocity behind the shock u_2 to that in front of the shock u_1 is

$$\frac{u_2}{u_1} = \frac{1}{2}\left(\frac{\gamma-1}{\gamma+1} + \frac{2}{(\gamma+1)M_1^2} + \frac{2\gamma h_1^2}{\gamma+1}\right) +$$

$$+\frac{1}{2}\left\{\left[\frac{\gamma-1}{\gamma+1}+\frac{2}{(\gamma+1)M_1^2}+\frac{2\gamma h_1^2}{\gamma+1}\right]^2+\frac{8\,(2-\gamma)\,h_1^2}{\gamma+1}\right\}^{\frac{1}{2}} \quad (6.84)$$

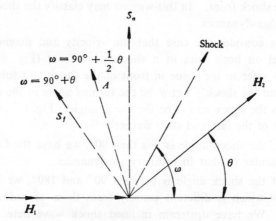

Figure 6.7 Magnetogasdynamic oblique shocks with $U \parallel H$

where $M_1 = u_1/a_1$ is the Mach number in front of the shock and $h_1 = H_1/(2\rho_1 u_1/\mu_e)^{\frac{1}{2}}$ is the nondimensional magnetic field in front of the shock. When there is no magnetic field, Eq. (6.84) reduces to the normal shock relation in ordinary gasdynamics. For a given value of M_1, the ratio u_2/u_1 increases and the strength of the shock decreases with the increase of magnetic field h_1. This occurs because the shock strength depends on the effective Mach number $M_{e1} = u_1/a_{e1}$ and

$$a_{e1} = (a_1^2 + V_y^2)^{\frac{1}{2}} \quad (6.85)$$

If $h_1^2 > h_c^2 = \frac{1}{2}\left(1-\frac{1}{M_1^2}\right)$ and $M_{e1}<1$, no shock will exist.

9.   Linearized Theory of Magnetogasdynamics

We may use the linearized theory to calculate the flow problems of a thin body in magnetofluid dynamics. Let us consider the case of an inviscid fluid in which the effects of viscosity and heat conductivity may be neglected. However in magnetofluid dynamics, there is a third transport phenomenon which is characterized by the electrical conductivity or magnetic diffusivity ν_H. The value of ν_H may be of the same order of magnitude as the coefficient of kinematic viscosity ν_g or much larger than ν_g. We have to treat them separately.

(1) Case of very large electrical conductivity.[27] In this case, we may consider an ideal plasma in which all the transport phenomena may be neglected, i.e., $\nu_g = \kappa = \nu_H = 0$. Let us consider a two-dimensional steady flow of a uniform stream with velocity U over a thin body under a uniform externally applied magnetic field. The velocity vector in the present problem has the following components:

$$u = U + u', \quad v = v', \quad w = 0$$

where $U \gg u'$, v', the perturbed velocity components. The magnetic field has the components

$$H_x = H_{0x} + h_x, \quad H_y = H_{0y} + h_y, \quad H_z = 0$$

where H_{0x} and H_{0y} are the applied x- and y-component of the magnetic field and h_x and h_y are the corresponding perturbed magnetic field components, and H_{0x}, H_{0y} are much larger than h_x and h_y.

For an ideal plasma, we may consider isentropic flow and then the pressure is a function of density only. We write:

$$\rho = \rho_0 + \rho', \quad \frac{dp}{d\rho} = a_0^2$$

where a_0 is the speed of sound of the undisturbed flow.

If we substitute the above variables into the fundamental equations of magnetogasdynamics (6.32) to (6.34) and neglect the higher order terms of the perturbed quantities, we have the following linearized equations:[27]

$$\left[(1 - M^2)(1 - R_{Hx}) + M^2 R_{Hy} \right] \frac{\partial^4 J}{\partial x^4} + \left[1 - (1 - M^2)(R_{Hx} + R_{Hy}) \right]$$

$$\times \frac{\partial^4 J}{\partial x^2 \partial y^2} - R_{Hy} \frac{\partial^4 J}{\partial y^4} - 2(R_{Hx} R_{Hy})^{1/2} \frac{\partial^2}{\partial x \partial y} \left(\frac{\partial^2 J}{\partial x^2} + \frac{\partial^2 J}{\partial y^2} \right) = 0$$

$$(6.86)$$

where

$$M = \frac{U}{a_0}, \quad R_{Hx} = \frac{\mu_e H_{0x}^2}{\rho_0 U^2} = \frac{V_x^2}{U^2}, \quad R_{Hy} = \frac{\mu_e H_{0y}^2}{\rho_0 U^2} = \frac{V_y^2}{U^2},$$

$$J_z = \frac{\partial h_y}{\partial x} - \frac{\partial h_x}{\partial y} = J$$

The solution of Eq. (6.86) over a thin airfoil depends on the parameters M, R_{Hx}, and R_{Hy} and in general, the results are similar as those given in Fig. 6.6.

For instance, if we consider the aligned field case in which H_0 and U are parallel, i.e., $R_{Hy}=0$, Eq. (6.86) reduces to

$$(1-M^2)(1-R_{Hx})\frac{\partial^2 J}{\partial x^2} + \left[1-(1-M^2)R_{Hx} \right]\frac{\partial^2 J}{\partial y^2} = 0 \quad (6.87)$$

This is the case corresponding to Fig. 6.6 (a) and (b). If $M>1$ and $R_{Hx}<1$, we have the case $U=OP_1$. Eq. (6.87) is of hyperbolic type, i.e., the two terms of Eq. (6.87) are of opposite sign. We have the ordinary downstream inclined wave pattern shown in Fig. 6.6 (a). If $M>1$ and $R_{Hx}>1$, we have the case $U=OP_2$, Eq. (6.87) is of elliptic type, i.e., the two terms of Eq. (6.87) are of the same sign. Similarly if $M<1$ and $R_{Hx}>1$, we also have the case $U=OP_2$ with $V_x<a_0$, while previously we have $V_x>a_0$. There will be no standing wave pattern. When $M<1$ and $R_{Hx}<1$, we have the hyperbolic equation (6.87) if $R_{Hx}<1/(1-M^2)$ which corresponds to the case $U=OP_3$ inside the curved triangle and the elliptical equation (6.87) if $R_{Hx}>1/(1-M^2)$ which is the case U is smaller than both a_0 and V_x.

It is a straightforward manner to discuss the nature of the solutions for a cross field case or a field with arbitrary direction with respect to U.

(2) Linearized equation of fluid of low electrical conductivity.[26] In practice, the electrical conductivity is low and the magnetic Reynolds number is of the order of unity, or less than unity. If the magnetic Reynolds number is small, less than unity, we may consider the flow of magnetogasdynamics as a perturbed flow from the corresponding gasdynamic problem with given external applied electro-

magnetic fields E_0 and H_0. The zeroth order electrical current density would be zero and the first order electrical current density J is proportional to electrical conductivity σ_e. We may write in the case of a uniform flow over a thin body with velocity vector $q = i(U+u') + jv' + kw'$ $(U \gg u', v', w')$, the electrical current density J as follows:

$$J = \sigma_e(E + q \times B) \cong \sigma_e(E_0 + Ui \times B_0) = \sigma_e G \qquad (6.88)$$

With finite electrical conductivity, the flow is anisentropic. The pressure p, density ρ, and entropy S of the fluid in the flow field may be written as

$$p = p_0 + p', \quad \rho = \rho_0 + \rho', \quad S = S_0 + S' \qquad (6.89)$$

where subscript 0 refers to the value of uniform flow. It is easy to show that

$$S' = \frac{\sigma_e}{\rho_0 T_0 U} \int_{-\infty}^{x} G^2 dx \qquad (6.90)$$

where we take $S_0 = 0$.

The linearized equation for the gasdynamic variables Q (u', v', w', or ρ') is of the following form:

$$\nabla_M^2 Q = \left[(1 - M^2) \frac{\partial^2}{\partial x^2} + \frac{\partial^2}{\partial y^2} + \frac{\partial^2}{\partial z^2} \right] Q = F_Q \qquad (6.91)$$

where $M = U/a_0$ and F_Q is a known function of the applied electromagnetic fields of the body, i.e., E_0 and B_0.

The general solution of Eq. (6.91) may be written as follows:

$$Q = Q_0 + Q_1 \qquad (6.92)$$

where Q_0 is the solution of the homogeneous equation of Eq. (6.91) with $F_Q = 0$ and the corresponding boundary condition on the body. Hence Q_0 is exactly the ordinary gasdynamic solution. Q_1 is a particular solution of Eq. (6.91) with homogeneous boundary condition on the body. Since F_Q is proportional to $R_H R_\sigma$, Q_1 is also proportional to $R_H R_\sigma$. Hence the contribution of the gasdynamic variables due to the electromagnetic fields E_0 and B_0 is proportional to $R_H R_\sigma$. We may use the electromagnetic fields E_0 and B_0 as a control device to change the gasdynamic forces on a magnetized thin body. The effectiveness of this control device is proportional to $R_H R_\sigma$.

The above perturbation theory may be applied to the case where the basic aerodynamic problem is nonlinear too.

10. Boundary Layer Flows and Wake Problems of Magnetofluid Dynamics

In this section, we are going to discuss some flow problems of a viscous and electrically conducting fluid. We shall assume that the Prandtl number of the fluid is of the order of unity and the Reynolds number is very large, but the magnetic Prandtl number (4.81) may be of the order of unity or very small. We are going to treat these two cases of magnetic Prandtl number separately, because when the magnetic Reynolds number is small, we do not have the magnetic boundary layer and when the magnetic Reynolds number is large, we shall have the magnetic boundary layer.

(1) Case of very small magnetic Reynolds number. First we consider the case of large Reynolds number but small magnetic Reynolds number. We have boundary layer flows of velocity and temperature but not for the magnetic field. Hence in the boundary layer flow, we may assume that the applied electromagnetic fields E_0 and H_0 or B_0 are unaffected by the flow field. For simplicity, let us consider the case where there is an externally applied transverse magnetic field H_{0y} in the direction normal to the main flow direction u. The boundary layer direction is y with $u \gg v$. The ponderamotive force has a component in the x-direction only, i.e.,

$$F_e = J \times B = -i\sigma_e u B_{0y}^2 \tag{6.93}$$

where $J = k\sigma_e u B_{0y}$ is in the z-direction only and i is the unit vector in the x-direction. The Joule heat in the present case is

$$\frac{J^2}{\sigma_e} = \sigma_e u^2 B_{0y}^2 \tag{6.94}$$

The two-dimensional boundary layer equation for the present case are:

$$\frac{\partial \rho}{\partial t} + \frac{\partial \rho u}{\partial x} + \frac{\partial \rho v}{\partial y} = 0 \qquad (6.95a)$$

$$\rho \left(\frac{\partial u}{\partial t} + u \frac{\partial u}{\partial x} + v \frac{\partial u}{\partial y} \right) = -\frac{\partial p}{\partial x} + \frac{\partial}{\partial y} \left(\mu \frac{\partial u}{\partial y} \right) - \sigma_e u B_{0y}^2$$
$$(6.95b)$$

$$\rho \left(\frac{\partial h}{\partial t} + u \frac{\partial h}{\partial x} + v \frac{\partial h}{\partial y} \right) = \frac{\partial p}{\partial t} + u \frac{\partial p}{\partial x} + \frac{\partial}{\partial y} \left(\kappa \frac{\partial T}{\partial y} \right)$$
$$+ \mu \left(\frac{\partial u}{\partial y} \right)^2 + \sigma_e u^2 B_{0y}^2 \qquad (6.95c)$$

$$p = \rho RT \qquad (6.95d)$$

where $h = c_p T$ is the enthalpy of the plasma and the pressure p is assumed to be a given function of x and t.

For steady flow, Eqs. (6.95) give the relation:

$$\rho \left(u \frac{\partial h_0}{\partial x} + v \frac{\partial h_0}{\partial y} \right) = \frac{\partial}{\partial y} \left(\mu \frac{\partial h_0}{\partial y} \right) + \left(\frac{1}{P_r} - 1 \right) \frac{\partial}{\partial y} \left(\mu \frac{\partial h}{\partial y} \right)$$
$$(6.96)$$

If the Prandtl number $P_r = c_p \mu / \kappa = 1$, we have

$$h_0 = h + \frac{1}{2} u^2 = \text{constant} = \text{stagnation enthalpy} \qquad (6.97)$$

as a particular integral of Eq. (6.96). This is known as the Busemann relation in ordinary gasdynamics which holds here too. But the corresponding Crocco relation of ordinary gasdynamics does not hold true here.

Eqs. (6.95) may be solved by the usual method of solution of boundary layer equations. For instance, if we consider the case of boundary layer flow over a flat plate for an incompressible steady fluid flow, we may use the power series method in x for the stream-function ψ as follows:

$$\psi = (U \nu_g x)^{1/2} [f_0(\eta) + mx f_2(\eta) + (mx)^2 f_4(\eta) + \cdots \cdots] \qquad (6.98)$$

where $m = \sigma_e B_0^2 y/(\rho U)$, $\eta = y[U/(\nu_g x)]^{1/2}$ and U is the freestream velocity. Substituting Eq. (6.98) into Eq. (6.95b), we have a set of total differential equations for f_n as follows:

$$2f_0''' + f_0'' f_0 = 0 \tag{6.99a}$$

$$2f_2''' = 2f_0' f_2' - f_0 f_2'' - 3f_2 f_0'' + 2f_0' \tag{6.99b}$$

$$2f_4''' = 4f_0' f_4' + 2f_2' f_2' - f_0 f_4'' - 3f_2 f_2'' - 5f_4 f_0'' + 2f_2' \tag{6.99c}$$

etc., with the boundary conditions:

$$\left. \begin{array}{l} \text{at } \eta = 0: \quad f_0 = f_2 = f_4 = \cdots = 0 \\ \eta = 0: \quad f_0' = f_2' = f_4' = \cdots = 0 \\ \eta \to \infty: f_0' = 1, \ f_2' = -1, \ f_4' = \cdots = 0 \end{array} \right\} \tag{6.100}$$

Eq. (6.99a) is the well known Blasius equation of ordinary boundary layer flow. Eqs. (6.99b), (6.99c), etc., are linear equations which can be numerically integrated from the boundary conditions (6.100). Up to the term f_2, the skin friction coefficient is[29]

$$c_f = \frac{\mu}{\frac{1}{2}\rho U^2} \left(\frac{\partial u}{\partial y} \right)_{y=0} = \frac{0.664 - 1.788 mx + \cdots}{(R_{ex})^{1/2}} \tag{6.101}$$

where $R_{ex} = Ux/\nu_g$.

After $u(x, y)$ is obtained, we may either use the relation (6.97) to find the temperature distribution if the wall is insulated or to solve Eq. (6.95c) by method of series expansion with the known values of velocity and streamfunction.

(2) Case of large magnetic Reynolds number. If the magnetic Reynolds number and the ordinary Reynolds number are large, we have the boundary layer equations for velocity, magnetic field and temperature which should be derived from Eqs. (6.32) to (6.34). For two-dimensional flow, we have the following boundary layer equations of a plasma with high magnetic Reynolds number:

$$p = \rho RT \tag{6.102a}$$

$$\frac{\partial \rho}{\partial t} + \frac{\partial \rho u}{\partial x} + \frac{\partial \rho v}{\partial y} = 0 \tag{6.102b}$$

$$\frac{\partial H_x}{\partial t} + u \frac{\partial H_x}{\partial x} + v \frac{\partial H_x}{\partial y} + H_x \frac{\partial v}{\partial y} - H_y \frac{\partial u}{\partial y} = \frac{\partial}{\partial y}\left(\nu_H \frac{\partial H_x}{\partial y}\right)$$

(6.102c)

$$\frac{\partial H_x}{\partial x} + \frac{\partial H_y}{\partial y} = 0 \qquad (6.102d)$$

$$\rho\left(\frac{\partial u}{\partial t} + u \frac{\partial u}{\partial x} + v \frac{\partial u}{\partial y}\right) - B_y \frac{\partial H_x}{\partial y} = -\frac{\partial p}{\partial x} + \frac{\partial}{\partial y}\left(\mu \frac{\partial u}{\partial y}\right)$$

(6.102e)

$$p + \frac{B_x H_x}{2} = p_0(x) = \text{given function of } x \qquad (6.102f)$$

$$\rho\left(\frac{\partial h}{\partial t} + u \frac{\partial h}{\partial x} + v \frac{\partial h}{\partial y}\right) = \frac{\partial p}{\partial t} + u \frac{\partial p}{\partial x} + \frac{\partial}{\partial y}\left(\kappa \frac{\partial T}{\partial y}\right)$$

$$+ \mu\left(\frac{\partial u}{\partial y}\right)^2 + \nu_H\left(\frac{\partial H_x}{\partial y}\right)^2 \qquad (6.102g)$$

In solving Eqs. (6.102), we have the new parameters of magnetic Reynolds number and the magnetic pressure number or the magnetic Mach number. There are many new phenomena. For instance, in the sub-Alfven flow, i.e., $M_m = U/V_x < 1$, we may have upstream influence of the boundary layer flow. Without going to very lengthy calculation, we shall consider the following simple wake problem to illustrate this phenomenon.

(3) Magnetohydrodynamic wakes behind a two-dimensional body.[14] We consider a two-dimensional flow of a uniform stream U over a body (Fig. 6.8) under a uniform magnetic field $H_0 = iH_{x0} + jH_{y0} + k0$. We assume that the compressibility effect is negligible and the deviation of the velocity and magnetic field from the uniform states are small. Hence we write:

$$u = U + u', \quad v = v', \quad H_x = H_{x0} + h_x, \quad H_y = H_{y0} + h_y$$

such that $u', v' \ll U$ and $h_x, h_y \ll H_{x0}, H_{y0}$.

If we neglect the higher order terms of the perturbed quantities u', v', etc., we may obtain the following nondimensional linearized

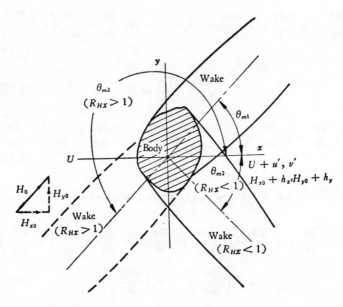

Figure 6.8   MHD wakes around a body in an electrically conducting liquid

equation:

$$\nabla^2\Big[\Big(\frac{\partial}{\partial x}-\frac{1}{R_e}\nabla^2\Big)\Big(\frac{\partial}{\partial x}-\frac{1}{R_\sigma}\nabla^2\Big)$$
$$-\Big(R_{Hx}\frac{\partial}{\partial x}-R_{Hy}\frac{\partial}{\partial y}\Big)^2\Big]u^*=0 \qquad (6.103)$$

where x and y are the nondimensional coordinates in terms of a reference length L, and

$$u^*=\frac{u'}{U}, \quad \nabla^2=\frac{\partial^2}{\partial x^2}+\frac{\partial^2}{\partial y^2}, \quad R_e=\frac{UL}{\nu_g}, \quad R_\sigma=\frac{UL}{\nu_H},$$
$$R_{Hx}=\frac{\mu_e H_{x_0}^2}{\rho\,U^2}, \quad R_{Hy}=\frac{\mu_e H_{y_0}^2}{\rho U^2}$$

The operator of the square bracket may be considered as a product of two Oseen-type operators. For instance, if we assume the $R_e=R_\sigma$, Eq. (6.103) becomes

$$\nabla^2 \left[(1 + R_{Hx}) \frac{\partial}{\partial x} + R_{Hy} \frac{\partial}{\partial y} - \frac{1}{R_e} \nabla^2 \right]$$

$$\times \left[(1 - R_{Hx}) \frac{\partial}{\partial x} - R_{Hy} \frac{\partial}{\partial y} - \frac{1}{R_e} \nabla^2 \right] u^* = 0$$

(6.104)

The two operators in the square bracket of Eq. (6.104) are the Oseen-type operator, which may be written in the general form:

$$\left(A_1 \frac{\partial}{\partial x} + B_1 \frac{\partial}{\partial y} - \frac{1}{R_e} \nabla^2 \right) u^* = 0 \qquad (6.105)$$

The solution of Eq. (6.105) is

$$u^* = \text{constant} \cdot \exp \left[\frac{(A_1 x + B_1 y) R_e}{2} \right] K_0 \left[\frac{(A_1^2 + B_1^2) R_e r}{2} \right] \quad (6.106)$$

where $r^2 = x^2 + y^2$ and K_0 is the Bessel function of zeroth order.

For a given radial distance from the origin, the maximum value of u^* occurs at

$$\tan \theta = \frac{y}{x} = \frac{B_1}{A_1} \qquad (6.107)$$

Thus we have a wake along the direction

$$\theta = \theta_m = \tan^{-1} \left(\frac{B_1}{A_1} \right) \qquad (6.108)$$

Since there are two Oseen operators in Eq. (6.104), we have two wakes which are respectively along the lines (Fig. 6.8):

$$\theta_{m1} = \tan^{-1} \left(\frac{R_{Hy}}{1 + R_{Hx}} \right); \quad \theta_{m2} = \tan^{-1} \left(\frac{-R_{Hy}}{1 - R_{Hx}} \right) \quad (6.109)$$

If $R_{Hx} \geqslant 1$, i.e., $M_m < 1$, we have an upstream inclined wake in sub-Alfven flow, which is impossible in ordinary hydrodynamics. There are two wakes instead of one wake in ordinary hydrodynamics.

From Eq. (6.106), we see that the rate of decrease of u^* with r increases with the Reynolds number R_e and we have boundary layer phenomena for large value of R_e, i.e., the spread of the wake is within a narrow region from the axis which make angles θ_{m1} and θ_{m2} with respect to the direction of the main flow U. Thus if $R_{Hx} > 1$, we have the upper stream influence of the boundary layer flow.

11. Ionization and Saha Relation

So far we consider the plasma as a single fluid of definite composition. Actually the composition, particularly the degree of ionization of the plasma, changes with temperature. If the temperature of the plasma varies through a large range in the flow field, the change of composition of the plasma may be important. The best way to take the variation of the composition into account is the multifluid theory which will be discussed in chapter IX section 4. We may also use the method discussed in chapter V by considering the ionization as a chemical process. The most important factor for the change of composition of a plasma is its degree of ionization. In a plasma, the temperature is high and the ionization is essentially due to thermal ionization which is a combination of ionization processes due to molecular collision, radiation, and electron collisions occuring in the plasma.[22] In plasma dynamics, we are not interested in the detailed processes of ionization but in the degree of ionization for a given flow condition. Some approximate relations may be used. If we use the approach of equilibrium flow of chapter V, we may assume that the degree of ionization corresponds to the equilibrium condition of the local temperature. This relation is known as the Saha relation.[4, 30]

Saha[30] was the first one to make a successful analysis of the degree of ionization of a plasma in an equilibrium state. He considered the dissociation of neutral particles into positive ions and electrons and applied the reaction equation of thermodynamics, known as the law of mass action to the ionization process, i.e.,

One mol of atoms + ionization energy for one mol of neutral atoms = One mol of ions + one mol of electrons

It gives the following equation:

$$\frac{p_i p_e}{p_a} = CT^{5/2} \exp\left(-\frac{eV_i}{kT}\right) \qquad (6.110)$$

where p_i is the partial pressure of ions, p_e is the partial pressure of electrons, p_a is the partial pressure of atoms, e is the elementary electric charge, V_i is the first ionization potential of the atom, k is the Boltzmann constant, T is the absolute temperature of the plasma, and C is a constant which is

$$C = \frac{(2\pi\mu_0)^{3/2} k^{5/2} Z_i Z_e}{h^3 Z_a} \qquad (6.111)$$

where $\mu_0 = m_e m_i / m_a = m_e =$ reduced mass = mass of the electron m_e, h is the Planck constant, Z is the internal partition function defined by

$$Z = \sum_n g_n \exp\left(-\frac{E_n}{kT}\right) \qquad (6.112)$$

where E_n and g_n are respectively the energy and the statistic weight of the nth state of the system. E_0 is taken as zero. The partition function of the electrons is 2 and that for the atoms is one. The partition function for ions may be taken as $4 + 2 \exp[-E_1/(kT)]$ as a first approximation.

Eq. (6.110) may be expressed in terms of the pressure p of the plasma as a whole and the degree of ionization α. For singly charged ions, Eq. (6.110) may be written as:

$$\log_{10}\left(\frac{\alpha^2 p}{1 - \alpha^2}\right) = 2.5 \log_{10} T - \frac{5040 V_i}{T} - 3.62 \qquad (6.113)$$

This is the well known Saha relation in which p is in terms of mm Hg, T is in degrees Kelvin and V_i is in volts.

Eq. (6.110) or (6.113) cannot be used to determine the degree of ionization of sth species in a mixture of different kinds of neutral atoms. This is due to the fact that in equilibrium condition, the electrons produced from one species depend also on the conditions of all the other species in the plasma. The actual degree of ionization depends on the equilibrium condition of the system as a whole. Hence the degree of ionization of sth species depends on the pressure, temperature as well as the mean degree of ionization of the plasma as a whole (see Problem 21).

For a flow problem with variable degree of ionization α, we may consider α as an extra unknown in addition to the other six gasdynamic variables, just as the concentration which we consider in chapter V and the Saha relation (6.113) or (6.110) serves as the additional relation. The influences of the degree of ionization are:

(1) The equation of state should be written as

$$p = (1+\alpha)\rho RT \qquad (6.114)$$

where R is the gas constant of the unionized gas.

(2) The internal energy of the ionized gas when the ions are singly charged is

$$U_m = \frac{RT}{\gamma-1}(1+\alpha) + \alpha R\left(\frac{eV_i}{k}\right) \qquad (6.115)$$

(3) The transport coefficients such as the coefficient of viscosity, the coefficient of thermal conductivity and the electrical conductivity are all functions of the degree of ionization α.

In general the effect of increasing degree of ionization will decrease the temperature of the flow field and increase the density in comparison with the case of constant degree of ionization. When the degree of ionization is high, the temperature of electrons may be different from these of heavy particles and the diffusion velocity of electrons may be very large so that the analysis based on transport coefficients may not be good. As a result, we should use the multifluid theory for such a case as we will discuss in chapter IX section 4.

12. Electrogasdynamics[6, 15]

In our previous discussions from sections 3 to 11, we consider the case where the plasma is essentially neutral, i.e., the number of ions (singly charged) is about the same as that of electrons and the excess electric charge ρ_e is approximately zero. There is another limiting case that the plasma consists of essentially one kind of charged particles, e.g., electrons alone. The study of motion of electron gas has been made by many authors from the point of view of electronics but by only a few authors from gasdynamical point

of view. Such a study would give us many interesting new phenomena about the motion of charged particles. One of the new features is the formulation of a sheath on the wall of a solid body and
the other is the appearance of a density spike in supersonic flow.
In this case, the electric field E is more important than the magnetic
field H. Hence we use the name electrogasdynamics (EGD). There
is also some practical applications, particularly the so-called electrogasdynamic power generation. The basic idea is that ions, forced
to move through a gas by electric field, tend to drag the gas along.
The induced motion of the gas is known as electric wind. We
shall discuss briefly the interaction between the electric field and
the flow of a dielectric gas containing a small percentage
of unipolar ions and some results of EGD power generation in this
section too.[11, 17, 18]

The fundamental equations of electromagnetofluid dynamics
discussed in section 2 still hold true for electrogasdynamics. However, we may simplify those equations of electromagnetofluid dynamics of section 2 by the electrogasdynamic approximations as
follows:

In the present case, the electric field is very large and the excess
electric charge is far from zero. For a gas of single kind of charged
particles, the excess electric charge is simply the electric charge
density of the gas, i.e., $\rho_e = Zen$ where n is the number density of
the gas, Ze is the charge of such particle in the gas. For instance,
for electrons, $Z = -1$ and for singly charged ion, $Z = 1$. The electrical current density $J = \rho_e q$, i.e., the convection electric current is
the main electric current. From Eq. (6.19), the order of magnitude of electric field must be as follows:

$$E \sim \frac{\rho_e L}{\epsilon} \qquad (6.116a)$$

where L is a reference length. The order of magnitude of the
convection current is then

$$J \sim \rho_e U \qquad (6.116b)$$

where U is a reference velocity. If we assume that the time parameter

R_t of Eq. (4.63) of the order of unity, the magnetic field has the following order of magnitude:

$$H \sim \rho_e UL \tag{6.116c}$$

If we use the results of Eqs. (6.116), it is easy to show that those terms with magnetic field H or B are negligibly small in comparison to those with excess electric charge ρ_e. For instance:

$$\rho_e E = \left(\frac{c^2}{U^2}\right) J \times B \tag{6.117}$$

Hence $J \times B$ is negligible in the equation of motion (6.4) in comparison with $\rho_e E$. We call this the electrogasdynamic approximation. Applying this approximation in the fundamental equations of electromagnetofluid dynamics of section 2, we have the following fundamental equations for electrogasdynamics:

In electrogasdynamics, we have ten unknowns: six gasdynamic variables p, ρ, T, q and four electrical variables E and ρ_e, because $J = \rho_e q$ and the terms with magnetic field are negligible. The fundamental equations for these ten variables are:

(1) Equation of state:

$$p = \rho RT \tag{6.118a}$$

(2) Equation of continuity:

$$\frac{\partial \rho}{\partial t} + \nabla \cdot (\rho q) = 0 \tag{6.118b}$$

(3) Equation of motion:

$$\rho \frac{Dq}{Dt} = -\nabla p + \nabla \cdot \tau' + \rho_e E \tag{6.118c}$$

(4) Equation of energy:

$$\rho \frac{DU_m}{Dt} = -p(\nabla \cdot q) + \Phi + \nabla \cdot (\kappa \nabla T) \tag{6.118d}$$

It is interesting to notice that there is no Joule heat term in electrogasdynamics because the electric conduction current is negligible or there is no electrical conduction current at all if we consider strictly the flow of a single species of a kind of charged particles.

(5) The electrical field equation from Eq. (6.7):

$$\nabla \times E = 0 \tag{6.118e}$$

(6) Poisson's equation:

$$\nabla \cdot E = \frac{\rho_e}{\epsilon} \tag{6.118f}$$

Even though Eqs. (6.118) are derived for an ionized gas of one kind of charged particles only, they may be used to a plasma in general if the electric field strength E is much larger than the induced electric field $q \times B$. In other words, when the conditions (6.116) are satisfied, we may use Eqs. (6.118) and the resultant problem may be called the electrogasdynamic problem.

(a) One-dimensional steady flow of an inviscid fluid in a nozzle in electrogasdynamics.[6, 11] We consider the one dimensional flow of an inviscid and nonheat conducting charged fluid in a nozzle of cross-sectional area $A(x)$. In the present case, we have four unknowns: p, n, u and $E_x = E$ where the temperature T may be expressed in terms of p and n, the number density of the gas. The equations governing these four unknowns are:

$$nAu = \text{constant} = A_0 N \tag{6.119a}$$

$$\frac{dp}{dx} + mnu\,\frac{du}{dx} = ZeEn \tag{6.119b}$$

$$\epsilon\,\frac{dE}{dx} = Zen \tag{6.119c}$$

$$p = p(n) \tag{6.119d}$$

where we replace the energy equation by the barotropic relation (6.119d), i.e., the pressure p is a function of the number density n only. Since the energy equation (6.118d) can be reduced to that

$$\frac{dS}{dx} = 0 \tag{6.120}$$

if the flow is adiabatic, we may consider the isentropic flow with $p = $ constant $(n)^\gamma$. The other simple case is the isothermal process, $T = $ constant and $p = $ (constant)n.

Eqs. (6.119) are a set of nonlinear differential equations and the well known method of nonlinear mechanics may be used to study this set of equations. For simplicity, let us consider the case of isothermal flow $T = $ constant in a channel of constant cross-sectional area $A(x) = A_0 = $ constant. For constant area channel, Eqs. (6.119) give

$$p + \frac{mN^2}{n} - \frac{1}{2}\epsilon E^2 = \text{constant} \qquad (6.121)$$

Since $p = p(n)$, Eq. (6.121) gives the phase plane relation between n and E. Clauser[6] discussed extensively the case of isothermal flow, i.e., $p = knT$. Clauser introduced the following nondimensional quantities for the isothermal case:

$$n^* = \frac{n}{n_a}; \quad E^* = \frac{E}{\left(\dfrac{2n_a kT}{\epsilon}\right)^{1/2}} \qquad (6.122)$$

where $n_a = [mN^2/(kT)]^{1/2}$ is the value of n at $u = a = (kT/m)^{1/2}$. Eq. (6.121) becomes

$$n^* + \frac{1}{n^*} - E^{*2} = \text{constant} \qquad (6.123)$$

Eq. (6.123) has been plotted in Fig. 6.9. There is a singular point at $E^* = 0$ and $n^* = 1$ and it is a saddle point. If we substitute the relation $E^* = E^*(n^*)$ obtained from Eq. (6.123) into Eq. (6.119c), we have a first order differential equation in $E(x)$ which can be numerically integrated. Some typical results are shown in Fig. 6.10. The most interesting point is that the integral curve of E^* has cusp at sonic line $n^* = 1$ or $M = u/a = 1$. This means that they cannot represent the real flow condition at the sonic point. The only curves which are reasonable are those integral curves which pass through the saddle point.

(b) Electrogasdynamic power generation and propulsion

device.[5, 17, 18] The basic analysis of electrogasdynamics may be used for electrogasdynamic power generation and for glow discharge as an advance propulsion device. In both cases, the gas is not a single species of one kind of charged particles in the whole system. The medium is essentially a neutral gas with a slight degree of ionization. However, a very large voltage of the order of tens of kilovolts is applied so that the electric field effect will be dominant in the flow field and the problem may be analyzed by the fundamental equations of electrogasdynamics.

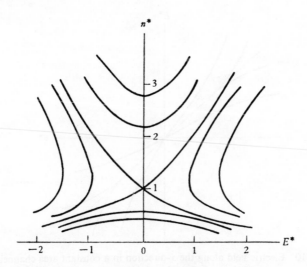

Figure 6.9   One dimensional flow in an isothermal constant area channel in electrogasdynamics

Fig. 6.11 shows the basic concept of electrogasdynamic generator which consists of three sections: an ionizer section where the charges are produced or injected into the flow, a converter section where the mechanical energy of the flow is converted into electrical energy, and the collector section where the transported charges are collected or neutralized.

The principle of the electrogasdynamic propulsion device is as follows: In the electric field, a neutral gas with a slight degree of ionization is known to move in the direction of the field. This is the electric wind. A massive ion is much more efficient than an electron in transferring to the colliding neutral molecule the momentum and energy that ions or electrons obtained from the electric field. Accordingly, the neutral molecules will have a net momentum in the direction of motion of the ions, even though the plasma is locally neutral. The structure carrying the electrodes across which the large electric field are applied will experience a thrust due to the gas motion. The advantages of such a device is that thrust may be derived directly

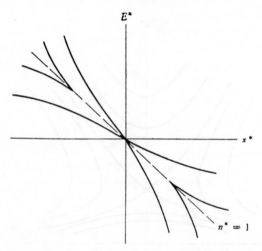

Figure 6.10   Electric field along the x-direction in a constant area channel of an isothermal flow in electrogasdynamics

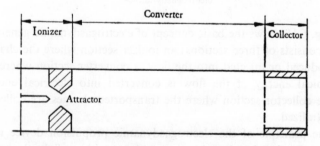

Figure 6.11   A sketch of electrogasdynamic generator

from the electric power at room temperature without any moving parts and that the ambient atmosphere is employed as the propellant.

Both of the above devices are still in the developing stage. At present time, it is difficult to say whether they are feasible for practical use.

13. Tensor Electrical Conductivity[23]

In our analysis of magnetogasdynamics or electromagneto-gasdynamics in previous sections, we use the simple generalized Ohm's law (6.11) for the electrical conduction current which represents the relative motion of charged particles or diffusion phenomena of charged particles. In Eq. (6.11), the electrical conductivity σ_e is assumed to be a scalar quantity. Hence the electrical conduction current I is in the direction of the electric field E_u. For such an electric current in a magnetic field H, there is an electromagnetic force $I \times B$ in the direction perpendicular to both H and I or E_u. Hence the charged particles will move in the direction of $I \times B$, and a Hall current will result. If we take this Hall effect into account, the resultant electrical conduction current will not parallel to E_u and we should consider the electrical conductivity σ_e as a tensor in the generalized Ohm's law. We shall show how to derive the generalized Ohm's law including the Hall effect in chapter IX section 4. Here we simply give the result and discuss its influence on the flow problems of magnetofluid dynamics.

We now consider a partially ionized monatomic gas which consists of electrons, singly charged ions, and atoms. If we assume that the gas is slightly ionized, the relation between the conduction current I and the electric field E_u is

$$I = \frac{1+\beta_i\beta_e+\beta_e^2}{(1+\beta_i\beta_e)^2+\beta_e^2}(\sigma_e E_u^{\|}) + \frac{(1+\beta_i\beta_e)\sigma_e E_n^{\perp}}{(1+\beta_i\beta_e)^2+\beta_e^2}$$

$$+ \frac{\beta_e}{(1+\beta_i\beta_e)^2+\beta_e^2}\left(\frac{B}{B} \times E_u\right) = \sigma_T E_u \qquad (6.124)$$

where σ_e is the scalar electrical conductivity of Eq. (6.11) and

$$E_u = E_u^{\|} + E_u^{\perp}$$

$E_u^{\parallel} = (E_u \cdot B)B/B^2 =$ component of E_u parallel to B and B is the magnitude of B

$E_u^{\perp} = [B \times (E_u \times B)]/B^2 =$ component of E_u perpendicular to B

$\beta_e = \dfrac{en_e B}{K_{ei}} = \dfrac{\omega_c}{f} = WB\left(\dfrac{\rho_0}{\rho}\right)$ with n_e as the number density of electrons

$K_{ei} =$ friction coefficient between electrons and ions

$f = K_{ei}/m_e =$ collision frequency between ions and electrons with m_e as the mass of an electron

$\sigma_e = n_e^2 e^2/K_{ei} =$ scalar electrical conductivity

$\omega_c = en_e B/m_e =$ cyclotron frequency of electron

$\beta_i = en_e B \Big/ \left[\left(1 + \dfrac{n_e}{n_a}\right) K_{ai}\right]$ with n_a as the number density of atoms

and K_{ai} as the friction coefficient between ions and atoms

$\sigma_T =$ is the tensor electrical conductivity defined by Eq. (6.124).

In Eq. (6.124), we neglect the effect of collision between the electrons and atoms. The factor β_e determines essentially the Hall effect while the factor β_i determines the ion slip effect. Fig. 6.12 shows some typical variation of the scalar electrical conductivity and the Hall factor W with the temperature and density of an ionized argon seeded with 1% potassium. If the magnetic field strength is small, we have both β_e and β_i are negligibly small in comparison with unity, Eq. (6.124) reduces to Eq. (6.11):

$$I = \sigma_e E_u \qquad (6.125)$$

Let us consider the case where the ion slip factor β_i is negligible but the Hall factor β_e is not. Eq. (6.124) may be written as follows:

$$I = \sigma_e E_u^{\parallel} + \sigma^{\perp} E_u^{\perp} + \sigma_h(E_u \times B_1) \qquad (6.126)$$

where B_1 is the unit vector in the direction of B. Eq. (6.126) shows that in the direction of the magnetic field H or B, the electrical conductivity of the plasma has the value of scalar electrical conductivity σ_e while the electrical conductivity in the other two directions perpendicular to B has a value less than the scalar electrical conductivity σ_e. In the direction of $E_u^{\perp}$, the electrical conductivity

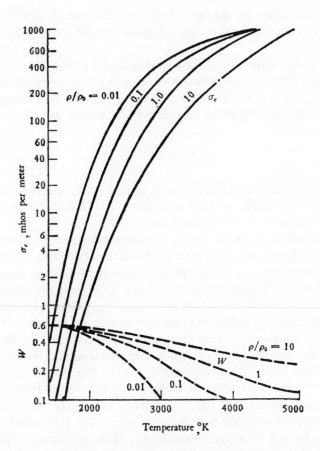

Figure 6.12 Tensor electrical conductivity in argon seeded with 1% potassium as function of density and temperature. ρ_0 is the density of air in standard atmosphere at sea level. (Figure 2 of reference [28], courtesy of AVCO)

is $\sigma^\perp$ which may be written as

$$\sigma^\perp = \frac{\sigma_e}{1+(\omega_c/f)^2} \qquad (6.127)$$

It is evident that $\sigma^\perp$ decreases with increase of magnetic induction B. Only when the ratio (ω_c/f) is negligibly small, $\sigma^\perp$ will be equal to σ_e, i.e., when the magnetic field strength is small so that ω_e is small and the density of the gas is large so that the collision frequency is large.

The last term on the right-hand side of Eq. (6.126) is the Hall current which is in the direction of $(E_u \times B_1)$, i.e., perpendicular to both E_u and B. The electrical conductivity of the Hall component of the electric conduction current may be written as

$$\sigma_h = \frac{\omega_c}{f} \sigma^\perp \tag{6.128}$$

Hence when ω_c/f is negligibly small, we may neglect the Hall current. If ω_c/f is not small, we have to consider the Hall current and the tensor character of the electrical conductivity.

The effect of the ion slip β_i is to decrease the electrical conductivity in all the three directions of the electrical conduction current.

It is possible to repeat all the analyses of classical magneto-fluid dynamics discussed in sections 4 and 10 by using the generalized Ohm's law (6.124) or (6.126) instead of Eq. (6.11). In order to show the essential difference of the flow pattern for the case of tensor electrical conductivity from those with scalar conductivity, we consider the Hartmann flow of section 7 by including the Hall current, i.e., we use Eq. (6.126) to replace Eq. (6.11). The general configuration of the channel is still that shown in Fig. 6.2. With Hall current, we cannot assume that the z-component of velocity w and the z-component of magnetic field H_z to be zero. The Hall current produces a force in the z-direction which causes the motion of the plasma in the z-direction even though there is no net flow in the z-direction. The pressure gradient in the z-direction p_z will not be zero either.

With Hall current, we have the following expressions for the nondimensional unknowns of the present problem.[32]

$$\left.\begin{aligned}
&u = U_0 u^*(y^*), \quad v = 0, \quad w = U_0 w^*(y^*), \\
&E_x = E_{x0}, \quad E_y = E_{y0}, \quad E_z = E_{z0} \\
&H_x = H_0 H_x^*(y^*), \quad H_y = H_0, \quad H_z = H_0 H_z^*(y^*) \\
&p = \rho U_0^2 p^*(x^*, y^*, z^*) = \rho U_0^2 [x^* p_x^* + z^* p_z^* + p^*(y^*)]
\end{aligned}\right\} \tag{6.129}$$

The definition of all the variables is the same as that in Eq. (6.56). The electric field components E_{x0}, E_{y0}, and E_{z0} are constant or zero. The pressure gradients p_x^* and p_z^* are constant. Here we have five unknowns u^*, w^*, H_x^*, H_z^*, and p_1^* which are governed by the following equations:

$$\frac{1}{R_e}\frac{d^2u^*}{dy^{*2}} + R_H\frac{dH_x^*}{dy^*} = p_x^* \tag{6.130a}$$

$$\frac{1}{R_e}\frac{d^2w^*}{dy^{*2}} + R_H\frac{dH_z^*}{dy^*} = p_z^* \tag{6.130b}$$

$$(1+\beta_e^2)\frac{d^2H_x^*}{dy^{*2}} + R_\sigma\frac{du^*}{dy^*} + \beta_e R_\sigma\frac{dw^*}{dy^*} = 0 \tag{6.130c}$$

$$(1+\beta_e^2)\frac{d^2H_z^*}{dy^{*2}} + R_\sigma\frac{dw^*}{dy^*} - \beta_e R_\sigma\frac{du^*}{dy^*} = 0 \tag{6.130d}$$

$$\frac{dp_1^*}{dy^*} = -\frac{1}{2} R_H\frac{d}{dy^*}(H_x^{*2}+H_z^{*2}) \tag{6.130e}$$

where R_e is the Reynolds number, R_σ is the magnetic Reynolds number, and R_H is the magnetic pressure number. All these non-dimensional parameters have the same definitions given in section 7 and $\beta_e = en_eB_0/K_{ei} = \omega c_0/f$. Strictly speaking, β_e depends on the resultant magnetic induction B instead of the applied magnetic induction B_0. Because the induced magnetic induction B_x and B_z are usually small in comparison with B_0 for small R_σ, we may use B_0 to replace B in β_e and β_i.

We consider here only the Poiseuille flow and the corresponding boundary conditions are:

$$\text{at } y^* = \pm 1: u^* = w^* = 0, \ H_x^* = H_z^* = 0$$
$$y^* = 1: p_1^* = p_0^* = \text{constant} \tag{6.131}$$

In solving Eqs. (6.130), it is convenient to introduce the following complex variables:

$$V^* = u^* - iw^*, \ H^* = H_x^* - iH_z^*, \ p_s^* = p_x^* - ip_z^* \tag{6.132}$$

where $i = (-1)^{\frac{1}{2}}$. Eqs. (6.130a) and (6.130b) may be combined and we have

$$\frac{1}{R_e} \frac{d^2 V^*}{dy^{*2}} + R_H \frac{dH^*}{dy^*} = p_s^* \qquad (6.133a)$$

Similarly, Eqs. (6.130c) and (6.130d) give

$$\frac{1}{R_\sigma^*} \frac{d^2 H^*}{dy^{*2}} + \frac{dV^*}{dy^*} = 0 \qquad (6.133b)$$

where

$$R_\sigma^* = \frac{R_\sigma}{1 - i\beta_e} = \text{complex magnetic Reynolds number} \qquad (6.134)$$

Eliminating H^* from Eqs. (6.133), we have

$$\frac{d^3 V^*}{dy^{*3}} - R_h^* \frac{dV^*}{dy^*} = 0 \qquad (6.135)$$

where

$$R_h^* = \frac{R_h}{(1 - i\beta_e)^{\frac{1}{2}}} = \frac{(R_e R_\sigma R_H)^{\frac{1}{2}}}{(1 - i\beta_e)^{\frac{1}{2}}} = \text{complex Hartmann number}$$

$$(6.136)$$

Eq. (6.135) has the same form as Eq. (6.57) except that the complex quantities are used in Eq. (6.135). Hence the general solution of Eq. (6.135) is of the same form as that of Eq. (6.57), i.e., Eq. (6.61):

$$V^* = A_1 \cosh R_h^* y^* + B_1 \sinh R_h^* y^* + C_1 \qquad (6.137)$$

where A_1, B_1, and C_1 are arbitrary complex constants to be determined by the boundary conditions. Some typical velocity distributions of MHD channel flow with Hall current as shown in Fig. 6.13. Since R_h^* is a complex number, in the expression V^*, we have both the sinusoidal terms such as $\sin a_1 y^*$ and $\cos a_1 y^*$ as well as the real hyperbolic terms occurred in the case of scalar electrical conductivity. Since we assume that there is no net flow in the z-direction, the velocity distribution of w is of sinusoidal form with the net flow equal

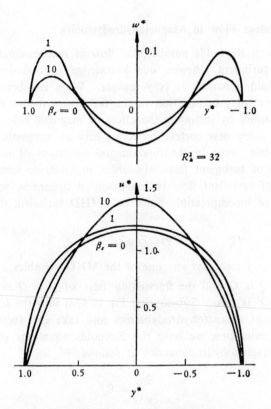

Figure 6.13 Velocity distribution in a MHD channel flow with Hall current to zero. The introduction of such sinusoidal z-velocity component is one of the new features of flow with Hall current.

Similarly, the Hall current or the effect of tensor electrical conductivity makes the flow pattern much more complicated than the corresponding case of scalar electrical conductivity. For instance,[8] if we consider the force on a symmetrical body in a uniform stream without angle of attack under applied transverse magnetic field. When β_e is small, the flow will be symmetrical and there is drag force but no lift force on the body. If β_e is not small, the flow will be asymmetrical and there will be both lift and drag on the body.

14. Turbulent Flow in Magnetohydrodynamics

At high Reynolds number, the flow of magnetofluid dynamics may be turbulent. Again, our knowledge of turbulent flow of magnetofluid dynamics is very meager. What has been done are straightforward generalization of the turbulence theory of ordinary fluid dynamics by including the effect of magnetic field. The new facts are many new correlation terms between magnetic fields and velocity field occur in the fundamental equations of magnetofluid dynamics of turbulent flow. In order to illustrate some essential features of turbulent flow in magnetofluid dynamics, we consider the case of incompressible fluid, i.e., MHD turbulent flow. First we write

$$Q = \bar{Q} + Q' \tag{6.138}$$

where Q may represent any one of the MHD variables. The mean value of Q is $\bar{Q}$ and the fluctuating part of Q is Q'. The mean value of Q' is zero. Substituting Eq. (6.138) into the fundamental equations of magnetohydrodynamics and take the average of the resultant equations, we have the Reynolds equations of turbulent flow of magnetohydrodynamics as follows:[22]

$$\frac{\partial \bar{u}_i}{\partial x_i} = 0 \tag{6.139a}$$

$$\rho \frac{\partial \bar{u}_i}{\partial t} + \frac{\partial}{\partial x_k}(\rho \overline{u_i u_k} - \mu_e \overline{H_i H_k}) = -\frac{\partial}{\partial x_i}(\bar{p} + \frac{1}{2}\mu_e \overline{H^2}) + \mu \frac{\partial^2 \bar{u}_i}{\partial x_k^2} \tag{6.139b}$$

$$\frac{\partial \bar{H}_i}{\partial t} + \frac{\partial}{\partial x_k}(\overline{H_i u_k} - \overline{H_k u_i}) = \nu_H \frac{\partial^2 \bar{H}_i}{\partial x_k^2} \tag{6.139c}$$

where

$$\overline{u_i u_k} = \bar{u}_i \bar{u}_k + \overline{u_i' u_k'}$$
$$\overline{H^2} = \overline{H_i H_i} = \bar{H}_i \bar{H}_i + \overline{H_i' H_i'}$$

and so forth, and the summation convention is used.

The seven equations of Eq. (6.139) may be used to determine the seven mean values of the pressure, three velocity components, and three magnetic field components. We still need some other relations for those turbulent correlation terms such as $\overline{u_i'u_k'}$, $\overline{H_i'H_k'}$, and $\overline{u_i'H_k'}$. Similar to the case of ordinary fluid dynamics, if we try to derive equations for these second order correlation terms, we introduce third order correlation terms. This is one of the main difficulties to study turbulent flow based on Reynolds equations. Usually for engineering problems, we make some empirical relation between these second order correlation terms and the mean flow variables with some empirical constant which should be determined experimentally. This is known as the semi-empirical theory of turbulent flow such as we have discussed for the chemical reaction case in chapter V section 16.

To illustrate this semi-empirical theory of turbulent MHD flow, let us reconsider the two-dimensional channel flow of Fig. 6.2. We assume that the two plates are at rest and insulated. There is no applied electric field. The boundary conditions for the mean and fluctuating quantities are as follows:

(1) Both the mean and the fluctuating velocity components vanish on the walls.

(2) We assume that the electrical conductivity is a scalar, the boundary conditions for the mean magnetic field are

$$\overline{H}_x = 0, \ \ \overline{H}_y = H_0 \ \text{ at } \ |y| \geqq L$$

(3) The curl and divergence of the fluctuating magnetic field on the walls must be zero and we have

$$\overline{H_x'H_y'} = 0, \ \ \overline{H_y'H_z'} = 0, \ \ \overline{H'^2} = \overline{H_x'^2} - \overline{H_z'^2} = 0 \ \text{ at } \ |y| \geqq L$$

Now we use the same type of nondimensional variables as those in Eq. (6.56), i.e., all the velocities are in terms of U_0 and all magnetic fields are in terms of H_0. The Reynolds equations (6.139) for the present case become

$$\frac{1}{R_e}\frac{d^2u^*}{dy^{*2}} + R_H \frac{d\overline{H_x^*}}{dy^*} - \frac{d}{dy^*}(\overline{u'^*v'^*} - R_H\overline{H_x'^*H_y'^*}) = \overline{p_x^*}$$

(6.140a)

$$\frac{1}{R_\sigma}\frac{d^2\overline{H_x^*}}{dy^{*2}} + \frac{d\overline{u^*}}{dy^*} - \frac{d}{dy^*}(\overline{H_x'^*v'^*} - \overline{H_y'^*u'^*}) = 0$$

(6.140b)

$$\overline{p_w^*} - \overline{p^*} = \overline{v'^{*2}} + \frac{1}{2}R_H(\overline{H_x^{*2}} - \overline{H_y'^{*2}} - \overline{H_z'^{*2}})$$ (6.140c)

Eqs. (6.140) have similar form as those of laminar flow case in section 7 except that we have three more Reynolds stresses here. We may consider each of the terms in the parenthesis as a new variable such as $(\overline{H_x'^*v'^*} - \overline{H_y'^*u'^*})$, etc. We have to make some reasonable assumptions to express these Reynolds stresses in terms of the mean flow variables $\overline{u^*}$ and $\overline{H_x^*}$ before we may solve for $\overline{u^*}$ and $\overline{H_x^*}$. Some preliminary attempt has been made in reference [12]. However, since we do not have the experimental results about the velocity distributions of the MHD channel, it is not possible to decide how good such an assumption is. Thus we will not discuss these assumptions here. In reference [12], some empirical formulas for the friction factor f are given:

$$f = 8\frac{\tau_0}{\rho\overline{v^2}} = 8\frac{u_c^2}{\overline{v^2}}$$

(6.141)

where τ_0 is the shearing stress at the wall, u_c defined in Eq. (6.141) is the characteristic velocity and $\overline{v}$ is the average velocity across the channel, and the mean velocity distribution has been proposed as follows:

$$\frac{1}{\sqrt{f}} = 2.0 \log_{10} R^* + 2.53 + \frac{R^*}{8R_h^2}\int_0^{R_h^2/R^*} F_1(x^*)\,dx^*$$

(6.142)

where $F_1(x^*)$ is an empirical function determined by experiments and for $x^* > 6$, we have approximately

$$F_1(x^*) = 2.07 - 5.657 \log_{10} x^*$$

(6.143)

R_h is the Hartmann number based on the mean values of the flow and $R^* = u_c L/\nu_g$.

The velocity distribution is

$$\frac{\overline{u}}{u_c} = 5.657 \log_{10}(R^*\xi) + 6.154 + F_1(R_h^2 \xi/R^*) \qquad (6.144)$$

where $\xi = (L-y)/L = 1 - y^*$.

Some of the general features of the turbulent MHD channel flow from the preliminary experimental results of Hartmann and Lazarus[13] and Murgatroyd[21] are as follows:

(1) The greatest decrease in friction factor is nearly proportional to the 3/2 power of the zero magnetic friction factor.

(2) The friction factor is the best parameter to describe the transition from laminar to turbulent flow. The critical value of f for transition is within the range 0.025 to 0.50.

(3) There is no sharp boundary between the laminar and turbulent MHD flows.

(4) The velocity profiles are constant over the major portion of the central part of the channel when the Hartmann number is very large.

(5) The maximum increase in the quantity $\overline{\nu}/u_c$ obtained by the application of the magnetic field is a constant for all configuration.

Since the experiments of Hartmann-Lazarus and Murgatroyd were performed in channels of very small cross section (0.60×3.72 mm² for Hartmann channel and 0.266×4.0 mm² for Murgatroyd channel) and no velocity profile was measured, there are many uncertainties in these experiments. More refined experimental investigations are needed, particularly for large channel in which the actual velocity profiles may be measured.

15. General Remarks on Plasma Dynamics

In this chapter, we consider only the dynamics of electrically conducting fluid based on continuum theory only. For many plasma dynamics problems, the density of the plasma is very low, the continuum theory may not hold. We shall discuss briefly the plasma dynamics based on kinetic theory in chapter VIII.

Even in the case of continuum theory, we discuss mostly the limiting case of magnetofluid dynamics in which the magnetofluid dynamic approximations of section 3 hold and briefly the electrogasdynamics in which the electric field is the most important electric quantity (see section 12). The general case where both the electrical field and the magnetic field are of same importance has not been studied extensively.

From the kinetic theory (chapter VIII) or from the multifluid theory (chapter IX), we shall show that there are three characteristic lengths (chapter IV) for plasma dynamics, i.e., (i) mean free path L_f [Eq. (4.22)], (ii) Larmor radius L_{Ls} [Eq. (4.27)], and (iii) Debye length L_D. There are two different Larmor radius: one is for electrons L_{Le} and the other is for ions L_{Li}. According to Kantrowitz and Petschek, we may divide the plasma dynamics into five domains according to the values of these characteristic lengths.[19]

(1) S-domain. In this domain, the mean free path of the plasma L_f is much smaller than the dimension of the flow field L ($L/L_f \gg 1$) but both the Larmor radii are larger than L ($L/L_{Ls} \ll 1$). In this domain, the electrical conductivity may be regarded as a scalar quantity and the continuum theory may be used. This domain also covers the flow of an electrically conducting liquid. This is the case where we study mainly in this chapter.

(2) T-domain. In this domain, the electron Larmor radius is of the same order of magnitude as the mean free path but $L/L_f \gg 1$. The ion Larmor radius L_{Li} is still much larger than L. In this domain, the Hall current becomes important and the electrical conductivity should be considered as a tensor quantity because as shown in section 13, ω_c/f is of the order of unity. However, the continuum theory still may be used in the study of the plasma dynamics.

(3) M-domain. In this domain, the ion Larmor radius L_{Li} becomes smaller than the mean free path. Now the magnetic terms become the most important terms in the fundamental equations. The ordinary magnetofluid dynamic equations discussed in section 2 may not be the proper equations and we should derive other fundamental equations from the Boltzmann equation, instead of those equations of section 2 which are derived by assuming that the mean

free path is the dominant factor in the Boltzmann equation. This case has not been extensively studied yet.

(4) EM-domain. In this domain, $L/L_{Li} \gg 1$ and $L_{Li} < L_f$. In this domain, the ions will move as free particles subjected to the electromagnetic forces. Both the electric and magnetic fields are important. The continuum theory will not be good.

(5) E-domain. In this domain, $L/L_{Le} \gg 1$ and $L_{Le} < L_f$. In this domain, the electric field is the dominant force. It is a region of very low density in which many glow discharges operate. If the density of the plasma is very low, we should not use the continuum theory for this case. However, if the density of the plasma is not very low, but the electric field strength is extremely high, we may use the continuum theory and this domain gives the results of electro-gasdynamics discussion in section 12.

Besides the five domain listed above, if the Debye length is larger than L, there will be no interaction between gas particles. One should treat the problem by considering individual particles instead of treating it as a continuum, i.e., the gas or the plasma as a whole. We shall discuss this case in chapter VIII.

From the point of view of continuum theory, we may divide the electromagnetofluid dynamics into four domains. In all the domains, we assume that the mean free path of the electrically conducting fluid is small so that the fundamental equations similar to those discussed in section 2 are applicable except the electrical current density equation (6.11).

(1) Classical magnetofluid dynamics. In this domain, the magnetofluid dynamic approximations of section 3 are satisfied and that the ion Larmor radius is much larger than the mean free path. In this case, the electrical current equation (6.11) is applicable and the electrical conductivity is a scalar quantity. The terms with excess electric charge such as convection electric current and the electrostatic force may be neglected. This is the case discussed in sections 3 to 10.

(2) Modified magnetogasdynamics. In this case, the magneto-gasdynamic approximations of section 3 are still satisfied but the density of the plasma is so low and the magnetic field strength is

so high that ω_c/f is not negligible. In this case, we have to consider the electrical conductivity as a tensor quantity, but the electric convection current and the electrostatic forces are still negligible.

If the density is sufficiently low and the magnetic field is sufficiently high, we should take into account not only the tensor behavior of the electrical conductivity but also the gasdynamic effects on the electric current density. We may either use a very complicated equation for the electrical current density or the multifluid theory. We shall discuss it in chapter IX section 4.

(3) Electrogasdynamics. When the electric field strength is very high so that the magnetofluid dynamic approximations (6.28) and (6.29) do not hold true, we can not neglect the excess electric charge nor the displacement current. If we consider the other extreme condition that the electrogasdynamic approximations of section 12 are satisfied, we may neglect the electric conduction current and all the magnetic terms. We have the case of electrogasdynamics discussed in section 12.

(4) Electromagnetofluid dynamics. This case lies between the magnetofluid dynamics and the electrogasdynamics so that none of the electromagnetic variables may be neglected, and we have to use the complete set of fundamental equations discussed in section 12 to study this problem. This class of flow problems of electrically conducting fluid has not been studied in any detail except the general discussion of the fundamental equations. It would be of interest to find out how the present case acts as a bridge between the magnetofluid dynamics on one side and electrogasdynamics on the other side.

Most of the magnetofluid dynamic problems published in the literature do not consider the ionization process as a chemical reaction. In fact, we may apply many concepts discussed in chapter V for the ionization process in the flow problems of a plasma. We may have the nonequilibrium flow of an ionized gas. In the nonequilibrium flow, one of the most interesting facts is that the electron temperature may be considerably different from those of the heavy particles. Since the electrical conductivity depends mainly on the motion of the electrons, we may have quite different results of the nonequilibrium flow from those of equilibrium flow. For instance, in MFD power generation, it is better to have nonequilibrium flow,[20] so that the electron temperature is higher than those of heavy particles and the electrical conductivity of the plasma will be much higher than the equilibrium flow with the same temperature of the plasma as a whole. The success of MFD power generation depends on whether the generator could work in this nonequilibrium condition. In such a case, we need at least two temperatures in the study of the flow of electrically conducting fluid which will be discussed in chapter IX section 4.

PROBLEMS

1. From Eqs. (6.3) and (6.5), derive an energy equation in terms of the stagnation enthalpy $h_0 = h + \frac{1}{2}q^2$.

2. What is the Gaussian system of units in electromagnetic theory? Write down the Maxwell's equations (6.6) and (6.7) in Gaussian units and the equation of vector and scalar potentials (6.22) in Gaussian units.

3. What is the system of electromagnetic units in electromagnetic theory? Write down Eqs. (6.6) and (6.7) in electromagnetic units.

4. Derive the boundary conditions of the electromagnetic fields (6.24) to (6.27).

5. Show that when the electrical conductivity is infinite, the magnetic field in an electrically conducting fluid follows the streamlines.

6. Define the following terms and discuss their physical significance:
   (a) Plasma frequency, (b) Cyclotron frequency,
   (c) Debye length, and (d) Larmor radius.

7. Derive the quasi-one-dimensional flow equations of magnetofluid dynamics by averaging the flow variables in a rectangular channel. Show that the correlations between various variables are important in the final equations.

8. Discuss the possible singularities for the flow in a nozzle of

magnetogasdynamics as given by Eqs. (6.43).

9. Discuss the properties of the strictly one-dimensional magneto-gasdynamic flow of a viscous and heat-conducting fluid with infinitely large electrical conductivity.

10. Calculate the velocity distributions for an incompressible, viscous, and electrically conducting fluid in a straight channel of rectangular cross section with an externally applied magnetic field perpendicular to one side of the rectangular section and the main flow (Fig. 6.4).

11. Calculate the velocity and magnetic field distributions near a two-dimensional stagnation point with an externally applied uniform magnetic field perpendicular to the surface of the body at the stagnation point. Assume that the fluid is incompressible.

12. Calculate the temperature distribution of a MHD Couette flow (Fig. 6.3) for the cases: $R_h = 4$, $R_E = 4$, and $T = T_1$ at $y = -1$ and $T = 4T_1$ at $y = 1$.

13. Draw in scale the Friedrichs diagram (Fig. 6.5) for the case
$$V_H = \frac{1}{2} a.$$

14. Find the velocity distributions for the case of a uniform flow over a magnetized sphere in an inviscid and electrically conducting fluid. Assume that both the magnetic field and the electrical conductivity of the fluid are small.

15. Find the velocity and magnetic field distributions over an infinite flat insulated plate setting impulsively into a uniform motion with a velocity U in its own plate in the presence of a transverse uniform magnetic field.

16. Show that if the proper coordinate system is used, an oblique shock in magnetogasdynamics with the velocity vector and magnetic field parallel in front of the shock has its velocity and magnetic field parallel behind the shock too.

17. Derive the fundamental equations of the linearized theory of magnetogasdynamics for the case of a uniform velocity over a thin body and an arbitrarily orientated externally applied magnetic field. Discuss the general solution of the resultant

equation.

18. Derive Eq. (6.84) and show that when the effective Mach number is unity, the strength of the normal shock is zero.

19. Calculate the velocity and the temperature distributions; the skin friction and the heat transfer at the wall over a flat plate in a uniform flow of velocity U parallel to the plate and an applied magnetic field perpendicular to the plate but fixed to the fluid. Assume that the magnetic Reynolds number is small, boundary layer approximations are applicable, and the fluid is incompressible. Both the temperature of the free stream T_e and that of the plate T_W are constant.

20. Discuss the Rankine-Hugoniot relations of a normal shock in an ionized gas where the degree of ionization changes considerably across the shock. Assume that Saha relation holds and the gas is argon.

21. Derive the Saha relation for a mixture of gases of N species in terms of degree of ionization of each species α_s.

22. Discuss the tensor character of electrical conductivity of an ionized gas with strong magnetic field and the collision between electron and atoms is not negligible (see reference [23]).

23. Discuss the flow in a nozzle of variable cross section for electrogasdynamics in some detail (see section 12).

24. Discuss the solution of steady radial flow in electrogasdynamics in (i) cylindrical case and (ii) spherical case by assuming that the gas is inviscid and nonheat-conducting.

25. Calculate the velocity and magnetic field distribution over an infinite flat plate moving impulsively into a uniform velocity U in its own plane in the presence of a transverse uniform magnetic field by assuming that the Hall current is not negligible but the ion slip may be neglected.

26. Derive the von Kármán-Howarth equation of isotropic turbulence in magnetohydrodynamics (see reference [22]).

27. Derive the stability equations of a two-dimensional parallel flow in magnetohydrodynamics (see reference [22]).

28. Discuss the turbulent boundary layer flow in magnetohydrodynamics by means of mixing length theory.

REFERENCES

[1]   Alfven, H. "On the existence of electromagnetic-hydrodynamic waves", *Arkiv f. Math. Astro. Ock. Fysik,* Vol. 29B, No. 2, 1943, pp. 1—7.

[2]   Bazer, J. and Ericson, W. B. "Oblique shock waves in a steady two dimensional hydromagnetic flow". Proc. of the Symp. on Electromagnetics and Fluid Dynamics of Gaseous Plasma, Vol. XI. Intersci. Publi., 1961.

[3]   Cabannes, H. "Attached stationary shock in ionized gases", *Rev. Mod. Phys.,* Vol. 32, No. 4, 1960, pp. 973—976.

[4]   Cambel, A. P., Duclos, D. P., and Anderson, T. P. Real Gases. Academic Press, 1963.

[5]   Cheng, S. I. "Glow discharge as an advanced propulsion device", *ARS Jour.,* Vol. 32, No. 12, 1962, pp. 1910—1916.

[6]   Clauser, F. H. "Plasma dynamics" in Aero. & Astro. International Ser. of Aero. Sci. & Space Flight, Div. IX, Vol. 4. Pergamon Press, 1960, pp. 305—343.

[7]   Faraday, M. Experimental Researches in Electricity, Vol. III. Richard & John Edward Taylor, 1835.

[8]   Fishman, F., Lothrop, J., Patrick, R., and Petschek, H. "Supersonic two dimensional MHD flow". Res. Report 39, AVCO, Res. Lab., 1959.

[9]   Friedrichs, K. O. and Kranzer, H. "Notes on MHD VIII, Non-Linear Wave Motion". N. Y. U. Report NYD 6486, July 1958.

[10] Goldsworthy, F. A. "On the dynamics of an ionized gas". Prog. in Aero. Sci., Vol. I. Pergamon Press, 1961, pp. 174—205.

[11] Gourdine, M. C. "Electrogasdynamic channel flow". Tech. Report 34—5, Jet Propulsion Lab., Calif. Inst. of Tech., 1960.

[12] Harris, L. P. Hydromagnetic Channel Flows. John Wiley, 1960.

[13] Hartmann, J. (1) "Hg-Dymanics I", Math. fys, Medd, Vol. 15, No. 6, 1937 and (2) with F. Lazarus, "Hg-Dynamics II", Math. fys, Medd, Vol. 15, No. 7, 1937.

[14] Hasimoto, H. "Magnetohydrodynamic wakes in a viscous conducting fluid", *Rev. Mod. Phys.,* Vol. 32, No. 4, 1960, pp. 860—866.

[15] Hasimoto, H. and Kuwabara, S. "Electrogasdynamics", *Jour. Phys. Soc. Japan,* Vol. 20, No. 5, 1965, pp. 859—868.

[16] Hughes, W. F. and Young, F. J. The Electromagnetodynamics of Fluids. John Wiley, 1966.

[17] Kahn, B. and Gourdine, M. C. "A basic study of slender channel electrogasdynamics". Report ARL 63—205, Aerospace Res. Lab., DAR, USAF, 1963.

[18] Kahn, B. "A continuation of the basic study of slender channel electrogasdynamics". Report ARL 65—4, Aerospace Res. Lab., DAR, USAF, 1965.

[19] Kantrowitz, A. R. and Petschek, H. E. "An introductory discussion of magnetohydrodynamics". Symp. on Magnetohydrodynamics, Stanford

Univ. Press, 1957, pp. 3—15.

[20] Kerrebrock, J. L. "Segmented electrodes losses in MHD generators with non-equilibrium ionization". I. AVCO Res. Report 179, April 1964; II. Paper in Eng. MHD Conference, Pittsburgh, Pa., 1965.

[21] Murgatroyd, W. "Experiments on magnetohydrodynamics channel flow", *Phil. Mag.* (7), Vol. 44, No. 359, 1953, pp. 1348—1354.

[22] Pai, S.-I. Magnetogasdynamics and Plasma Dynamics. Springer-Verlag, 1962.

[23] Pai, S.-I. "A critical survey of magnetofluid dynamics: Part I. General discussion". Oct. 1963; "Part II. Magnetohydrodynamic channel flows". May 1964; "Part III. Quasi-one dimensional analysis of magnetogasdynamics". Dec. 1965; "Part IV. Modern aspects of magnetofluid dynamics (Tensor electrical conductivity and multifluid theory)". March 1966. Reports ARL 63—175 and ARL 66—0060, Aerospace Res. Lab., OAR, USAF, 1963, 1964, 1965, and 1966.

[24] Pai, S.-I. "Quasi-one-dimensional analysis of magnetogasdynamics". Proc. Electricity from MHD, Vol. 1, Symp. MHD Power Gen., Salzburg. International Atomic Energy Agency, July 1966, pp. 283—294.

[25] Pai, S.-I. "Magnetohydrodynamics of channel flow", in Adv. in Hydro-Science, Vol. 3, V. T. Chow (ed.). Academic Press, 1966, pp. 63—110.

[26] Pai, S.-I. "Linearized theory of airfoils in fluids of low electrical conductivity". Mizzellaneen der Ang. Mach. Acad. Berlin, Verlag, 1962, pp. 232—243.

[27] Resler, E. L., Jr. and McCune, J. E. "Compressibility effects in magnetoaerodynamic flows", *Jour. Aero. Sci.*, Vol. 27, No. 7, 1960, pp. 493—503.

[28] Rose, R. J. "Physical principle of MHD power generator". AVCO Res. Report No. 69, 1960.

[29] Rossow, V. J. "On the flow of electrically conducting fluids over a flat plate in the presence of a transverse magnetic field". NACA Report No. 1358, 1958.

[30] Saha, M. N. "Ionization in the solar chromosphere", *Phil. Mag.*, Vol. 40, No. 238, 1920, pp. 472—488.

[31] Sears, W. R. and Resler, E. L., Jr. "Sub- and Super-Alfvenic flows past bodies", in Adv. in Aero. Sci., Vol. 4. Pergamon Press, 1961, pp. 657—673.

[32] Sherman, A. and Sutton, G. W. "The combined effect of tensor conductivity and viscosity on MHD generator with segmented electrodes". Magnetohydrodynamics. Proc. 4th Biennial Gasdyn. Symp., Northwestern Univ. Press, Chapter 12, 1962.

[33] Steg, L. and Sutton, G. W. "The prospects of MHD power generation", *Astronautics*, Vol. 5, No. 8, 1960, pp. 22—25, 82—86.

[34] Stratton, J. A. Electromagnetic Theory. McGraw-Hill, 1941.

[35] Yen, J. T. "Viscous Hall and ion slip effect on MHD channel flow and power generation". Ses. 3b., Paper 46, Intern. Symp. on MHD Power Generation, Paris, July, 1964.

RADIATION GASDYNAMICS

1. Thermal Radiation Effects

At low temperature such as $T < 10^4 \,^\circ K$ and moderate pressure such as one atmosphere, the thermal radiation effects are negligible in the analysis of fluid dynamics. Thus in the last two chapters, we did not consider the thermal radiation effects. However, in this space age, we are concerned with many technological developments, in space hypersonic flight, fission and fusion reactions, etc., in which the temperature is very high and the density of the gas is very low. As a result, the thermal radiation effects are no longer negligible. We use the name of Radiation Gasdynamics[9] for that branch of fluid dynamics in which the thermal radiation effects must be considered simultaneously with all other fluid dynamic effects.

The study of radiation in high temperature gases has been made by physicists for a long time. At the turn of the present century, Planck[16] found the correct theory of radiation. The radiative heat transfer has been extensively studied by astrophysicists,[1, 6, 18] but the interaction between the radiation field and the gasdynamic field has been studied only recently. The main purpose of this chapter is to study the influence of thermal radiation on the velocity and temperature distributions of a very high temperature gas.

There are three main thermal radiation effects on the flow field of a high temperature gas which are the radiation energy density, the radiation stresses, and the heat flux of radiation. Let us first show how and when these radiation effects become important in a flow field. For simplicity, let us estimate these thermal radiation terms for an optically thick medium and compare them with the corresponding gasdynamic terms. We shall show later that the radiation energy density E_R is of the same order of magnitude as the radiation stresses τ_R, particularly the radiation pressure p_R.

Hence we should consider these two effects simultaneously. For an optically thick medium, the radiation energy density per unit mass of the gas is

$$E_R/\rho = a_R T^4/\rho \qquad (7.1)$$

where a_R is the Stefan-Boltzmann constant which is 7.67×10^{-15} erg$\cdot$cm$^{-3}\cdot{}^{\circ}$K^{-4} and T is the absolute temperature in $^{\circ}$K and ρ is the density of the gas. We may compare this radiation energy density with the internal energy of the gas $U_m = c_v T$. As a rough estimate, we take $\rho = 1.23 \times 10^{-6}$ gr/cm^3 which is 10^{-3} of the value of air at standard sea level conditions and $c_v = 7 \times 10^6$ erg/gr$\cdot{}^{\circ}$K which is the value of air. If we take $T = 10^{5\,\circ}$K, we find that the radiation energy density and the internal energy are of the same order of magnitude. Since the radiation energy density is proportional to T^4 while the internal energy is proportional to T, at a temperature of $T = 10^{4\,\circ}$K, the radiation energy density will be three orders smaller than the internal energy and may be neglected. Since $T = 10^{4\,\circ}$K is a rather high temperature in many engineering problems, we may neglect the radiation energy density in many engineering problems. For instance, for the problem of reentry from Mars, the temperature is of the order of $10^{4\,\circ}$K, thus in many current space flight problems we may neglect the radiation energy density and the radiation stresses. However, in a fusion reaction, the temperature $10^{+5\,\circ}$K may be considered as a low temperature. Thus in a fusion reaction, the radiation energy density and radiation stresses may not be neglected.

For the radiation heat flux, we may estimate its value by the simple equilibrium formula:

$$q_R = \frac{\sigma T^4}{\rho} \qquad (7.2)$$

where $\sigma = ca_R/4 = 5.75 \times 10^{-5}$ erg$\cdot$cm$^{-2}\cdot$sec$^{-1}\cdot{}^{\circ}$K^{-4} is the Stefan-Boltzmann constant for radiative transfer and c is the velocity of light in vacuum ($c = 3 \times 10^8$m/sec). The heat flux by convection may be written as

$$q_v = Uc_v T \qquad (7.3)$$

where U is a typical flow velocity.   If we take $U=6.5\times10^4$ m/sec which is an average speed of a satellite, we find that at $T=10^4$ °K, the radiative heat flux q_R and the convective heat flux q_v are of the same order of magnitude.   At present time, many reentry problems have a temperature of the order of 10^4°K.   It is one of the reasons that aerospace engineers study the radiation gasdynamics extensively by including only the radiative heat flux.

When the radiation terms are much smaller than the corresponding terms of ordinary gasdynamics, we may calculate the temperature distribution without the radiation term and then estimate the radiation terms from the known temperature distribution.   If the radiation terms are of the same order of magnitude as those of ordinary gasdynamics, we should solve the radiation field simultaneously with the gasdynamic field.

Generally speaking, thermal radiation is a far more complicated phenomenon than the conduction of heat or convection of heat in ordinary fluid dynamics.   Heat convection depends on the velocity field of the fluid, and heat conduction depends essentially on the temperature distribution of the medium.   Thermal radiation depends on the temperature of the medium in a very complicated manner. There are cases in which the radiation of heat is apparently independent of the temperature of the medium through which it passes. For instance, we may concentrate the solar rays at a focus by passing them through a converging lens of ice which remains at a constant temperature of 0 °C and ignite an inflammable body.

There are a great number of radiation phenomena, such as fluorescence, phosphorescence, etc., besides the thermal radiation. We consider here only the thermal radiation, particularly its interaction with gasdynamic field.

2.   Specific Intensity of Radiation

We may either consider the thermal radiation as a stream of photons or as electromagnetic waves.   When we consider the thermal radiation as a stream of photons, we should use the relativistic Boltzmann equation to study the motion of these particles (see chapter IX section 6).   When we consider the thermal radiation as electromagnetic waves, we may use the geometric optics to study the behavior of thermal radiation from the macroscopic point of view. In this chapter, we consider only the case of continuum theory in which the method of geometric optics will be used and in which the thermal radiation may be expressed in terms of a specific intensity I_ν which is defined as follows:

$$I_\nu = \lim_{d\sigma_0, d\nu, dt, d\omega \to 0} \left(\frac{dE_\nu}{d\sigma_0 \cdot \cos \theta \cdot d\nu \cdot dt \cdot d\omega} \right) \tag{7.4}$$

where I_ν is a function of time t, spatial coordinates, direction θ which is the angle between the direction of the ray L and the normal of the elementary area $d\sigma_0$ (Fig. 7.1), and the frequency of the wave ν. The amount of radiative energy flowing through the area $d\sigma_0$ in the frequency range ν and $\nu + d\nu$, in the direction of the ray L which

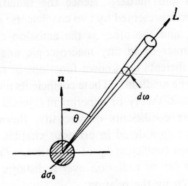

Figure 7.1 A heat ray in a radiation field

makes an angle θ with the normal of the area $d\sigma_0$ within the solid angle $d\omega$ in the time interval dt is dE_ν. The total amount of energy radiated over the whole spectrum is

$$dE = \int_0^\infty \left(\frac{dE_\nu}{d\nu} \right) d\nu = I \cos \theta \cdot d\sigma_0 \cdot d\omega \cdot dt \tag{7.5}$$

where

$$I = \int_0^\infty I_\nu d\nu = \text{integrated intensity of thermal radiation} \tag{7.6}$$

The main interesting point for the specific intensity I_ν is that it depends on the frequency ν. In general, the specific intensity tends to be zero for both very high and very low frequencies. There is a maximum in the intermediate frequency. Hence the radiation effects would depend on the frequency too.

3.   Absorption and Emission Coefficients of Radiation

The specific intensity is the result of the interaction between radiation and matter in the atoms of the medium which should be studied by quantum mechanics from the microscopic point of view. However for radiation gasdynamics, we are interested in the problem from macroscopic point of view only. In other words, we are interested in the average situation due to a great number of interactions between radiation and matter. Hence the radiative properties of the medium may be represented by two coefficients: one is the absorption coefficient k_ν and the other is the emission coefficient j_ν. In radiation gasdynamics or in any macroscopic analysis, we assume that these two coefficients are known functions of frequency and the states variables of the medium. These coefficients must be determined by either microscopic theory or experiments, just as those transport coefficients such as coefficients of viscosity, thermal conductivity, etc., which we have considered in previous chapters.

(1)   Absorption coefficient k_ν of radiation.   The loss of specific intensity along the ray of radiation over a distance ds in a medium of density ρ is given by the relation:

$$dI_\nu = -\rho k_\nu I_\nu ds \tag{7.7}$$

where k_ν is known as the mass absorption coefficient or simply absorption coefficient of the medium and ρk_ν is known as linear absorption coefficient of the medium because its dimension is $1/L$. Sometimes, we introduce the term of mean free path of radiation as

$$L_{R\nu} = \frac{1}{\rho k_\nu} \tag{7.8}$$

The mean free path of radiation represents the average distance between collision of a photon and a molecule and plays a similar role in radiative transfer as ordinary mean free path in gasdynamics. This similarity will be further discussed in later sections. The integration of Eq. (7.7) gives

$$I_\nu(s) = I_\nu(s_0) \exp\left(-\int_{s_0}^{s} \rho k_\nu \, ds \right) = I_\nu(s_0) \exp\left(-\tau_\nu \right) \tag{7.9}$$

where τ_ν defined in Eq. (7.9) is known as optical thickness of radiation of the layer $(s-s_0)$ and s_0 is a reference point where the specific intensity is $I_\nu(s_0)$. The optical thickness τ_ν is a nondimensional distance which shows the effective length in absorption of radiation. For a given physical length $L = s - s_0$, if τ_ν is large, the medium is said to be optically thick, while if τ_ν is small, the medium is said to be optically thin.

The absorption coefficient k_ν is a function of temperature and pressure of the medium as well as the frequency ν. The absorption coefficient k_ν consists of two parts: one is the true absorption which is the absorption of photons by atoms or molecules exposed in the radiation field; and the other is due to scattering which represents that part of radiation energy scattered by optical obstacles in the medium to other directions from the original direction. At present stage of radiation gasdynamics, we may neglect the scattering phenomena even though in many radiative transfer processes such as in meteorology, the scattering is one of the most important processes. In radiation gasdynamics, we consider only the homogeneous medium, and the scattering by molecules is taken into account by calculating the true coefficient of absorption. Hence we may neglect scattering as a first approximation. The absorption coefficient in radiation gasdynamics has a similar position as other coefficients such as that of viscosity, thermal conductivity, etc., in ordinary gasdynamics. It may be considered as a new transport coefficient.

(2) Emission coefficient j_ν of radiation. The radiation energy emitted from a mass dm may be written as

$$dE_e = j_\nu \, dm \, d\omega \, d\nu \, dt \tag{7.10}$$

where dE_e is the radiant energy emitted by the medium of mass dm in the solid angle $d\omega$ in the time interval dt and in the frequency range ν and $\nu + d\nu$.

The emission coefficient j_ν is a function of the temperature and pressure of the medium as well as the frequency ν. The emission coefficient j_ν also consists of two parts: one is due to the creation of photons from the matter, and the other is due to the contribution of photons from scattering from all other directions into the direction of the radiation ray under consideration. One of the most difficult problems in radiation gasdynamics is to determine the proper emission coefficient. We shall discuss it in the next section.

4. Equation of Radiative Transfer

If we have the absorption coefficient k_ν and the emission coefficient j_ν of a medium, the conservation of radiative energy through an elementary cylinder of density ρ, base area $d\sigma_0$ and length ds gives the radiative transfer equation which governs the specific intensity I_ν, i.e.;

$$dE_0 - dE_i = dE_e + dE_a - dE_t \qquad (7.11)$$

The difference between the outgoing radiative energy dE_0 and the incoming radiative energy dE_i must be equal to the sum of the energy emitted dE_e and the energy absorbed dE_a minus the net change of the radiative energy in this volume with time dE_t. Eq. (7.11) in terms of k_ν and j_ν is

$$\frac{1}{c}\frac{\partial I_\nu}{\partial t} + n_i \frac{\partial I_\nu}{\partial x_i} = \rho k_\nu (J_\nu - I_\nu) \qquad (7.12)$$

where

$$J_\nu = \frac{j_\nu}{k_\nu} = \text{source function of radiation} \qquad (7.13)$$

and n_i is the direction cosine of the ray considered with respect to the axis x_i.

One of the most difficult problems in radiation gasdynamics is to determine the source function of radiation J_ν. At the present stage of investigation, we usually use the assumption of local thermodynamic equilibrium condition which is similar to the equilibrium flow case in the flow with chemical reaction.   After we know more about the results of radiation gasdynamics under the local thermodynamic equilibrium condition, we should study the source function in the nonequilibrium condition.

If we assume that the gas is in local thermodynamic equilibrium, i.e., the emission is determined by the local temperature, Eq. (7.12) becomes

$$\frac{1}{c}\frac{\partial I_\nu}{\partial t}+\frac{\partial I_\nu}{\partial s}=\rho k'_\nu(B_\nu-I_\nu) \tag{7.14}$$

where s is the distance along the radiation ray, and

$$k'_\nu=k_\nu\left[1-\exp\left(-\frac{h\nu}{kT}\right)\right]=\text{reduced absorption coefficient}$$

$$\tag{7.15}$$

$$B_\nu=\frac{2h\nu^3}{c^2}\frac{1}{\exp\left[h\nu/(kT)\right]-1}=\text{Planck radiation function}$$

$$\tag{7.16}$$

where $h=6.62\times10^{-27}$ erg·sec is the Planck constant and $k=1.379\times10^{-16}$ erg/°K is the Boltzmann constant.

The reduction of absorption coefficient in Eq. (7.14) is due to the induced emission of radiation.   In many fluid dynamics problems, the first term on the left-hand side of Eq. (7.14), i.e., $(1/c)(\partial I_\nu/\partial t)$, is negligible because the value of light speed c is very large in comparison with the flow velocity.   We shall neglect this term in the remaining sections of this chapter.

5.   Radiation Energy Density and Radiation Stresses

The energy density of radiation U_ν is the amount of radiant energy per unit volume in the state of frequency interval ν and $\nu+d\nu$ which is on course of transit in the neighborhood of the point considered.   It is easy to show that[9]

$$U_\nu=\frac{1}{c}\int_{4\pi}I_\nu d\omega \tag{7.17}$$

The total energy density of radiation for the whole spectrum at any point in space at any time is then

$$E_R = \int_0^\infty U_\nu \, d\nu = \frac{1}{c} \int_{4\pi} I \, d\omega \qquad (7.18)$$

The energy density of radiation per unit mass is then E_R/ρ which should be added to the internal energy due to molecular motion so that we may have the total or effective internal energy of a radiating gas.

A quantum of energy $h\nu$ is associated with a momentum $h\nu/c$. Hence we may calculate the radiation stress tensor from the rate of change of momentum associated with the energy of photons. The ijth component of the radiation stress tensor τ_R is

$$\tau_{Rij} = -\frac{1}{c} \int_{4\pi} I \, n_i n_j \, d\omega \qquad (7.19)$$

We may define a radiation pressure p_R as

$$p_R = -\frac{1}{3}(\tau_{R11} + \tau_{R22} + \tau_{R33}) = \frac{1}{3c} \int_{4\pi} I \, d\omega = \frac{E_R}{3} \qquad (7.20)$$

Hence we see that the radiation pressure and the radiation energy density are always of the same order of magnitude.

For isotropic radiation, we have

$$p_R = -\tau_{R11} = -\tau_{R22} = -\tau_{R33} = \frac{1}{3} E_R, \ \tau_{Rij} = 0 \text{ if } i \neq j \qquad (7.21)$$

6. Radiative Heat Flux

The flux of heat energy by thermal radiation is q_{Ri} which is given by the following formula:

$$q_{Ri} = \int_{4\pi} I \, n_i \, d\omega \qquad (7.22)$$

We should add the divergence of this radiation flux in the energy equation of radiation gasdynamics, i.e.,

$$Q_R = \nabla \cdot q_R \qquad (7.23)$$

In spherical coordinates, Eq. (7.22) may be written as

$$q_{RL}(\theta, \phi, r, t) = \int_0^\infty \int_0^{2\pi} \int_0^\pi I_\nu(\theta, \phi, r, t) \sin\theta \cos\theta \, d\theta \, d\phi \, d\nu$$

$$(7.24)$$

where the ray of radiation L is specified by the angular variables θ $(0 \le \theta = \pi)$ and the azimuth angle ϕ $(0 \le \phi \le 2\pi)$. In many engineering problems and astrophysical problems, we consider the radiative transfer with respect to a straight plane (Fig. 7.2). It is convenient to split the net radiative flux q_{RL} into two parts: one part q_{RL}^+ represents the contribution coming from the side of the

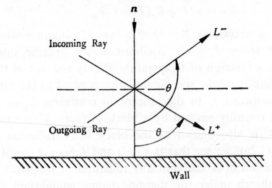

Figure 7.2 Incoming and outgoing rays of radiation with respect to a straight wall

unit normal vector n, the incoming rays, and the other part q_{RL}^- represents the contribution from the opposite side, the outgoing rays, i.e.,

$$q_{RL}^+ = \int_0^\infty \int_0^{2\pi} \int_0^{\pi/2} I_\nu(\theta, \phi, r, t) \sin\theta \cos\theta \, d\theta d\phi \, d\nu \qquad (7.25a)$$

$$q_{RL}^- = \int_0^\infty \int_0^{2\pi} \int_{\pi/2}^\pi I_\nu(\theta, \phi, r, t) \sin\theta \cos\theta \, d\theta \, d\phi \, d\nu \qquad (7.25b)$$

and the net radiative flux through a unit area of the straight wall with normal n is

$$q_{RL} = q_{RL}^+ - q_{RL}^- \qquad (7.25c)$$

Customarily, we choose the normal **n** toward the wall and q_{RL} is the net radiative heat flux toward the wall.

7. Radiative Equilibrium Conditions

Many of the properties of a radiation field under the thermodynamic equilibrium condition have been known before Planck found the correct law of radiation. In complete equilibrium, the variations of specific intensity with respect to time and space are zero. From Eqs. (7.12) and (7.14), we have

$$I_\nu = B_\nu(T) = j_\nu/k_\nu \qquad (7.26)$$

Eq. (7.26) is known as Kirchhoff's law of radiation which states that under thermodynamic equilibrium, the specific intensity of radiation is a function of temperature T only and not of the nature of the medium, and the specific intensity is equal to the ratio of the emission coefficient j_ν to the absorption coefficient k_ν or equal to the specific intensity emitted by a black surface. Hence the Planck function B_ν is also known as the black body radiation function. From Eq. (7.26), we see that if $k_\nu = 0$ and $I_\nu \neq \infty$, j_ν must be zero. Thus a medium does not emit any radiation of frequency which it does not absorb under the thermodynamic equilibrium condition. Thus in a medium which is transparent for a certain frequency of radiation, thermodynamic equilibrium can exist for any finite intensity of that frequency.

Wien found the displacement law:[5]

$$\frac{U_\lambda}{T^5} = G\left(\frac{c}{\lambda T}\right) \qquad (7.27)$$

where $U_\lambda \, d\lambda = U_\nu \, d\nu$ and λ is the wave length with $c = \lambda\nu$. Eq. (7.27) shows that if we plot energy density U_λ divided by T^5 vs. $c/(\lambda T)$, we have a single curve for all temperature under the thermodynamic equilibrium condition. The maximum energy density occurs at a constant value of $c/(\lambda T)$, i.e.,

$$\lambda_m T = 0.290 \text{ cm} \cdot {}^\circ C \qquad (7.28)$$

Eq. (7.28) shows that the wave length λ_m at the maximum energy density decreases with increase of temperature. In other words, the position of the maximum energy density is displaced as the temperature changes. Thus we call Eq. (7.27) the displacement law.

Before the Planck's law of radiation (7.16) was found, we had the Rayleigh-Jeans radiation law:

$$U_\nu = \frac{8\pi\nu^2}{c^2} kT \tag{7.29}$$

By means of Eq. (7.29), the total radiation energy density E_R will be infinite. This fact is referred to as ultraviolet catastrophe. However, we may show that Eq. (7.29) is a good approximation for Eq. (7.16) only if the frequency is small. For high frequency, Eq. (7.29) does not hold true.

8. Fundamental Equations of Radiation Gasdynamics

For simplicity, we consider the case of a neutral gas including thermal radiation effects. The analysis may be easily generalized in a straightforward manner for an electrically conducting fluid by what we may call the radiation magnetogasdynamics.[15] For radiation gasdynamics, we have seven variables, i.e., the temperature T, the pressure p, and the density ρ of the gas; the velocity vector q with components u_i and the specific intensity of radiation I_ν. The following equations govern these variables:

(1) Equation of state

$$p = \rho RT \tag{7.30}$$

This is the same as that in ordinary gasdynamics.

(2) Equation of continuity

$$\frac{\partial \rho}{\partial t} + \nabla \cdot (\rho q) = 0 \tag{7.31}$$

This is also the same as that in ordinary gasdynamics.

(3) Equation of motion

$$\rho \frac{Dq}{Dt} = -\nabla p + \nabla \cdot \tau_v + \nabla \cdot \tau_R + F \tag{7.32}$$

where τ_v is the viscous stress tensor given by Eq. (5.34), τ_R is the radiation stress tensor given by Eq. (7.19), and F is the body forces such as gravitational force, electromagnetic force, etc. Since the radiation stress tensor is in general an integral expression, Eq. (7.32) is an integro-differential equation.

(4)   Energy equation

$$\rho\frac{D\bar{e}_m}{Dt}=\nabla\cdot(pq)+\nabla\cdot(q\cdot\tau_v)+\nabla\cdot(q\cdot\tau_R)+\nabla\cdot(\kappa\nabla T)+\nabla\cdot q_R+Q$$

(7.33)

where in the total energy of the gas $\bar{e}_m$, we should add the radiation energy, i.e., $\bar{e}_m=U_m+\frac{1}{2}q^2+K+E_R/\rho$. In addition, we have the dissipation due to radiation stress tensor and the energy loss due to radiative heat flux. The rest are the same as those in ordinary gas-dynamics [see Eq. (3.95)].

(5)   Equation of radiative transfer

$$l\frac{\partial I_\nu}{\partial x}+m\frac{\partial I_\nu}{\partial y}+n\frac{\partial I_\nu}{\partial z}=\frac{\partial I_\nu}{\partial s}=\rho k'_\nu(B_\nu-I_\nu)$$ (7.34)

where we neglect the term with $1/c$, i.e., the unsteady term; l, m. and n are respectively the direction cosines of the radiation ray s with respect to x-, y-, and z-axis and s is the distance along the radiation ray. The reduced absorption coefficient k'_ν is considered as a given function of the state variables T and p or T and ρ as well as the frequency ν. Eq. (7.34) is under the assumption of local thermodynamic equilibrium and is a first order total differential equation of distance s which may be integrated as follows:

$$I_\nu(s)=I_\nu(s_0)\exp[-\tau_\nu(s,s_0)]+\int_{s_0}^{s}B_\nu(s_1)\{\exp[-\tau_\nu(s,s_1)]\}\rho k'_\nu ds_1$$

(7.35)

where $\tau_\nu(s, s_1)$ is the optical thickness defined by the formula:

$$\tau_\nu(s, s_1)=\int_{s_1}^{s}\rho k'_\nu\, ds$$ (7.36)

and s_0 is an initial point on the radiation ray whose specific intensity

$I_\nu (s_0)$ is known. The determination of $I_\nu (s_0)$ depends on the boundary condition of radiation intensity which will be discussed in next section. With the help of Eq. (7.36), all the radiation terms such as q_R, E_R, etc., may be expressed as integrals of the state variables T, p and frequency ν, and our fundamental equations of radiation gasdynamics (7.32) and (7.33), etc., are a system of integro-differential equations. Since such a system of integro-differential equations is too complicated to solve for any practical problem, we have to make reasonable approximations so that the fundamental equations may be simplified into a form which can be analyzed. The following are some of these approximations:

(A) All the Well Known Approximations of Gasdynamics May be Used

For instance, we may consider

(1) The inviscid and nonheat-conducting flow. Outside the boundary layer or other transition regions, the viscosity and heat conductivity may be neglected when the Reynolds number of the flow is large. In radiation gasdynamics, we should add the proper radiation terms to the equations of an inviscid and nonheat-conducting fluid, and

(2) Boundary layer flow. In the boundary layer region, the well known Prandtl boundary layer approximations may be used.

(B) Some Approximations on the Radiation Terms

(1) Optically thick medium. When the mean free path of radiation $L_{R\nu}$ is very small in comparison with the typical dimension of the flow field, the solution of Eq. (7.34) may be written as

$$I_\nu = B_\nu - L_{R\nu} \left(n_i \frac{\partial B_\nu}{\partial x_i} \right) + O(L_{R\nu}^2) \qquad (7.37)$$

where the summation convention in i is used. If we neglect the second and higher order terms of $L_{R\nu}^2$, the radiation terms in gasdynamics can be easily evaluated, i.e.,

$$E_R = a_R T^4 = 3 p_R \qquad (7.38)$$

and all shearing stresses of radiation vanish. The formula (7.38) is the one we used in Eq. (7.1) to estimate the radiation energy density.

The radiative heat flux (7.22) with the help of Eq. (7.37) gives for first approximation:

$$q_R = D_R \nabla E_R = \kappa_R \nabla T \qquad (7.39)$$

where

$$D_R = c/(3\rho K_R) = \text{Rosseland diffusion coefficient}$$
$$\text{of radiation} \qquad (7.40)$$

and

$$K_R = \left(\int_0^\infty \frac{\partial B_\nu}{\partial T} d\nu \right) \Big/ \left(\int_0^\infty \frac{1}{k_\nu'} \frac{\partial B_\nu}{\partial T} d\nu \right)$$

$$= \text{Rosseland mean absorption coefficient} \qquad (7.41)$$

For optically thick medium, the integro-differential equations of radiation gasdynamics are reduced to differential equations as follows:

$$p = \rho RT \qquad (7.42a)$$

$$\frac{\partial \rho}{\partial t} + \nabla \cdot (\rho q) = 0 \qquad (7.42b)$$

$$\rho \frac{Dq}{Dt} = -\nabla(p + p_R) + \nabla \cdot \tau_v + F \qquad (7.42c)$$

$$\rho \frac{D\bar{e}_m}{Dt} = -\nabla \cdot [q(p + p_R)] + \nabla \cdot (q \cdot \tau_v) + \nabla \cdot [(\kappa + \kappa_R)\nabla T] + Q \qquad (7.42d)$$

where

$$\kappa_R = 4D_R a_R T^3 = \text{coefficient of heat conductivity by}$$

$$\text{thermal radiation} \qquad (7.43)$$

and p_R, E_R, and K_R are given by Eqs. (7.38) and (7.41) respectively. Eqs. (7.42) are very similar to those of ordinary gasdynamics but the radiation terms introduce many new features of the flow as we shall discuss later.

(2) Low radiation pressure number R_p [Eq. (4.73)]. If the radiation pressure number R_p of Eq. (4.73) is negligibly small, the

radiation energy density E_R and the radiation stress tensor τ_R are negligible in comparison with the corresponding terms of ordinary gasdynamics. This is the case for most of current reentry problems in which the temperature is of the order of $2 \times 10^4\,°K$ and pressure of 0.01 atmosphere (see Fig. 4.1). However, in this condition, the radiative heat flux may not be neglected because the radiative flux number (4.88) or (4.89) depends on the product of radiation pressure number and the velocity ratio (c/U) which is a very large quantity in reentry problems. In Eqs. (4.86), (4.88), and (4.89), we define the radiative flux number R_F as the ratio of radiative heat flux to heat conduction flux. Similarly we may also define another radiative flux number as the ratio of radiative heat flux to the convective flux which will be useful in the study of inviscid flow of radiation gasdynamics. We have then

$$R'_F = \frac{\text{radiative heat flux}}{\text{convective heat flux}} = \frac{R_F}{P_r R_e} \qquad (7.44)$$

where R_F is given by Eq. (4.88) or (4.89). We have two forms for R'_F: one for optically thick medium and the other for optically thin medium and the case of finite mean free path of radiation. These two forms can be easily obtained from Eqs. (4.88) and (4.89). For instance, for the optically thick medium we have from Eqs. (4.88) and (7.44):

$$R'_{F_1} = \frac{\kappa_R T}{c_p \rho U T L} = 4\left(\frac{\gamma - 1}{\gamma}\right) K_r\, R_p / R_r \qquad (7.45)$$

(3) Some approximations for the radiative heat flux formula. In general, we should substitute the expression of specific intensity I_ν of Eq. (7.35) into the radiative heat flux expression (7.24). For simplicity, let us consider the radiative heat flux in the y-direction. The evaluation of the integral (7.24) depends on the geometrical configuration of our problem. We consider here the two-dimensional problem of radiative heat transfer over a flat plate (Fig. 7.2) and all variables are independent of the azimuth angle ϕ. Hence the integration of the radiation terms with respect to ϕ can be easily carried out. The y-wise radiative heat flux is then:

$$q_{Ry} = 2\pi \int_0^\infty \int_0^\pi I_\nu \cos\theta \sin\theta \, d\theta \, d\nu$$

$$= 2\pi \int_0^\infty \left\{ \int_0^{\pi/2} \left[\int_{\tau_\nu}^\infty B_\nu(t,\theta) \right. \right.$$

$$\left. \times \exp\left[-m(t-\tau_\nu)\right] dt \right] \sin\theta \, d\theta$$

$$+ \int_{\pi/2}^\pi \left[I_{\nu 0}(0,\theta) \cos\theta \exp(m\tau_\nu) \right.$$

$$\left. \left. - \int_0^{\tau_\nu} B_\nu(t,\theta) \exp[-m(t-\tau_\nu)] \, dt \right] \sin\theta \, d\theta \right\} d\nu \qquad (7.46)$$

where $m = \sec\theta$. and $\tau_\nu = \int_0^y \rho k'_\nu \, dy$ is the optical thickness, t is the dummy τ_ν in the integration. The specific intensity on the plate is $I_{\nu 0}$ and the other boundary is at infinity. The specific intensity at infinity will not affect q_{Ry}. In general, the Planck function B_ν depends on two spatial coordinates x and y or τ_ν and θ and k'_ν is also a function of y and θ. Hence we cannot further simplify the integral expression (7.46) because we have to solve it simultaneously with other gasdynamic equations for T and ρ and then k'_ν. We may simplify Eq. (7.46) by some approximations as follows:

(a) Grey gas approximation. The absorption coefficient k'_ν is in general a function of frequency ν. As a result, the optical thickness τ_ν is a function of frequency ν. Thus we may carry out the integration with respect to frequency in Eq. (7.46) if we know the variation of k'_ν with frequency ν. Such a integration may be carried out numerically but seldom analytically because k'_ν is usually a very complicated function of frequency ν. If we assume that the gas is grey so that the absorption coefficient k'_ν is independent of frequency ν, we can integrate Eq. (7.46) with respect to frequency ν because

$$\int_0^\infty B_\nu \, d\nu = B(T) = \frac{\sigma}{\pi} T^4 \qquad (7.47)$$

It is easy to show that the radiative heat flux over a half plane, i.e., q_{RL}^{+} or q_{RL}^{-} with the help of Eq. (7.47) gives the result of Eq. (7.2) as we used to estimate the radiative heat flux in section 1. By means of Eq. (7.47), Eq. (7.46) becomes

$$q_{Ry} = 2 \left\{ \int_0^{\pi/2} \int_{\tau}^{\infty} \sigma\, T^4(t, \theta) \exp\left[-m(t-\tau)\right] dt \sin\theta\, d\theta \right.$$

$$+ \int_{\pi/2}^{\pi} \left[B_0(0, \theta) \cos\theta \exp(m\tau) \right.$$

$$\left. - \int_0^{\tau} \sigma T^4(t, \theta) \exp\left[-m(t-\tau)\right] dt \right] \sin\theta\, d\theta \left. \right\} \qquad (7.48)$$

where $\tau = \tau_{\nu}$ because we assume that τ is independent of frequency ν. Actually, we should use a Planck mean absorption coefficient K_p in Eq. (7.48) which is defined as

$$K_p = \frac{1}{B(T)} \int_0^{\infty} k'_{\nu} B_{\nu}\, d\nu \qquad (7.49)$$

and

$$\tau = \int_0^{y} \rho K_p\, dy$$

and

$$B_0(0, \theta) = \int_0^{\infty} I_{\nu 0}(0, \theta)\, d\nu$$

A second approximation over Eq. (7.48) may be made by assuming that the absorption coefficient k'_{ν} is replaced by several step functions, i.e., it has constant value for certain frequency intervals. We may evaluate the integral with respect to frequency in Eq. (7.46) for various frequency intervals.[2] Since the temperature T is a function of both τ and θ, we cannot simplify Eq. (7.48) any further except by solving it simultaneously with other fundamental equations.

(b) One-dimensional approximation. However, Eq. (7.48) may be further simplified if we assume that the temperature T depends only on τ but independent of θ. This is known as one-dimensional approximation. In other words, we assume that the variation of temperature T with respect to y is much larger than that with respect to x. Such an approximation is good for boundary layer flow and other one-dimensional flow such as flow near a stagnation point and it has been extensively used in the current literature. If we use this approximation, we may carry out the integration with respect to θ in Eq. (7.48) and obtain:

$$q_{Ry} = 2\pi \int_{\tau}^{\infty} B(t)\, \mathscr{E}_2(t-\tau)dt - 2\pi \int_{0}^{\tau} B(t)\, \mathscr{E}_2(\tau-t)\, dt$$

$$-q_R(0)\, \mathscr{E}_3(\tau) \tag{7.50}$$

where the exponential integral $\mathscr{E}_n$ is defined by the formula:

$$\mathscr{E}_n(t) = \int_{1}^{\infty} m^{-n} \exp(-mt)\,dm = \int_{0}^{1} z^{n-2} \exp(-t/z)\,dz \tag{7.51}$$

and $q_R(0) = B_0(0)$.

From Eqs. (7.48) or (7.50), we may say that the order of magnitude of the radiative heat flux is as follows:

$$q_{Ry} \sim \sigma T^4 \frac{L}{L_R} \sim \frac{ca_R T^4 L}{L_R} \tag{7.52}$$

where $L_R = 1/(\rho K_p)$ is the Planck mean free path of radiation. As a result, for the case of finite mean free path, we have the radiative flux number Eq. (7.44) as follows:

$$R'_{F2} = \frac{ca_{RT}4L}{L_R c_p \rho UT} = 3\left(\frac{\gamma-1}{\gamma}\right) R_p/(K_r\, R_r) \tag{7.53}$$

We see that for a large or finite mean free path of radiation, the radiative heat flux is inversely proportional to the mean free path of radiation, while that for a very small mean free path of radiation, the radiative heat flux is proportional to the mean free path of radiation. Hence the radiative heat flux tends to be zero for both very large and very small mean free path of radiation.

(c) Differential approximation. Eq. (7.50) is much simpler than that of Eq. (7.46) or (7.48). But it is still an integral expression and the fundamental equations of radiation gasdynamics, with the expression of Eq. (7.50), is still a system of integro-differential equa-

ions which is difficult to solve. If we approximate the exponential
ntegrals in Eq. (7.50) by an exponential such that

$$\mathscr{E}_2(t) = \frac{m^2}{3} \exp(-mt)$$ (7.54)

vhere m is a constant and apply the conditions that the limiting
/alues of optically thick medium and optically thin medium are
:atisfied, Eq. (7.50) gives the following differential equation for the
:adiative heat flux q_{Ry}:

$$\frac{d^2 q_{Ry}}{d\tau^2} - 3 q_{Ry} + 16\sigma T^3 \frac{dT}{d\tau} = 0$$ (7.55)

[f we consider q_{Ry} as a new variable in the energy equation, we may
:olve Eq. (7.55) with the other fundamental equations of radiation
gasdynamics simultaneously for the variables T, p, ρ, q, and q_{Ry}.
[n this case, we neglect the radiating energy density and radiation
stresses.

For a three dimensional case of a grey gas, Eq. (7.55) may be
replaced by the formula[7]

$$\frac{1}{\rho K} \frac{\partial}{\partial x_i} \left(\frac{1}{\rho K} \frac{\partial q_{Rj}}{\partial x_j} \right) - 3q_{Ri} + 16\sigma T^3 \frac{1}{\rho K} \frac{\partial T}{\partial x_i} = 0$$ (7.56)

where K is the mean absorption coefficient of the medium.

If we use the differential approximation (7.55) or (7.56), the
resultant differential equation of radiation gasdynamics is one order
higher than that without radiative flux because the occurrence of the
second order derivatives in Eq. (7.55) or (7.56). Recently, Shen[20]
proposed an alternative differential approximation such that

$$q_{Ri} = a_i + b_i B + c_{ij} \frac{\partial B}{\partial x_j}$$ (7.57)

where a_i, b_i, and c_{ij} are coefficients to be integrated at each point
in space, essentially geometrical factors except that they depend on
the mean free path of radiation and the specific intensity on the
boundary. The advantage of Shen's approximation is that the
resultant equation is of the same order as that of ordinary gasdy-

namics and the disadvantage is that these coefficients are not known a priori and have to be evaluated for each case.

(d) Optically thin approximation. For an optically thin medium, the optical thickness is usually small. It should be noticed that the upper limit ∞ in the first integral of Eq. (7.50) should be replaced by a finite value of τ_2 in the optically thin medium, because in actual flow field, the region which we are interested in is small compared with the mean free path of radiation. For the optically thin medium, we may expand the exponential integral as power series of τ. If we keep only the lowest term, we have

$$\frac{dq_{Ry}}{dy} = \frac{4\sigma}{L_R}\left(- T^4 + \frac{1}{2} T_w^4\right) \tag{7.58}$$

where $L_R = 1/(\rho K)$ and T_w is the temperature of the wall. In Eq. (7.58), we neglect completely the absorption of radiation. However, if we keep the terms up to $(1/L_R^2)$ some integral terms will remain in the radiative heat flux term.[15] The formula (7.58) is one of the most popular formulae used several years ago when the radiative heat flux just became important and the medium is optically thin. However, we shall show later, it may give very wrong results by neglecting completely the absorption of radiation.

9. Initial and Boundary Conditions of Radiation Gasdynamics

For every specific problem of radiation gasdynamics, we have certain initial and boundary conditions. Our problem is to find proper solution of the fundamental equations of radiation gasdynamics which satisfies these initial and boundary conditions. Similar to all other fluid dynamical problems, we require that the initial conditions be consistent with the boundary conditions at time $t = 0$ and the fundamental equations. Hence we need to examine the boundary conditions only. In radiation gasdynamics, we have to consider the boundary conditions of both the gasdynamic field and the radiation field. The gasdynamic boundary conditions in radiation gasdynamics are the same as those of ordinary gasdynamics, as we have discussed in chapter III section 12. Hence we discuss only the boundary conditions of radiation field in this section.

The boundary conditions of the radiation field depend on the mean free path of radiation. When the mean free path of radiation is very small, i.e., the optically thick medium, all the radiation terms can be expressed in terms of temperature by the Rosseland approximations, i.e., Eqs. (7.42), if we assume that the local thermodynamic equilibrium is attained. Under this condition, we do not need explicitly the boundary condition of the specific intensity of radiation and we need only the boundary condition of temperature. Hence the consideration of gasdynamic conditions only is sufficient for this case of radiation gasdynamics.

When the mean free path of radiation is small but not negligible, we find that the fundamental equations of radiation gasdynamics for optically thick medium, Eqs. (7.42), are sufficient to describe the flow field away from the boundary, but the no-slip condition of temperature will not be satisfied. Hence in analogy to the slip flow of rarefied gasdynamics (chapter VIII section 11), we have a temperature jump in this case. This temperature jump is proportional to the mean free path of radiation and depends on the boundary condition of specific intensity of radiation.

For the finite mean free path of radiation, we have to consider the boundary conditions of the radiation field by studying the interaction of radiation at the interface of two media. When a ray of radiation strikes a surface, part of the radiative energy may penetrate into the second medium and part of the radiative energy may reflect back into the first medium. Hence an incident ray may result in a transmitted ray and a reflected ray. The transmitted ray and reflected ray depend on the smoothness of the surface separating the two media.

(1) Smooth surface.[5] For an optically smooth surface separating two media which have respectively the index of refraction n_1 and n_2 where

$$n_i = \frac{c}{q_i} \tag{7.59}$$

and q_i is the velocity of light in the ith medium (Fig. 7.3), the Snell-Fresnel laws hold, i.e.,

$$\sin \theta_1 = \sin \theta'_1 \tag{7.60a}$$

$$\frac{\sin \theta_1}{\sin \theta_2} = \frac{q_1}{q_2} = \frac{n_2}{n_1} \tag{7.60b}$$

where θ'_1 is the angle between the normal of the surface and the incoming ray, θ_1 is the angle between the normal of the surface and the reflected ray and θ_2 is the angle between the normal of the surface and the refracted ray (Fig. 7.3).

Let r be the coefficient of reflection of the surface S which is the fraction of radiative energy reflected by the surface and r' be the coefficient of reflection of a ray coming from medium 2 and incident on the surface S. It is easy to show that

$$r = r' \tag{7.61}$$

and

$$q_1^2 I_{\nu 1} = q_2^2 I_{\nu 2} \tag{7.62}$$

For the thermodynamic equilibrium condition, we have

$$I_\nu = n^2 B_\nu \tag{7.63}$$

Hence in Eq. (7.34), we assume implicitly that $n = 1$. If the index of refraction n is different from unity, the equation of radiative transfer (7.34) should be written as follows:

$$\frac{\partial I_\nu}{\partial s} = \rho k'_\nu (n^2 B_\nu - I_\nu) \tag{7.64}$$

(2) **Rough surface.** In all the practical problems, we have rough surfaces. Furthermore, we do not study each ray of radiation individually but a large number of rays statistically. Hence with a beam of incoming rays in the direction of L'_1, we may have distributed reflected and refracted rays shown in Fig. 7.4. Thus we have to use statistical averages to consider the actual surface conditions in engineering problems. The properties of a surface in radiation gasdynamics may be expressed in terms of the following three overall coefficients:

(a) The absorption coefficient of the surface a_ν which is equal to the emissivity coefficient of the surface e_ν.

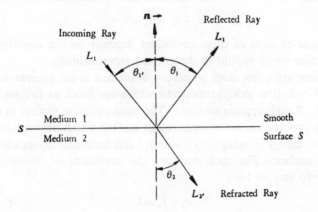

Figure 7.3 Reflection and refraction of a radiation ray on a smooth surface

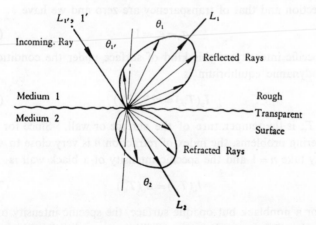

Figure 7.4 Reflected and refracted rays on a rough transparent surface

In general, these coefficients are functions of frequency ν of the radiation ray as well as the properties of the surface.

(b) the reflection coefficient of the surface r_ν.

(c)   the transparency coefficient of the surface tr_ν.

These three coefficients are connected by the following relation:

$$a_\nu + r_\nu + tr_\nu = 1 \qquad (7.65)$$

The value of each of these coefficient depends on the condition of the surface which should be determined experimentally.

There are a few cases which are of interest in the general discussion of radiation gasdynamics and which are listed as follows:

(i)   Rough opaque surface.   The most common surface in engineering problems is an opaque rough surface which absorbs all the radiative energy entering into it within a few molecular layers adjacent to the surface.   For such surfaces, the coefficient of transparency tr_ν is zero and we have

$$a_\nu + r_\nu = 1 \qquad (7.65a)$$

(ii)   Grey surface.   For surfaces, all the three coefficients a_ν, r_ν, and tr_ν are independent of the frequency ν.

(iii)   Black surface.   For such a surface, both the coefficients of reflection and that of transparency are zero and we have

$$a_\nu = e_\nu = 1 \qquad (7.66)$$

The specific intensity of this kind of surface under the condition of thermodynamic equilibrium is

$$I_\nu(T_w) = n^2 B_\nu(T_w) \qquad (7.67)$$

where T_w is the temperature of the surface or wall.   Since for most engineering problems, the index of refraction n is very close to unity, we may take $n = 1$ and the specific intensity of a black wall is

$$I_\nu(T_w) = B_\nu(T_w) \qquad (7.67a)$$

For a nonblack but opaque surface, the specific intensity on the wall under the thermodynamic equilibrium condition with $n = 1$ is

$$I_\nu(T_w) = e_\nu B_\nu(T_w) = (1 - r_\nu) B_\nu(T_w) \qquad (7.68)$$

Eqs. (7.67) and (7.68) have been widely used to determine the boundary condition of the specific intensity of radiation.

10. Mean Free Path of Radiation

In the radiative heat flux formula, one of the most important physical properties of the gas is the mean free path of radiation L_{R_ν} or the absorption coefficient k_ν. This property represents a new diffusive phenomenon in the medium in addition to the momentum diffusivity by viscosity and thermal diffusivity by heat conductivity. Many new and interesting phenomena may be introduced by this radiative diffusivity. The value of the mean free path of radiation should be determined either by experiments or from some microscopic theory.

In general the mean free path of radiation L_{R_ν} or absorption coefficient k_ν is a function of frequency ν of the radiation ray and

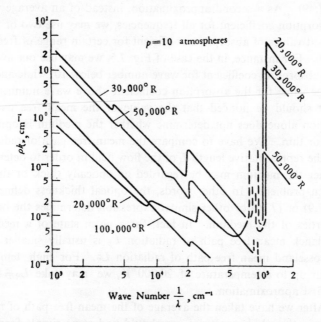

Figure 7.5 Variation of absorption coefficient of hydrogen gas with wave number at a pressure of 10 atmospheres and various temperatures. (Figure 4 of reference [8] by N.L. Krascella, courtesy of NASA)

the state variables of the medium. For instance[8], Fig. 7.5 shows the variation of mean free path of radiation with frequency ν or wave length λ and the temperature of hydrogen gas. If we use the actual variation of the mean free path of radiation with frequency and state variables, the gas is said to be nongrey and the radiative terms such as radiative heat flux should be expressed in integral form. The solution of the radiation gasdynamic equations would be very complicated and extensive numerical computations are required. In engineering problems, it is advisable to use some simple expression of the average value of the mean free path of radiation over the whole spectrum of frequency. For an optically thick medium, we use the Rosseland mean absorption coefficient given by Eq. (7.41). For an optically thin medium or finite mean free path of radiation case, we may use the Planck mean absorption coefficient given by Eq. (7.49). As a second approximation, instead of an average value of absorption coefficient for all frequencies, we may use two or more constant values of absorption coefficient for certain range of frequencies.[3] For instance, in the case of Fig. 7.5, we may use one average mean absorption coefficient for wave number below 10^5 while another average value for the absorption coefficient above wave number 10^5.

It should be noticed that the value of the mean free path of radiation alone does not determine whether the medium is optically thick or thin. We have to compare the mean free path of radiation with the representative length L of the flow field in order to determine whether the medium may be regarded as optically thick or thin in a given problem. In other words, the optical thickness defined by Eq. (7.9) or (7.36) or other similar expression determines the optical properties of the medium. In fact, for a given state of a medium, the Planck mean free path of radiation L_p is usually smaller than the Rosseland mean free path of radiation L_R. For high temperature air up to a temperature of 20,000°K, we may take $L_R = 8.3 L_p$ as a first approximation.

After we have taken the average of the mean free path of radiation over the whole spectrum, we should find some simple formulas for the variation of the average mean free path of radiation with the temperature and density or pressure of the medium. Since the absorption coefficient increases with density, the mean free path of radiation decreases with the increase of density. The variation of the mean free path of radiation with temperature is very complicated. At low temperature, say $T < 10,000\,°K$, the mean free path of radiation decreases with increase of temperature, while at very high temperature, $T > 100,000\,°K$, the mean free path of radiation increases with the temperature and there is a minimum in the intermediate temperature range depending on the density and composition of the medium. Fig. 7.6 shows some typical variations of absorption coefficients of air.[3] If the temperature range in the flow field is not too large, a power law for the average values of the mean free path of radiation may be used, i.e.,

$$\frac{L_R}{L_{R_0}} = \frac{L_p}{L_{p_0}} = \left(\frac{T_0}{T}\right)^{m_1} \left(\frac{\rho_0}{\rho}\right)^{m_2} \tag{7.69}$$

where the subscript 0 refers to the values at some reference conditions. The powers m_1 and m_2 should be so chosen that formula (7.69) gives the best fit with the opacity data over the temperature range considered. For instance, in the temperature range of $7000\,°K$

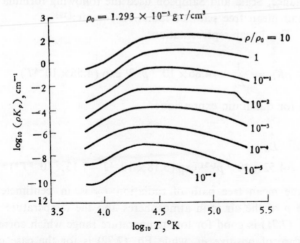

Figure 7.6 Planck mean absorption coefficient of air. (Figure 11 of reference [3] by B.H. Armstrong, J. Sokoloff, R.W. Nicholls, D.H. Holland, and R.E. Meyerott, courtesy of Pergamon Press, Ltd.)

to 12,000°K, the following values may be used for air:

$$m_1 = 4.4, \quad m_2 = 1, \quad \rho_0 = 1.23 \times 10^{-3} \text{ gr/cm}^3,$$

$$T_0 = 10,000°K, \quad L_{P0} = 0.5 \text{ meter}$$

For higher temperature in the neighborhood of 20,000°K, we may take $m_1 = 2.5$ and $m_2 = 1$. At very high temperature in some astrophysical problems, we may take $m_1 = -7/2$ and $m_2 = 2$.

If we consider a large range of temperature including the minimum value of the mean free path of radiation, the following formula may be used

$$\frac{L_{R0}}{L_R} = \frac{L_{p_0}}{L_p} = \frac{L_{p_{01}}}{L_{p_1}} + \frac{L_{p_{02}}}{L_{p_2}} \tag{7.70}$$

where L_{p_1} is the mean free path of radiation given by Eq. (7.69) which is good for low temperature range, i.e., m_1 is positive while L_{p_2} is the corresponding value for high temperature range, i.e., m_1 is negative.

There are some other formulas for the mean free path of radiation with the state variables of the medium which have been used. For instance, Scala and Sampson used the following formula for the Rosseland mean free path of radiation for air:[19]

(i) For linear representation:

$$\rho K_R = \frac{1}{L_R} = 4.86 \times 10^{-7} p^{1.31} \exp (4.56 \times 10^{-4} T) \tag{7.71}$$

and (ii) for quadratic expression:

$$\rho K_R = \frac{1}{L_R}$$
$$= 4.52 \times 10^{-7} p^{1.31} \exp (5.18 \times 10^{-4} T - 7.13 \times 10^{-9} T^2) \tag{7.72}$$

where the mean free path of radiation L_R is in centimeters, the pressure p of the air is in atmospheres and the temperature T is in °K. Eq. (7.71) is good for low temperature range which corresponds to the case of positive m_1 while Eq. (7.72) is for the case of large variation of temperature including the minimum value of L_R which corresponds to the formula (7.70).

For more accurate numerical calculations, the variations of mean free path of radiation with the state variables should be taken into account. On the other hand, for qualitative examination, we may take the mean free path of radiation as a constant in order to show the essential features of the radiation effects. We shall use both approaches in the next few sections when we discuss some flow problems in radiation gasdynamics.

11. Wave Motion in a Radiating Gas[4, 10]

The first problem of radiation gasdynamics which we are going to discuss is the wave motion in a viscous heat conducting and radiating gas. The wave motion will bring out many new characteristic features in radiation gasdynamics which may differ considerably from those of ordinary gasdynamics. The properties of a wave in a gas depend on the amplitude of the wave. The simplest type of wave is the wave of infinitesimal amplitude. For waves of infinitesimal amplitude, we may linearize the fundamental equations of radiation gasdynamics. Since the resultant equations for the wave motion are linear, we may use the method of superposition and the sum of two solutions of them is also a solution of these equations. Thus we may study any typical solution of these wave equations which will give the general features of the wave propagation.

For waves of finite amplitude, the shape of the wave will distort as the wave propagates. When the distortion of the wave is large, ordinary compression wave will develop into shock wave in which large change of physical variables occurs in a very thin region. We shall discuss the shock wave in a radiating gas in the next section.

Now we consider the waves of infinitesimal amplitude in an optically thick medium in which Rosseland approximations may be used. Since we consider an infinite domain, it is a good approximation. We assume that originally the gas is at rest with a pressure p_0, a temperature T_0, and a density ρ_0. The gas is perturbed by a small disturbance so that in the resultant disturbed motion, we have

$$u = u(x, t), \ v = v(x, t), \ w = w(x, t), \ p = p_0 + p'(x, t),$$

$$T = T_0 + T'(x,t), \ \rho = \rho_0 + \rho'(x, t) \tag{7.73}$$

where u, v, and w are respectively the perturbed x-, y-, and z-components of velocity, prime refers to the perturbed quantities of the state variables of the gas. All the perturbed quantities are functions of one spatial coordinate x and time t; thus we consider the wave motion in the direction of x. Substituting Eq. (7.73) into the fundamental equations of radiation gasdynamics (7.42) and neglecting the higher order terms of perturbed quantities, we have the following linear equations for the wave motion in radiation gasdynamics:

$$\frac{p'}{p_0} = \frac{T'}{T_0} + \frac{\rho'}{\rho_0} \tag{7.74a}$$

$$\frac{\partial \rho'}{\partial t} + \rho_0 \frac{\partial u}{\partial x} = 0 \tag{7.74b}$$

$$\rho_0 \frac{\partial u}{\partial t} = -\frac{\partial p'}{\partial x} + 4 R R_p \rho_0 \frac{\partial T'}{\partial x} + \frac{4}{3} \mu \frac{\partial^2 u}{\partial x^2} \tag{7.74c}$$

$$\rho_0 \frac{\partial v}{\partial t} = \mu \frac{\partial^2 v}{\partial x^2} \tag{7.74d}$$

$$\rho_0 \frac{\partial w}{\partial t} = \mu \frac{\partial^2 w}{\partial x^2} \tag{7.74e}$$

$$c_p^* \rho_0 \frac{\partial T'}{\partial t} - 4 R R_p T_0 \frac{\partial \rho'}{\partial t} = \frac{\partial p'}{\partial t} + \kappa^* \frac{\partial^2 T'}{\partial x^2} \tag{7.74f}$$

where

$R_p = a_R T_0^3 / (3 R \rho_0) =$ radiation pressure number of undisturbed flow

$c_p^* = c_p + 12 R R_p =$ effective specific heat at constant pressure including

the radiation effect (7.75)

$\kappa^* = \kappa + 12 R R_p \rho_0 D_R = \kappa + \kappa_R =$ effective coefficient of heat conductivity including radiation effect (7.76)

Examining the linearized equations (7.74), we see that the perturbed quantities may be divided into two groups:

(1) Transverse wave: v and w. In our problem, there is no distinction between the y- and z-directions. Hence both v and w are governed by similar equations (7.74d) and (7.74e). It shows that the wave propagation of variables in the direction perpendicular to the direction of wave propagation, i.e., x-axis.

(2) Longitudinal waves: u, p', T', and ρ'. These four variables are coupled and governed by the rest four equations of Eqs. (7.74). The wave motion due to these variables may be called longitudinal waves which include the ordinary sound wave as a special case.

We are looking for a periodic solution in which all the perturbed quantities are proportional to

$$\exp\left[i(\omega t - \lambda x)\right] = \exp\left[-i\lambda_R(x - Vt)\right]\exp\left(\lambda_i x\right) \qquad (7.77)$$

where ω is the given real angular frequency of the wave, $\lambda = \lambda_R + i\lambda_i$ is the complex wave number, $i = \sqrt{-1}$ and

$$V = \frac{\omega}{\lambda_R} = \text{speed of wave propagation} \qquad (7.78)$$

Substituting the perturbed quantities in the form of (7.77) into Eq. (7.74), we obtain the dispersion relation $\lambda(\omega)$ for both the transverse and longitudinal waves.

(a) Transverse wave. The dispersion relation for the transverse wave v and w is

$$\nu_g\omega^2 + i\lambda = 0 \qquad (7.79)$$

where $\nu_g = \mu/\rho_0$ is the coefficient of kinematic viscosity of the gas. It is the well known damped wave in a viscous fluid as we have discussed in chapter VI section 8. This wave is independent of the compressibility effect of the medium. Since the radiation effect is essentially the compressibility effect, it can be shown that radiation effect has no influence on the transverse magnetofluid dynamic waves even when we include the thermal radiation effects in the fluid of an electrically conducting fluid.

(b) Longitudinal waves. The dispersion relation of the longitudinal waves obtained from Eqs. (7.74a), (7.74b), (7.74c) and (7.74f) is

$$\kappa^*\left(\frac{1}{\rho_0}+\frac{4}{3}\frac{i\omega\nu_g}{\rho_0}\right)\lambda^4-\left\{\frac{\omega^2\kappa^*}{\rho_0}+\frac{4}{3}\frac{\nu_g\omega^2}{T_0(\gamma-1)}\left[1+12(\gamma-1)R_p\right]\right.$$

$$\left.-i\omega(c_p+20RR_p+16RR_p^2)\right\}\lambda^2-\frac{i\omega^3}{T_0(\gamma-1)}\left[1+12\,(\gamma-1)\,R_p\right]$$

$$=0 \tag{7.80}$$

Eq. (7.80) is similar to the first square bracket of Eq. (6.80) which represents two modes of the longitudinal waves in a radiating gas: one is the sound wave in a viscous, heat-conducting, and radiating gas and the other is the heat wave of this medium. Both the sound wave and the heat wave are affected by the thermal radiation effects. There are two types of thermal radiation effects: one is due to the radiation pressure and radiation energy density which is characterized by the radiation pressure number R_p, and the other is due to the radiation heat flux which is characterized by radiation heat conductivity κ_R.

Let us first consider the effects of R_p. For an inviscid and nonheat-conducting fluid without radiative heat flux, i.e., $\nu_g=0$ and $\kappa^*=0$, Eq. (7.80) gives the sound wave in a radiating gas with the speed of propagation C_R given by the formula:

$$C_R^2=\frac{\omega^2}{\lambda^2}=a_0^2\frac{1+20\left(\frac{\gamma-1}{\gamma}\right)R_p+16\left(\frac{\gamma-1}{\gamma}\right)R_p^2}{1+12\,(\gamma-1)\,R_p} \tag{7.81}$$

where $a_0=(\gamma RT_0)^{1/2}$ is the ordinary sound speed without radiation effects. For a given temperature, the radiation sound speed increases with the radiation pressure number as shown in Fig. 7.7.

The effects of κ^* are twofold: First it introduces some damping in the sound wave discussed above. Of course, if the value of κ^* is not small, it will affect the sound speed too. Second it introduces another mode of the longitudinal wave which corresponds to the heat wave in ordinary gasdynamics. In ordinary gasdynamics, if the heat conductivity κ is different from zero, we have a heat wave besides the sound wave. Since the thermal radiation increases the effective value of heat conductivity, we will have this heat wave even if the ordinary heat conductivity is negligible but the radiative heat conductivity is not negligible. When we consider the case of an inviscid and nonheat-conducting gas with a small amount of radiative heat flux so that κ^* is a small quantity, Eq. (7.80) gives the following two roots of λ as a first approximation.

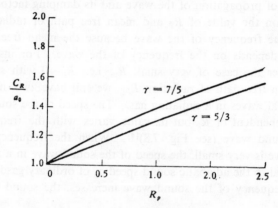

Figure 7.7   Sound speed in a radiating gas as a function of radiation pressure number R_p

(i)   Radiation sound wave

$$\lambda_1 = \pm \frac{\omega}{C_R}[1 - i\omega f(R_p)D_R] \qquad (7.82)$$

and (ii) radiation heat wave

$$\lambda_2 = \pm \frac{\omega}{C_R}\left[\frac{g(R_p)}{\omega D_R}\right]^{1/2}(-1+i) \qquad (7.83)$$

where $f(R_p)$ and $g(R_p)$ are functions of R_p. Eq. (7.82) shows that the first order effects of radiative heat flux are the introduction of damping in the sound wave without change of its speed of propagation C_R. Eq. (7.83) shows that the speed of propagation of the heat wave is proportional to C_R and $(\omega D_R)^{1/2}$, i.e., its speed of propagation increases with the square root of frequency and D_R and the damping factor of the heat wave is inversely proportional to $D_R^{1/2}$. Hence when the diffusion coefficient of radiation tends to zero, the heat wave is highly damped and we have only one undamped radiation sound wave in the medium.

For optically thin medium or medium with finite mean free path of radiation, the waves are much more complicated than that for the case of optically thick medium. One of the new features is that the speed of propagation of the wave and its damping factor depend not only on the value of R_p and mean free path of radiation but also on the frequency of the wave because the mean free path of radiation depends on the frequency of the wave. For instance, if we consider the case of very small R_p, i.e., $R_p \to 0$ with small but finite mean free path of radiation L_{R_ν}, we still have two modes of longitudinal waves in a radiating gas. The speed of the sound wave C_R is independent of R_p but its value varies with the frequency ω of the sound wave (see Fig. 7.8)[4]. When the frequency of the sound wave is very small, the speed of the sound wave in a radiating gas is equal to the adiabatic sound speed a_0 of ordinary gasdynamics. As the frequency of the sound wave increases, the sound speed of

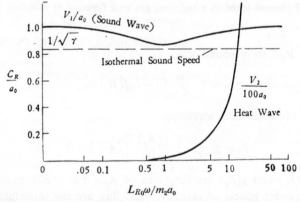

Figure 7.8 Longitudinal waves in a nongrey radiating gas with $R_p = 0$. (Figure 2 of reference [4] by B.S. Baldwin, Jr., courtesy of NASA)

the radiating gas decreases until it reaches a value a little higher than the isothermal sound speed $(RT)^{1/2}$ of ordinary gasdynamics. After the minimum of this radiation sound speed is reached, further increase of frequency will cause an increase of the speed of sound. At very high frequency, the speed of sound in the radiating gas becomes equal to that of adiabatic sound speed a_0 again. The speed of heat wave increases with the frequency in a similar manner as that of the optically thick medium.

If we include the radiation effects in magnetofluid dynamics, we find that the most important effect is the change of sound speed from a_0 to C_R. For instance, for the case of an ideal plasma, i.e., $\nu_q = 0$, $\nu_H = 0$, and $\kappa^* = 0$, we will have the fast and slow waves as given in Eq. (6.81) except that the ordinary sound speed a_0 should be replaced by the radiation sound speed C_R.

12. Shock Wave in a Radiating Gas[11]

It is well known that waves of finite amplitude may develop into shock waves across which large changes of velocity and state variables occur. Since shock waves are important in high speed flow where thermal radiation is also important, it is interesting to see what are the effects of thermal radiation on the shock waves. We first consider a normal shock wave in an optically thick medium. We choose the coordinate system such that the shock wave is stationary. In our system, the gas is moving in the direction of x-axis and has a velocity component u in the x-direction only. Both the velocity and the state variables in front of the shock are uniform and there is a sharp transition region in which a large variation in velocity and state variables occurs. Finally far behind the shock, the velocity and state variables become uniform again but at different values from those in front of the shock. The fundamental equations which govern the flow field with this normal shock are as follows:

$$\rho u = \text{constant} = m \qquad (7.84a)$$

$$mu + p_t - \frac{4}{3}\,\mu\,\frac{du}{dx} = \text{constant} = mC_1 \qquad (7.84b)$$

$$mh_R + up_t - \frac{4}{3}\,\mu u\,\frac{du}{dx} - \kappa^*\,\frac{dT}{dx} = \text{constant} = mC_2 \quad (7.84c)$$

where

$$p_t = p + p_R,\ h_R = \frac{1}{2}\,u^2 + c_v\,T + E_R/\rho\ \text{and}\ \kappa^* = \kappa + 4D_R a_R\,T^3$$

In the analysis of the shock wave, we want to study two items: one is the Rankine-Hugoniot relations across the shock wave and the other is the shock transition region.

(1) Rankine-Hugoniot relations in radiation gasdynamics. The Rankine-Hugoniot relations connect the values of the two uniform states in front of the shock with subscript 1 and behind the shock with subscript 2. In these uniform states, we have

$$\frac{du}{dx} = \frac{dT}{dx} = 0 \tag{7.85}$$

Now we introduce the following nondimensional variables:

$$\xi = \frac{u}{u_1}, \quad T^* = \frac{RT}{u_1^2}, \quad T_1^* = \frac{1}{\gamma M_1^2}, \quad M_1 = \frac{u_1}{a_1}, \quad a_1 = (\gamma RT_1)^{1/2},$$

$$Q = \frac{a_R u_1^6}{\rho_1 R^4}, \quad R_p = \frac{p_R}{p} = \frac{1}{3} Q \xi T^{*3} \tag{7.86}$$

Substituting Eqs. (7.85) and (7.86) into Eqs. (7.84), we have

$$\xi^2 + (1 + R_p) - [1 + T_1^*(1 + R_{p_1})]\xi = 0 \tag{7.87a}$$

$$-\frac{1}{2}\xi^2 + \left(\frac{1}{\gamma - 1} + 3R_p\right)T^* + [1 + T_1^*(1 + R_{p_1})]\xi$$

$$-\left[\frac{1}{2} + \frac{\gamma T_1^*}{\gamma - 1} + 4T_1^* R_{p_1}\right] = 0 \tag{7.87b}$$

If we eliminate T^* from Eqs. (7.87), we have

$$(\xi - 1)\left[\xi - \frac{8R_p + r^2 + 1}{7R_p + r^2}(1 + R_{p_1})fT_1^* + (1 + R_{p_1})\right]$$

$$= (\xi - 1)(\xi - \xi_2) = 0 \tag{7.88}$$

where

$$r^2 = \frac{\gamma + 1}{\gamma - 1}, \quad f = \frac{\xi - g(R_p)}{\xi - 1}$$

$$g(R_p) = \frac{(R_p + 1)(8R_{p_1} + r^2 + 1)}{(R_{p_1} + 1)(8R_p + r^2 + 1)}$$

There are two roots of Eq. (7.88). The root $\xi = 1$ represents the velocity of the original flow, i.e., no shock. The other root $\xi = \xi_2$ represents the velocity behind a normal shock. The formal expression for ξ_2 is

$$\xi_2 = \frac{\gamma_e - 1}{\gamma_e + 1} + \frac{2\gamma_e p_i^*}{\gamma_e + 1} \tag{7.89}$$

where

$$\gamma_e = \frac{4(\gamma - 1)R_{p_2} + \gamma}{3(\gamma - 1)R_{p_2} + 1} = \begin{array}{l}\text{effective ratio of specific heats}\\ \text{in radiation gasdynamics}\end{array} \tag{7.90}$$

$$p_i^* = (1 + R_{p_1})f(R_{p_2})T_1^* = \begin{array}{l}\text{effective value of } T_1^* \text{ in}\\ \text{radiation gasdynamics}\end{array} \tag{7.91}$$

when $R_{p_2} \cong 0$, $\gamma_e = \gamma$ and $p_i^* = T_1^*$, we have the Rankine-Hugoniot relation across a normal shock in ordinary gasdynamics. When R_{p_2} is very large, $\gamma_e = 4/3$ for all values of γ.

Since both γ_e and p_i^* are functions of R_{p_2} and R_{p_2} depends on ξ_2, we have to find ξ_2 for a given set of initial conditions T_1^* and R_{p_1} by the method of successive approximations.

It is interesting to find the values of ξ_2 for a few limiting cases:

(a) Low temperature case. If the temperatures both in front of and behind the normal shock are not too high, we have $R_{p_1} \cong R_{p_2} \cong 0$. Hence $\gamma_e = \gamma$ and $p_i^* = T_1^*$, Eq. (7.89) is identical to the normal shock relation in ordinary gasdynamics.

(b) Weak shock in a high temperature gas. If the temperature of the gas is initially very high, R_{p_1} is then not negligible. If in addition, the shock wave strength is weak, R_{p_2} will be approximately equal to R_{p_1}. Hence in Eq. (7.89), we may write $\gamma_e \cong \gamma_{e_1}$ and $p_i^* \cong p_{i_1}^*$. The effects of radiation on the uniform state behind a weak shock in this case are:

(i) The value of γ is replaced by the effective value γ_{e_1}, i.e.,

$$\gamma_{e_1} = \frac{4(\gamma - 1)R_{p_1} + \gamma}{3(\gamma - 1)R_{p_1} + 1} \tag{7.92}$$

and (ii) the value of T_1^* is replaced by $p_{i_1}^*$, i.e., the gas pressure is replaced by the total pressure which is the sum of the gas pressure and the radiation pressure. When the shock strength is infinitesimally small, we have

$$u_1^2 = \gamma_{e_1} \frac{p_1 + p_{R1}}{\rho_1} = C_R^2 \tag{7.93}$$

This formula (7.93) is another way to define a radiation sound speed C_R which is practically identical to that given by Eq. (7.81).

(c) Very strong shock in a cold gas. In this case, $R_{p_1} \ll 1$ but $R_{p_2} \gg 1$, we have then

$$\xi_2 = \frac{1}{7} + \frac{27}{70 M_1^2} \tag{7.94}$$

where we take $\gamma = 5/3$. In the limit of $M_1 = \infty$, $\xi_2 = 1/7$ and $\gamma_{e_2} = 4/3$.

(d) Shock wave in a very hot plasma. $R_{p_1} \gg 1$ and we have

$$\xi_2 = \frac{1}{7} + \frac{1}{6 M_{e\,1}^2} \tag{7.95}$$

where $M_e = u/C_R$.

It is interesting to notice that the effect of radiation on the drop in velocity ξ_2 is usually not very large. For instance, in the limit of infinite Mach number M_1, $\xi_2 = 1/7$ with radiation effect and $\xi_2 = 1/6$ when $\gamma = 7/5$ or $\xi_2 = 1/4$ when $\gamma = 5/3$ without radiation effect. However, the effect on temperature by thermal radiation may be very large. Let us consider the case of a very hot gas. Eq. (7.87b) with the help of Eq. (7.86) may be written as

$$T^{*4} + A^{-1} T^* - A^{-1} B = 0 \tag{7.96}$$

where

$$A^{-1} = \frac{T_1^{*3}}{R_{p_1} \xi_2} > 0$$

and

$$B = [(R_{p_1} + 1) T_1^* + 1] \xi_2 - \xi_2^2 > 0$$

For $R_{p_1} \gg 1$, $A^{-1} T^* \ll T^{*4}$, we have

$$\frac{T_2}{T_1} = \frac{T^*}{T_1^*} = \frac{(A^{-1}B)^{1/4}}{T_1^*} \cong \left[1 + \frac{8}{7}(M_{e_1}^2 - 1) \right]^{\frac{1}{4}} \quad (7.97)$$

For very large M_{e_1}, Eq. (7.97) becomes

$$\frac{T_2}{T_1} = 1.033 M_{e_1}^{\frac{1}{2}} \quad (7.98)$$

Without radiation effect, it is well known that at very high shock Mach number, the temperature jump across a normal shock is proportional to the square of the Mach number. Now with radiation effect, the temperature jump is proportional to the square root of the Mach number. Hence there is a very large difference in temperature with and without the radiation effect.

(2) Shock wave structure. In previous paragraphs, we consider the shock as a surface of discontinuity with two uniform states in front of and behind it. Actually there is a transition region in which the flow variables change gradually from the values in front of the shock to those behind it. For optically thick medium, we may obtain the transition region by solving Eqs. (7.84). We have to consider viscosity, heat conductivity, and radiative heat transfer. The main feature of the present problem is the introduction of a new diffusive transport property, the mean free path of radiation L_{R_ν}, in addition to the viscosity and heat conductivity in shock structure problems of ordinary gasdynamics in which the mean free path of gas L_f plays an important role. Since the relative values of L_{R_ν} and L_f may vary greatly, new phenomena would occur, particularly when L_{R_ν} is much larger that L_f. The present problem is very similar to the shock structure in a chemically reacting medium in which the relaxation phenomena are important and in which a relaxation length L_r may be introduced to represent the distance to attain the chemical equilibrium condition in the shock transition region (see chapter V section 13). If L_r is much larger than L_f, we have the inviscid, nonheat conducting tail of the shock transition region which is determined by L_r. We may have a partially dispersed shock or fully dispersed shock depending on the value of the relaxation length as well as the shock strength.

We have similar situations in the shock transition region in a radiating gas. For simplicity, let us consider the case where R_p is negligible but R_F is not negligible. In this case, we have the new feature of mean free path of radiation L_p, where we use the Planck mean free path of radiation as a representative value in our problem. If L_p is much larger than L_f, we would have some portion of the flow in the shock transition region in which the flow field will be determined mainly by L_p. Hence we would have a phenomena similar to the partly dispersed shock and fully dispersed shock in a chemically reacting gas. The chemical reaction is important only behind the normal shock while the radiative heat transfer may be important both in front of and behind the shock. Hence in a radiating gas, we may have the inviscid and nonheat conducting tails in front of and behind the shock. For a partly dispersed shock, the front and the rear inviscid and nonheat conducting tails may join together by a surface of discontinuity, the ideal shock in which we should consider the viscosity and heat conductivity effects. For a fully dispersed shock, the front and the rear inviscid and nonheat conducting tails would merge together without any discontinuity between them. In other words, the radiation effect may increase the transition region so that the viscosity and heat conductivity may be neglected in the whole shock transition region.[21]

A complete analysis of shock structure including the effects of radiation energy, radiation stresses, and frequency effects has not been carried out yet.

13. Inviscid Flow in Radiation Gasdynamics

In this section, we study the effects of radiation energy density and radiation pressure on the flow problem. We consider an optically thick medium with such a high temperature and low density that the radiation pressure number R_p is not negligible. We shall assume that the mean free path of radiation is so small that, except in the boundary layer region, the term of radiative heat transfer is negligibly small. If we restrict ourselves to the flow field outside the boundary layer or other transition regions, we may consider an ideal radiating gas in which viscous, heat-conducting, and radiative transfer terms are all negligible. The fundamental equations of this ideal radiating gas are:

(1) Equation of state

$$p = \rho R T \qquad (7.99a)$$

(2) Equation of continuity

$$\frac{\partial \rho}{\partial t} + \nabla \cdot (\rho q) = 0 \qquad (7.99b)$$

(3) Equation of motion

$$\rho \frac{Dq}{Dt} = -\nabla(p + p_R) \qquad (7.99c)$$

(4) Equation of energy

$$\rho \frac{D\bar{e}_m}{Dt} = -\nabla \cdot [q(p + p_R)] + Q \qquad (7.99d)$$

Now we define an effective enthalpy of a radiating gas H_R and its stagnation effective enthalpy H_{R0} as follows:

$$H_R = U_m + \frac{E_R}{\rho} + \frac{p + p_R}{\rho} \qquad (7.100)$$

$$H_{R0} = H_R + \tfrac{1}{2} q^2 \qquad (7.101)$$

From the fundamental equations (7.99), we have

$$\frac{DH_{R0}}{Dt} = T\frac{DS}{Dt} + \frac{\partial(p + p_R)}{\partial t} = Q + \frac{\partial(p + p_R)}{\partial t} \qquad (7.102)$$

and

$$T\frac{DS}{Dt} = \frac{DU_m}{Dt} + \frac{D}{Dt}\left(\frac{E_R}{\rho}\right) + (p + p_R)\frac{D}{Dt}\left(\frac{1}{\rho}\right) = Q \qquad (7.103)$$

where S is the entropy of the radiating gas. Eqs. (7.102) and (7.103) are similar to those corresponding equations of ordinary gasdynamics except that the corresponding terms of radiation effects should be added.

If we substitute the relation (7.38) into Eq. (7.103), we have

$$\frac{D}{Dt}\left(\frac{S}{c_v}\right)=\frac{D}{Dt}\left[\ln\left(\frac{T}{\rho^{\gamma-1}}\right)+4(\gamma-1)R_p\right] \qquad (7.104)$$

where

$$R_p=\frac{p_R}{p}=\frac{a_R T^4}{3p}=\frac{a_R T^3}{3R\rho} \qquad (7.105)$$

Integration of Eq. (7.104) gives the equation of state of a radiating gas in equilibrium as follows:

$$\frac{1}{c_v}(S-S_0)=\ln\left(\frac{T}{\rho^{\gamma-1}}\right)+4(\gamma-1)R_p+\text{constant} \qquad (7.106)$$

where subscript 0 refers to values at a reference point.

By the help of Eq. (7.99a), Eq. (7.106) becomes

$$\frac{p}{p_0}=\left(\frac{\rho}{\rho_0}\right)^\gamma\exp\left(\frac{S-S_0}{c_v}\right)\exp\left[4(\gamma-1)(R_{p0}-R_p)\right] \qquad (7.107)$$

If both R_p and R_{p_0} are negligibly small, Eq. (7.107) reduces to the formula of ordinary gasdynamics, i.e.,

$$\frac{p}{p_0}=\left(\frac{\rho}{\rho_0}\right)^\gamma\exp\left(\frac{S-S_0}{c_v}\right) \qquad (7.108)$$

If R_p is very large so that the term $\ln(T/\rho^{\gamma-1})$ in Eq. (7.104) is negligible, we have

$$S-S_0=\frac{4}{3}a_R\left(\frac{T^3}{\rho}-\frac{T_0^3}{\rho_0}\right) \qquad (7.109)$$

For isentropic flow $S=S_0$, Eq. (7.109) may be considered as a special case of (7.108) with $\gamma_e=4/3$ but the difference of behavior of Eqs. (7.108) and (7.109) should be noticed.

For isentropic flow $S=S_0=$ constant, the following relations of the variation of state variables of a radiating gas hold:

$$\frac{dT}{d\rho}=\frac{T}{\rho}\frac{(\gamma-1)(1+4R_p)}{1+12(\gamma-1)R_p} \qquad (7.110)$$

$$\frac{dp}{d\rho} = \frac{p}{\rho} \frac{\gamma + 16(\gamma - 1)R_p}{1 + 12(\gamma - 1)R_p} \qquad (7.111)$$

$$\frac{dp_R}{dT} = 4\frac{p_R}{T} \qquad (7.112)$$

$$\frac{d(p + p_R)}{d\rho} = \frac{p}{\rho} \frac{\gamma + 20(\gamma - 1)R_p + 16(\gamma - 1)R_p^2}{1 + 12(\gamma - 1)R_p} = C_R^2 \qquad (7.113)$$

Eq. (7.113) is identical to Eq. (7.81) which gives the sound speed of an ideal radiating gas. It is interesting to notice that R_p is a function of temperature and density of the gas. Hence C_R is a function of both T and ρ. Of course for isentropic flow, the density ρ may be considered as a function of temperature T, i.e., $\rho = \rho(T)$ but the function $\rho(T)$ still depends on some reference density and we cannot say that the sound speed C_R depends only on temperature because $\rho(T)$ depends on some reference density. Hence we conclude that the sound speed of a radiating gas depends on both temperature and density of the gas. Only in the limiting case, R_p approaches to zero, $C_R = a = (\gamma RT)^{1/2}$ which is a function of temperature only. For the other limiting case of very large R_p, i.e., $R_p \to \infty$, we have

$$C_R = \frac{2}{3} a_R \frac{T^2}{\sqrt{\rho}} = \left(\frac{4}{3} RT\right)^{1/2} (R_{p_0})^{1/2} \qquad (7.114)$$

where $R_{p_0} = p_{R_0}/p_0$ is a reference value of the radiation pressure number. Since R_{p_0} is a constant, at a first glance of Eq. (7.114), one might attempt to draw the conclusion that C_R depends on the temperature only which behaves like a gas with effective value of the ratio of specific heats $\gamma_e = 4/3$. However, the constant R_{p_0} depends on both the temperature T_0 and the density ρ of the gas. Hence even when $R_p \to \infty$, we cannot draw the conclusion that C_R is independent of the density of the gas, because we may have different value of C_R for given values of T and T_0 but various values of ρ_0.

Now let us consider the adiabatic and steady flow of a radiating gas. Eqs. (7.102) and (7.103) become respectively:

$$q \cdot \nabla H_{R0} = 0 \qquad (7.115)$$

and

$$q \cdot \nabla S = 0 \qquad (7.116)$$

Hence both the effective stagnation enthalpy and entropy are constant along a streamline for steady and adiabatic flow. For isoenergetic flow, integration of Eq. (7.115) gives

$$c_p T + \frac{4}{3} \frac{E_R}{\rho} + \frac{1}{2} q^2 = \frac{\gamma}{\gamma - 1} \frac{p}{\rho} + 4 \frac{p_R}{\rho} + \frac{1}{2} q^2 = \text{constant} \qquad (7.117)$$

Eq. (7.117) may be written in the following form:

$$\frac{1}{2} q^2 + \frac{C_R^2}{\gamma_e - 1} = \frac{C_{R0}^2}{\gamma_{e0} - 1} \qquad (7.118)$$

where

$$\gamma_e = \frac{\gamma + 4(\gamma - 1) R_p}{1 + 3(\gamma - 1) R_p} = \text{effective value of } \gamma \qquad (7.119)$$

and

$$C_R = \left(\gamma_e \frac{p + p_R}{\rho} \right)^{\frac{1}{2}} = \begin{array}{l} \text{effective sound speed of} \\ \text{radiation gasdynamics} \end{array} \qquad (7.120)$$

Eq. (7.119) is a general expression for Eqs. (7.90) or (7.92) and Eq. (7.120) is identical to Eq. (7.93). In Eq. (7.118), subscript 0 refers to the value at stagnation point $q = 0$. It should be noticed that the effective ratio of specific heats γ_e varies with the velocity q.

The maximum possible velocity q_m from a given stagnation condition T_0 and ρ_0 is

$$q_m = \left(\frac{2}{\gamma_{e0} - 1} \right)^{\frac{1}{2}} C_{R0} \qquad (7.121)$$

The maximum possible velocity q_m is a function of both stagnation temperature T_0 and stagnation density ρ_0 instead of a function of T_0 only in the case of ordinary gasdynamics, i.e., $R_p \to 0$, $q_m = \left(\frac{2}{\gamma - 1} \gamma R T_0 \right)^{\frac{1}{2}}$. For the other limiting case, $R_p \to \infty$, we have

$$q_m(R_p \to \infty) = \left(\frac{8 a_R}{3} \right)^{\frac{1}{2}} \frac{T_0^2}{\sqrt{\rho_0}} \qquad (7.122)$$

14. Two-Dimensional Unsteady Laminar Boundary Layer on an Infinite Plate[12]

Now we consider an unsteady uniform flow $U(t)$ of a compressible and radiating gas of constant pressure p_∞ and constant temperature T_∞ over an infinite long flat plate. The plate is kept at a constant temperature T_w. Fluid may be injected or sucked from the surface of the plate. In order to bring out some essential features of radiation gasdynamics, we use the following assumptions in our analysis:

(1) The pressure in the whole flow field is a constant.

(2) The gas is an ideal gas for which the perfect gas law holds:

$$\rho T = \rho_\infty T_\infty \tag{7.123}$$

where subscript ∞ refers to the value in the free stream.

(3) The coefficient of viscosity μ of the gas is proportional to its absolute temperature

$$\frac{\mu}{\mu_\infty} = C\left(\frac{T}{T_\infty}\right) \tag{7.124}$$

where C is a constant.

(4) Both the specific heat at constant pressure c_p and the Prandtl number P_r are constant.

(5) The normal velocity on the surface of the plate is small so that boundary layer approximations hold.

(6) The gas is optically thick so that Rosseland approximations hold. The radiation pressure number is small and we may neglect the radiation energy density and radiation pressure but not the radiative heat flux.

(7) All variables are functions of time t and normal coordinate y from the plate only.

Under the above assumptions, our fundamental equation (7.42) becomes

$$\frac{\partial \rho}{\partial t} + \frac{\partial \rho v}{\partial y} = 0 \tag{7.125}$$

$$\rho \frac{\partial u}{\partial t} + \rho v \frac{\partial u}{\partial y} = \rho \frac{dU}{dt} + \frac{\partial}{\partial y}\left(\mu \frac{\partial u}{\partial y}\right) \tag{7.126}$$

$$\rho c_p \left(\frac{\partial T}{\partial t} + v\,\frac{\partial T}{\partial y} \right) = \frac{\partial}{\partial y}\left[(\kappa + \kappa_R)\,\frac{\partial T}{\partial y}\right] + \mu \left(\frac{\partial u}{\partial y} \right)^2 \quad (7.127)$$

with the boundary conditions:

$$\left.\begin{array}{l} y=0: \ u=0, \ v=v_w(t), \ T=T_w = \text{constant} \\[4pt] y\rightarrow\infty: \ u\rightarrow U(t), \ T=T_\infty = \text{constant} \end{array}\right\} \quad (7.128)$$

where subscript w refers to the value on the surface of the plate, and v_w is rate of injection or suction of the gas on the surface of the plate which is a given function of time t.

We introduce a new variable Y such that

$$Y = \int_0^y \frac{\rho}{\rho_\infty}\,dy \quad (7.129)$$

Eq. (7.125) gives

$$\rho v = \rho_w v_w - \frac{\partial Y}{\partial t} \quad (7.130)$$

We also introduce the nondimensional variables:

$$u^* = \frac{u}{U}, \quad t^* = \frac{u_0 t}{L}, \quad \eta = \left(\frac{Y}{L} \right)\left[\frac{R_0}{C(1-\alpha t^*)} \right]^{\frac{1}{2}} \quad (7.131)$$

where u_0 is a reference velocity, L is a reference length, α is a constant and

$$R_0 = \frac{\rho_\infty u_0 L}{\mu_\infty} \quad (7.132)$$

is a reference Reynolds number in the freestream. For steady flow $\alpha=0$ and for unsteady flow, we take $\alpha=-1$ without loss of generality.

If the function $U(t)$ and $v_w(t)$ have the following form, we have similar solution for our boundary layer equations (7.125) to (7.127).

$$U(t) = u_0(1-\alpha t^*)^{-\beta} \quad (7.133)$$

and

$$\frac{\rho_w v_w}{\rho_\infty u_0}\left(\frac{R_0}{C} \right)^{\frac{1}{2}} = \gamma_0 (1-\alpha t^*)^{-\frac{1}{2}} \quad (7.134)$$

where β and γ_0 are constant and γ_0 may be called the injection parameter.

With the expressions (7.133) and (7.134), the velocity $u^* = u^*(\eta)$ is a function of η only, i.e., similar solution, which is governed by the total differential equation:

$$u^{*\,\prime\prime} - \left(\frac{1}{2}\alpha\eta + \gamma_0\right)u^{*\,\prime} + \alpha\beta(1 - u^*) = 0 \qquad (7.135)$$

where prime refers to differentiation with respect to η. We may integrate Eq. (7.135) for various values of β and γ_0. Even though the velocity $u^*(\eta)$ does not depend on temperature T explicitly, the velocity distribution in the physical plane $u(y,t)$ depends on the temperature distribution and the radiation flux number

$$R_{F1} = \frac{(\gamma - 1)^3 M_0^6 16\sigma T^3 L_R}{3 P_r C \kappa_\infty} \qquad (7.136)$$

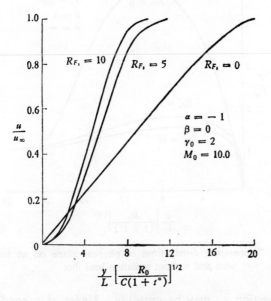

Figure 7.9 Velocity distributions on an infinite plate
with suction and radiation effects

where $M_0 = u_0/(\gamma R T_\infty)^{1/2}$ is the Mach number of the flow. Fig. 7.9 shows the effect of radiative flux number on the velocity distribution in the physical plane.

The nondimensional equation of energy (7.127) is

$$\left(\frac{1}{P_r} + R_{F1}\,\zeta^3\right)\frac{d^2\zeta}{d\eta^2} + \left(R_{F1}\,\frac{d\zeta^3}{d\eta} - \frac{1}{2}\alpha\eta - \gamma_0\right)\frac{d\zeta}{d\eta}$$
$$+ \left(\frac{du^*}{d\eta}\right)^2 = 0 \tag{7.137}$$

where $\zeta = c_p T/U^2$ and the boundary conditions are

$$\eta = 0:\ \zeta = \zeta_w;\ \eta \to \infty,\ \zeta = \zeta_\infty \tag{7.138}$$

In general the energy equation does not have similar solution. Eq. (7.137) holds true only for the case where the freestream is steady

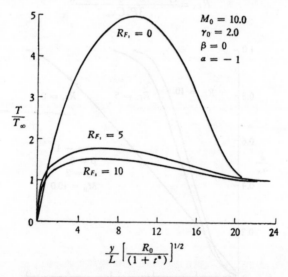

Figure 7.10 Temperature distributions in physical plane on an infinite plate with and without radiation heat flux

but the injection v_w may be unsteady. Under this condition, we have a similar solution for temperature, i.e., $\zeta(\eta)$. We have integrated Eq. (7.137) for various R_{F1}, M_0, and γ_0. Some typical results are shown in Figs. 7.10 and 7.11. The general conclusions are:

(1) The peak temperature in the boundary layer due to viscous dissipation at high Mach number is suppressed if the radiative flux number is high enough.

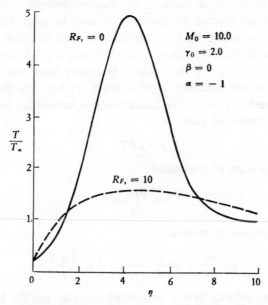

Figure 7.11   Similar solution in temperatures on an infinite plate for an optically thick radiating gas with and without radiative flux term

(2) This suppression of peak temperature may lead to a calculation of radiative heat flux that is too high if one neglects the thermal radiation effects in solving the boundary layer equations.

(3) The thermal boundary layer thickness increases slightly, while the velocity boundary layer thickness decreases with increase of R_F in the present case because of its effect on density.

15.   A Uniform Flow of a Radiating Gas over a Semi-Infinite Plate[13]

Now we are going to improve the results of the last section by two aspects: first we consider the case of a semi-infinite plate so that the variation of x-coordinate, distance along the plate, will be considered; second, we consider the finite mean free path of radiation. However, we shall consider only the steady boundary layer flow without injection or suction on the surface of the plate. Since the radiative heat flux may be regarded as an increase of the effective heat conductivity or a decrease in effective Prandtl number, the thickness of the thermal boundary layer may be much larger than the velocity boundary layer thickness. We shall consider both the case that the thermal boundary layer thickness is of the same order of magnitude as that of the velocity boundary layer, and the case where the thermal boundary layer is much thicker than that of the velocity boundary layer. Our fundamental equations are as follows:

(1)   Equation of state:

$$p = \rho RT \tag{7.139}$$

(2)   Equation of continuity:

$$\frac{\partial \rho u}{\partial x} + \frac{\partial \rho v}{\partial y} = 0 \tag{7.140}$$

(3)   Equation of motion:

$$\rho u \frac{\partial u}{\partial x} + \rho v \frac{\partial u}{\partial y} = \frac{\partial}{\partial y} \left(\mu \frac{\partial u}{\partial y} \right) \tag{7.141}$$

As far as the velocity field is concerned, we may use the boundary layer approximations in our problem. We consider only the case where the pressure is a constant over the whole flow field.

(4)   Equation of energy:

$$\rho u \frac{\partial c_p T}{\partial x} + \rho v \frac{\partial c_p T}{\partial y} = \mu \left(\frac{\partial u}{\partial y} \right)^2 + \frac{\partial}{\partial y} \left(\kappa \frac{\partial T}{\partial y} \right)$$
$$+ \frac{\partial q_{Rx}}{\partial x} + \frac{\partial q_{Ry}}{\partial y} \tag{7.142}$$

where we apply the boundary layer approximations to the heat conduction terms but not the thermal radiative heat flux q_{Rx} and q_{Ry}. Even though when the radiative flux number is not very large, we may use boundary layer approximations and neglect the term $\partial q_{Rx}/\partial x$, we should retain the term $\partial q_{Rx}/\partial x$ when the radiative flux number is very large because the thermal boundary layer is so thick that the boundary layer approximations are not good for the radiative transfer terms.

First we consider the case when the boundary layer approximations may be applied to the radiative flux terms as well as the velocity and heat conduction terms. Hence we may use the one-dimensional approximation for the radiative heat flux q_{Ry}. If we further assume that the gas is a grey gas and local thermodynamic equilibrium is applicable, we have

$$\frac{\partial q_{Ry}}{\partial y} = 2\sigma\rho K_p \left[\int_0^\tau T^4 \mathscr{E}_1(\tau - t')\, dt' + \int_\tau^\infty T^4 \mathscr{E}_1(t' - \tau)dt' \right.$$
$$\left. -2T^4 + T_w^4 \, \mathscr{E}_2(\tau) \right] \tag{7.143}$$

where K_p is the Planck mean absorption coefficient and τ is the optical thickness defined by the formula:

$$\tau = \int_0^y \rho K_p dy \tag{7.144}$$

The boundary conditions of our problems are

$$\left. \begin{array}{l} x > 0,\ y = 0:\ u = v = 0,\ T = T_w,\ \rho = \rho_w \\ y \to \infty:\ u \to U,\ T \to T_\infty,\ \rho \to \rho_\infty \end{array} \right\} \tag{7.145}$$

where U is the free stream velocity which is a constant and T_w is the temperature of the plate which is also a constant. Subscript ∞ refers to the values in the freestream.

Since the thermal radiation has a larger influence on temperature distribution than on velocity distribution, we consider the similar solution of the velocity profile by using Eq. (7.124) for the coefficient of viscosity and Eq. (7.129) for the modified y-coordinate. We have the following similar solution for the velocity profile:

$$\frac{u}{U} = \frac{df(\xi)}{d\xi} = f'(\xi) \tag{7.146}$$

and

$$w=-\left(\frac{\partial \psi}{\partial x}\right)_Y=\frac{1}{2}\left(\frac{C\mu_\infty U}{x}\right)^{\frac{1}{2}}(f'-f) \qquad (7.147)$$

where

$$\xi=\frac{Y}{x}\left(\frac{\rho_\infty Ux}{\mu_\infty}\right)^{\frac{1}{2}} \qquad (7.148)$$

and ψ is the streamfunction such that

$$\frac{\partial \psi}{\partial y}=\frac{\rho}{\rho_\infty}u; \quad \frac{\partial \psi}{\partial x}=-\frac{\rho}{\rho_\infty}v \qquad (7.149)$$

and $f(\xi)$ is the well known Blasius function of an incompressible boundary layer flow which has been tabulated in a standard text-book.[14]

In terms of Y and x, the energy equation (7.142) without the x-wise radiative heat flux becomes

$$c_p u \frac{\partial T}{\partial x}+c_p w \frac{\partial T}{\partial Y}=\frac{C\mu_\infty}{\rho_\infty}\left(\frac{\partial u}{\partial Y}\right)^2+\frac{C\mu_\infty c_p}{\rho_\infty P_r}\frac{\partial^2 T}{\partial Y^2}$$

$$+\left[T_w^4\, \mathscr{E}_2(\tau)-2T^4\right]2\rho\sigma K_p$$

$$+2\sigma\rho K_p\left[\int_0^\tau T^4\, \mathscr{E}_1(\tau-t')dt'\right.$$

$$\left.+\int_\tau^\infty T^4\, \mathscr{E}_1(t'-\tau)dt'\right] \qquad (7.150)$$

with the boundary conditions:

$$x>0: \ Y=0, \ T=T_w; \ Y\to\infty, \ T\to T_\infty \qquad (7.151)$$

where

$$\tau=\int_0^y \rho K_p dy$$

Even though we have similar solution for velocity $u(\xi)$, there will be no similar solution for temperature T when the absorption coefficient is finite or large. Only when the absorption coefficient is very small and Rosseland approximation (7.42) may be used, we may have a similar solution in temperature $T(\xi)$. In Eq. (7.150), we assume that both c_p and P_r are constant.

We integrate Eq. (7.150) numerically with the expression (7.72) for K_p; $P_r = 0.74$; $T_w = 2000°K$ and T_∞ at several temperatures. The results are shown in Figs. 7.12 to 7.15.

Fig. 7.12 is the case of low Mach number $M = 1.25$. Since there is no similar solution in $T(\xi)$, the temperature profiles change with x-distance. We use the following definition for the radiative flux number:

$$R_f = \frac{\sigma T_\infty^3 x^2}{\kappa_\infty L_p} = \frac{\sigma T_\infty^3 x^2 \rho_\infty K_p}{\kappa_\infty} \qquad (7.152)$$

when $R_f = 0$, i.e., without radiative heat flux, we have a similar solution for temperature $T(\xi)$ which is shown by the dotted curve in Figs. 7.12 to 7.15.

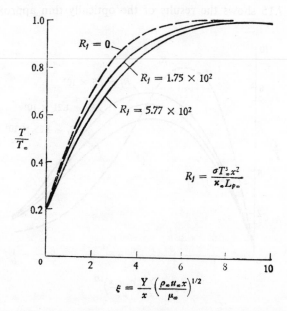

Figure 7.12 Temperature distributions in a boundary layer over a semi-infinite plate at $M_\infty = 1.25$, $T_\infty = 10,000°K$, $p_\infty = 5.0$ atmospheres

Fig. 7.13 shows a corresponding case for high Mach number $M_\infty = 15$.

In the numerical calculations of Figs. 7.12 and 7.13, we use the approximate formula (7.54) with $m = 1.562$.

The main effects of thermal radiation are (i) to decrease the maximum temperature in the boundary layer at high Mach number case, (ii) to increase the boundary layer thickness of temperature and (iii) to decrease the slope of $[d(T/T_\infty)/d\xi]_w$ at the wall.

It is interesting to compare the results of the integral expression of radiative heat flux with those approximate expressions of optically thick and thin media.   Fig. 7.14 shows the case of optically thick gas.   In comparison with the results of integral expression, it was found that the general trend of the results of the optically thick approximation is the same as that of the integral expression but quantitatively, the optically thick medium approximation overestimates the effect of thermal radiation.

Fig. 7.15 shows the results of the optically thin approximation

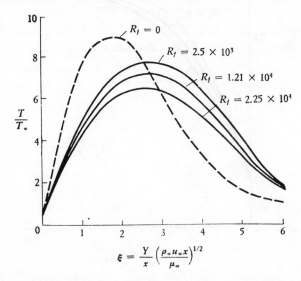

Figure 7.13   Temperature distributions in a boundary layer over a semi-infinite plate at $M_\infty = 15.0$, $T_\infty = 20{,}000°K$, $p_\infty = 1.0$ atmosphere

(7.57). Since the wall temperature T_w is always much smaller than the local temperature T except in the neighborhood of the wall, Eq. (7.57) shows that the radiative heat flux acts as a heat source in the boundary layer. In our numerical results of Fig. 7.15, the optically thin medium approximation gives an entirely wrong result in comparison with the results of the integral expression. Since the optically thin approximation (7.57) was a popular expression in many literatures, one should be very careful to use such an optically thin approximation when the temperature range is very high.

When the temperature of the flow field is very high, the thickness of the thermal boundary layer will be so large that the x-wise radiative heat flux will not be negligible. When we include the x-wise radiative heat flux term in the energy equation (7.142), there will be an upstream influence.[13] For instance, when the wall temperature is much smaller than that of the free stream, we have an upstream wake of temperature. At a given x-station upstream, the temperature is

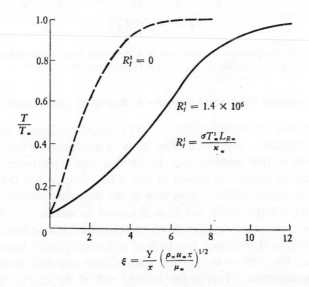

Figure 7.14 Temperature distribution for optically thick medium in a boundary layer. $T_\infty = 40,000°K$, $p_\infty = 5.0$ atmospheres, $M_\infty = 0.65$ (similar solution)

lowest at $y=0$ and increases with y. The defect of temperature along the x-axis increases from zero at minus infinity to the value of $T_\infty - T_w$ at $x=0$.

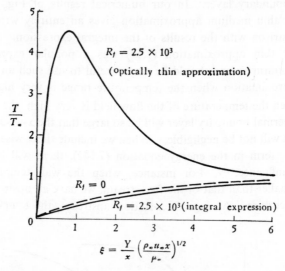

$$\xi = \frac{Y}{x}\left(\frac{\rho_\infty u_\infty x}{\mu_\infty}\right)^{1/2}$$

Figure 7.15 Temperature distributions in a boundary layer of an optically thin gas. T_∞ = 20,000°K, p_∞ = 1.0 atmosphere, M_∞ = 0.9

16. Stagnation Point Heat Transfer in Radiation Gasdynamics

Another interesting radiative heat transfer problem is the flow near a stagnation point of a blunt body in a hypersonic flow. The flow field in this problem may be divided into three more or less distinguished regions as shown in Fig. 7.16. Region I is the shock transition region which is very thin if the Reynolds number of the flow field is large which has been discussed in section 12. Region II is the inviscid shock layer in which the radiative heat flux should be considered if the mean free path of radiative is much larger than the mean free path of the gas but viscous force and heat conduction may be neglected. This is the inviscid tail of the shock. Region III is the boundary layer flow near the stagnation point. When the Reynolds number is small, or the Knudsen number is large, these three regions may merge into one. We shall discuss here only the case when Reynolds number of the flow field is large and that these three regions can be treated separately as follows:

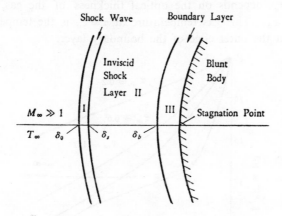

Figure 7.16 Flow regions near a stagnation point of a blunt body in a hypersonic stream

We assume that the temperature in the free stream is not high enough so that the thermal radiation effects in the free stream can be completely neglected. Hence we may use the ordinary Rankine-Hugoniot relation to determine the shock strength and the flow conditions immediately behind the shock which may be considered as a surface of discontinuity.

Secondly, we determine the flow in the inviscid shock region II from the initial conditions immediately behind the shock wave. In region II, we assume that the temperature is so high that radiative heat flux should be considered but the radiation energy density and radiation pressure are still negligible. Since we are interested in the region near the axis of the body of revolution and near the stagnation point, a one-dimensional approximation of radiative heat flux may be used as a first approximation. The most important coordinate is the distance along the axis of symmetry. For a first approximation, we may even neglect the curvature effect. Some typical temperature distributions in the inviscid shock layer region are shown in Fig. 7.17 where $p_2 = 10$ atmospheres and $T_2 = 15,000°K$ which is the temperature immediately behind the shock.[22]   For a given T_2, the temperature of the gas at the outer edge of the boundary layer, i.e., $\tau = \tau_w$, depends on the optical thickness of the gas, i.e., the value of τ_w.   This gas temperature at $\tau = \tau_w$ is the temperature of the gas at the outer edge of the boundary layer.

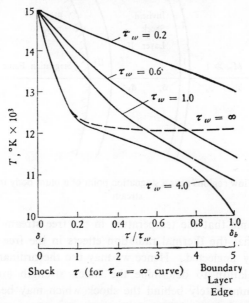

Figure 7.17   Temperature distributions in inviscid shock layer near a stagnation point. (Figure 11 of reference [22] by K.K. Yoshikawa and D.R. Chapman, courtesy of NASA)

Finally we may solve the stagnation point boundary layer flow with the free stream temperature determined in the solution of the inviscid shock layer.   The thermal radiation effect is to increase the thickness of the thermal boundary layer and to decrease the peak temperature in the boundary layer.

17. Two-Dimensional Channel Flows of an Ionized and Radiation Gas[15]

At very high temperature when the thermal radiation effects

are important the gas will be ionized. We should consider simultaneously the radiation effects and the electromagnetic effects on the flow problem. We may call such flow problems as radiation magnetogasdynamics. Now we reexamine the two dimensional channel flow of magnetogasdynamics of chapter VI section 7. All the dimensions, velocity and electromagnetic fields are the same as those given in chapter VI section 7. The temperatures of the two plates are kept at constant values T_0 and T_1 respectively. In the present problem we have two radiation terms, the radiation pressure p_R and the y-wise radiative heat flux q_{Ry}. We shall assume that the gas is a grey gas under the local thermodynamic equilibrium condition. The radiation pressure p_R is given by the following formula:

$$p_R = \frac{2\pi}{c} B(T_0)\, \mathscr{E}_4(\tau) + \frac{2\pi}{c} \int_0^\tau B(t)\, \mathscr{E}_3(\tau - t)\, dt$$

$$+ \frac{2\pi}{c} B(T_1)\, \mathscr{E}_4(\tau_2 - \tau) + \frac{2\pi}{c} \int_\tau^{\tau_2} B(t)\, \mathscr{E}_3(t - \tau)\, dt$$

$$(7.153)$$

where

$$\tau = \int_0^y \rho K_p\, dy \qquad (7.154)$$

and τ_2 is the value of τ at $y = L$. We assume that the lower plate is at $y = 0$ and the upper plate is at $y = L$. If the optical thickness τ is large, Eq. (7.153) reduces to the Rosseland expression (7.38). If the optical thickness τ is small, we obtain the following expression when we keep the terms up to the order of:

$$p_R = \frac{1}{2}[p_{Rb}(T_1) + p_{Rb}(T_0)] - \frac{3}{4}[p_{Rb}(T_0)\,\tau - p_{Rb}(T_1)\,(\tau_2 - \tau)]$$

$$+ \frac{1}{2} a_R \int_0^{\tau_2} T^4(t)\, dt \qquad (7.155)$$

where $p_{Rb}(T) = (1/3)a_R T^4$ is the Rosseland approximate value.
The radiative heat flux with black plate is

$$q_{Rv} = 2\pi B(T_1) \, \mathscr{E}_3(\tau_2 - \tau) + 2\pi \int_\tau^{\tau_2} B(t) \, \mathscr{E}_2(t - \tau) \, dt$$

$$- 2\pi B(T_0) \, \mathscr{E}_3(\tau) - 2\pi \int_0^\tau B(t) \, \mathscr{E}_2(\tau - t) dt \qquad (7.156)$$

Since the thermal radiation terms can be expressed in terms of the state variables T and ρ, we have five unknowns in our problem, i.e., the x-component of velocity u, the x-component of magnetic field H_x and the pressure p, density ρ, and temperature T of the gas. The fundamental equations for these five unknowns are as follows:

We express all the variables in nondimensional form: the velocity is expressed in terms of a reference velocity U, all the lengths are in terms of L and all the state variables are in terms of their values at the lower wall and the magnetic field is in terms of the applied magnetic field H_0. Hence we have the following nondimensional fundamental equations (see chapter VI section 7):

(1) Equation of state:

$$p^* = \rho^* T^* \qquad (7.157)$$

(2) x-wise equation of motion:

$$\mu^* \frac{du^*}{dy^*} + R_e \, R_H \, H_x^* = \left(\frac{du^*}{dy^*}\right)_0 = \text{constant} \qquad (7.158)$$

where $R_e = LU\rho_0/\mu_0$ is the Reynolds number and $R_H = \mu_e H_0^2/(\rho_0 U^2)$ is the magnetic pressure number. Star refers to the nondimensional quantity.

(3) The y-wise equation of motion:

$$p^* + R_{p_0} p_R^* + \frac{1}{2}\gamma M^2 R_H \, H_x^* = 1 + R_{p_0} \qquad (7.159)$$

where $M = U/a_0$ is the Mach number and $R_{p_0} = p_{R_0}/p_0$ is the radiation pressure number.

(4) Electrical current equation:

$$\frac{dH_x^*}{dy^*} = R_\sigma(-u^* + R_E) \qquad (7.160)$$

where $R_\sigma = UL/\nu_H$ is the magnetic Reynolds number and $R_E = E_0/(UB_0)$ is the electric field number.

(5) Energy equation:

$$M^2 R_E R_e R_H H_x^* + M^2 \mu^* \frac{d}{dy^*}\left(\frac{1}{2} u^{*2}\right)$$

$$+ \frac{\kappa^*}{(\gamma-1)P_r}\frac{dT^*}{dy^*} + R_{p_0} R_F R_e q_R^*(T^*)$$

$$= \text{constant} \tag{7.161}$$

where $R_F = cL/(UL_p)$ for L_p/L is large and $R_F = cL_R/(UL)$ for small L_R/L and the nondimensional radiative heat flux $q_R^* = q_{Ry}/(\gamma T_0^4/L)$.

In general, we have to solve Eqs. (7.157) to (7.161) simultaneously. In order to bring out the essential features, we consider the case of fluid of constant transport properties so that $\mu^* = 1$ and $\kappa^* = 1$. In this case, we solve u^* and H_x^* from Eqs. (7.158) and (7.160) which give exactly the same results as we consider in chapter VI section 7, i.e., the solution of ordinary magnetohydrodynamics. After we obtain u^* and H_x^* we solve Eq. (7.161) for temperature T^*. Fig. 7.18 shows some typical temperature distributions for the case of plane Couette flow. The most important result of Fig. 7.18 is that the temperature drops enormously due to the radiation effect.

It should be noticed that Eq. (7.158) is true only when $d(p+p_R)/dx = 0$, i.e., the plane Couette flow case. If the pressure gradient is not zero, strictly speaking we cannot assume that the state variables are functions of y only. They must be functions of both y and x and then for compressible fluid, all the variables including u and H_x are functions of x and y. Hence we can only treat this problem approximately by assuming that the velocity u and the magnetic field H_x are function of y only.

For a plane Poiseuille flow where both walls are at rest and there is a pressure gradient, all the variables should be functions of both x and y if the radiation pressure p_R is not negligible. Eq. (7.158) should be replaced by the following equation as a first ap-

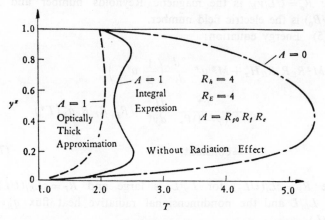

Figure 7.18   Temperature distributions of plane Couette flow in radiation magnetogasdynamics

proximation:

$$\frac{d}{dy^*}\left(\mu^*\frac{du^*}{dy^*}\right)+R_e\,R_H\frac{dH_x^*}{dy^*}=\frac{d}{dx^*}\left(p^*+R_{p_0}p_R^*\right)\left(\frac{R_e}{\gamma M^2}\right)$$

(7.162)

When the total pressure gradient is zero, Eq. (7.162) reduced to Eq. (7.158). In general, we do not have the fully developed flow in this problem and all variables are functions of both x and y and many other terms should be added. Fully developed flow may be approximately obtained only in the case when p_R, E_R, and $\partial q_{Rx}/\partial x$ are negligibly small. In this case, $\partial T/\partial x$ is not zero because $\partial p/\partial x$ is not zero. But we may assume that

$$\frac{\partial T}{\partial x}=\frac{T_w-T}{T_w-T_m}\frac{dT_m}{dx}$$

(7.163)

where the mean temperature over the cross section of the channel T_m is defined as

$$T_m=\left(\int_0^L Tu\,dy\right)\bigg/\left(\int_0^L u\,dy\right)$$

(7.164)

We may approximately take the variation of temperature along the channel by the heat balance over a section of length dx of the channel as follows:

$$c_p L u_m dT_m = 2q_t dx \qquad (7.165)$$

where u_m is the mean velocity u across the cross section and q_t is the total heat flux at the section far away from the entrance and the exit sections of the channel, i.e.,

$$q_t = -\left(\kappa\,\frac{\partial T}{\partial y}\right)_{y=0} - q_{Ry}\bigg|_{y=0} = \left(\kappa\,\frac{\partial T}{\partial y}\right)_{y=L} + q_{Ry}\bigg|_{y=L}$$

$$(7.166)$$

Substituting Eqs. (7.163) to (7.166) into the energy equation, we have

$$\frac{d}{dy}\left(\kappa\,\frac{dT}{dy}\right) + \frac{dq_{Ry}}{dy} = \frac{2q_t}{L}\left(\frac{T_w - T}{T_w - T_m}\right)\frac{u}{u_m} - \frac{d}{dy}\left(\mu u\,\frac{du}{dy}\right)$$

$$(7.167)$$

Eq. (7.167) is a total differential equation of y only and we have approximately the fully developed temperature distribution $T(y)$ for a plane Poiseville flow by solving Eq. (7.167).

PROBLEMS

1. Discuss the absorption coefficient in a medium due to scattering.
2. Discuss the emission coefficient in a medium due to scattering.
3. Derive the equation of radiative transfer including scattering as well as creation and true absorption of photons.
4. Derive the expression of radiation energy density in the frequency interval ν and $\nu + d\nu$ of Eq. (7.17) by considering the transient radiation energy in an elementary volume around a point in space from the macroscopic point of view.
5. Derive the Planck's radiation function (7.16).
6. Show that from Planck's radiation function, the maximum energy of radiation occurs at a wave length given by Eq. (7.28).
7. Calculate the specific heat at constant volume and that at

constant pressure of a mixture of an ideal gas and photons in thermodynamic equilibrium condition and express the ratio of these specific heats in terms of radiation pressure number.

8. Derive the expression of radiative heat flux between parallel infinite plates with different emission and reflection coefficients but opaque. Assume that the gas is grey and the plates are grey too. Discuss various limiting cases for this radiative heat flux.

9. Show that the case of finite but small mean free path of radiation, there is a radiation slip at the surface of the plate.

10. Consider that the radiating gas is in a semi-infinite domain bounded one side by an infinite plane black wall. The temperature of the gas is so high that radiative heat flux is not negligible but the radiation energy density and radiation stresses are still negligible. Initially, the gas is at rest with a uniform state. Assume that the plane wall moves sinusoidally with a small amplitude. Find the disturbed motion in the gas by assuming that the gas is an inviscid, nonheat-conducting but nongrey gas.

11. Derive the three dimensional formula of a grey gas (7.56).

12. Prove the relation (7.62) by considering the radiative energy in the two sides of a smooth surface.

13. Discuss the wave motion of small amplitude in an ideal radiating ionized gas under the action of an externally applied uniform magnetic field.

14. Calculate the shock structure at shock Mach number of 10 in an optically thick, viscous and heat-conducting gas in which the radiation pressure number is negligibly small but the radiative flux number is not negligible.

15. Discuss the shock structure at shock Mach number of 10 in an inviscid and nonheat-conducting but radiating gas in which the radiation pressure number is negligibly small but the radiative flux number is not negligible.

16. Calculate the hypersonic flow near the stagnation point of a cylinder at Mach number of free stream of 40 in an inviscid and nonheat-conduction gas in which the radiation pressure

number is negligible but the radiative flux number is not negligible.

17. By linearized theory, calculate the flow over a two-dimensional wedge with attached shock in an inviscid and nonheat-conducting gas in which the radiation pressure number is negligibly small but the radiative flux number is not negligible.

18. By linearized theory, calculate the steady flow over a two-dimensional wavy wall in an inviscid, nonheat-conducting gas in which the radiation pressure number is large but the radiative flux number is negligibly small.

19. Discuss the one-dimensional steady flow in a nozzle of variable cross section of an inviscid, nonheat-conducting gas in which the radiation pressure number is large but the radiative flux number is negligibly small.

20. Calculate the velocity and temperature distributions over an infinite flat plate setting into motion in its own plane impulsively. The fluid is assumed to be viscous, heat-conducting, and radiating for the cases (i) $R_p = 0$, $R_F \neq 0$, (ii) $R_p \neq 0$, but $R_F = 0$. For simplicity, we assume that the transport properties are constant and the gas is optically thick.

21. Repeat the case (i) of the above problem by assuming that the gas is grey and of finite mean free path of radiation.

22. Calculate the heat transfer near the stagnation point of a two-dimensional blunt body in a viscous, heat conducting, and radiating gas in which $R_p = 0$ and $R_F \neq 0$.

23. By the similar approach of nonequilibrium flow of chemical reaction, discuss how we could improve the study of radiation gasdynamics by discarding the assumption of local thermodynamic equilibrium.

REFERENCES

[1] Ambartsumyan, V. A. Theoretical Astrophysics. Pergamon Press, 1958.
[2] Anderson, J. D., Jr. "A simplified analysis for reentry stagnation point heat transfer from a viscous nongrey radiating shock layer". AIAA Paper No. 68—164, AIAA 6th Aerospace Sciences Meeting, Jan. 1968.
[3] Armstrong, B. H., Sokoloff, J., Nicholls, R. W., Holland, D. H., and

Meyerott, R. E. "Radiative properties of high temperature air", *Jour. Quant. Spec. Rad. Transfer*, Vol. 1, No. 2, Pergamon Press, 1961, pp. 143—162.

[4]   Baldwin, B. S., Jr. "The propagation of plane accoustic waves in a radiating gas". NASA TR R—139, 1962.

[5]   Born, M. and Wolf, E. Principles of Optics. (5th ed.) Pergamon Press, 1975.

[6]   Chandrasekhar, S. Stellar Structure. Dover Publications, 1957.

[7]   Cheng, P. "Dynamics of a radiating gas with application to flow over a wavy wall", *AIAA Jour.*, Vol. 4, No. 2, 1966, pp. 238—245.

[8]   Krascella, N. L. "Tables of composition, opacity and thermodynamic properties of hydrogen at high temperature". NASA SP—3005, 1963.

[9]   Pai, S.-I.   Radiation Gasdynamics. Springer-Verlag, 1966.

[10]   Pai, S.-I. and Speth, A. I. "The wave motions of small amplitude in radiation electromagnetogasdynamics". Proc. 6th Midwestern Conf. on Fluid Mech., Univ. of Texas Press, 1959, p. 466.

[11]   Pai, S.-I. and Speth, A. I. "Shock waves in radiation magnetogasdynamics". *Phys. Fluids.*, Vol. 4, No. 1, 1961, pp. 1232—1237.

[12]   Pai, S.-I. and Scaglione, A. P. "Unsteady laminar boundary layers on an infinite plate in radiation gasdynamics". Report SID 66—553, North American Avaiation, Inc., Space & Inf. Div., 1966.

[13]   Pai, S.-I. and Tsao, C. K. "A uniform flow of a radiating gas over a flat plate". Proc. 3rd Intern. Conf. of Heat Transfer. ASME, 1966, pp. 129—137.

[14]   Pai, S.-I. Viscous Flow Theory, I. Laminar Flow. D. Van Nostrand, 1956, p. 178.

[15]   Pai, S.-I. "Plane Couette flow in radiation magnetogasdynamics". Proc. 6th Intern. Symp. in Ionization Phenomena of Gases. Paris, 1963, pp. 431—436.

[16]   Planck, M. The Theory of Heat Radiation. Dover Publications, 1959.

[17]   Proko'fev, V. A. "Propagation of forced plane compression waves of small amplitude in a viscous gas when radiation is taken into account", *ARS Jour.*, Vol. 31, No. 7, 1961, pp. 988—997.

[18]   Rosseland, S. Theoretical Astrophysics. Oxford Press, 1936.

[19]   Scala, S. M. and Sampson, D. H. "Heat transfer in hypersonic flow with radiation and chemical reaction", in Supersonic Flow, Chemical Processes and Radiative Transfer. Pergamon Press, 1964, pp. 319—354.

[20]   Shen, S. F. "On the differential equation approximation in radiative gas dynamics". NOL TR 67—23, U. S. Naval Ord. Lab. Aero. Res. Report 282, 15 May, 1967.

[21]   Traugott, S.C. "Shock structure in a radiation, heat-conducting and viscous gas". Martin Co. Res. Report RR57, 1964.

[22]   Yoshikawa, K. K. and Chapman, D. R. "Radiative heat transfer and absorption behind a hypersonic normal shock wave". NASA TN—1424, 1962.

RAREFIED GASDYNAMICS

1. Introduction

Most of the previous discussions in this book are concerned with dense fluids and we may consider the fluid as a continuous medium and the fundamental equations are of the type of Navier-Stokes equations. However, if the fluid is rarefied, one would expect that the coarse molecular structure would affect the flow phenomena and the fundamental equations for rarefied gas may be different from those of Navier-Stokes equations. In the extremely rarefied gas, the gas may behave as individual particles. The degree of rarefication of a gas may be expressed by the nondimensional parameter known as the Knudsen number (4.76), defined as

$$K_f = \frac{\text{mean free path of a gas}}{\text{characteristic length}} = \frac{L_f}{L} \qquad (8.1)$$

where L_f is the mean free path of the gas (4.22) and L is the characteristic length of the flow field. When the Knudsen number is not negligibly small, the flow should be considered as a rarefied gas flow. When the Knudsen number is negligibly small, the fluid may be considered as a continuous medium. The choice of the characteristic length L depends on the problems considered. Hence we may choose the typical dimension of the body as L when we study the forces on this body in a gas flow. We may use the boundary layer thickness on a body as L when we are interested in the skin friction and heat transfer through the boundary layer. When we investigate the transition region in a shock wave, the thickness of the shock wave may be used as L. Because of the various choice of characteristic length L, whether a gas flow should be considered as rarefied or not, depends on the particular problems considered. Once the value of L is chosen, the Knudsen number K_f tells us the degree of rarefication.

As we shall show in section 7, the mean free path of a gas is closely associated with the coefficients of viscosity and heat conduction [see Eqs. (1.52) and (1.57)]. For a plasma, besides the ordinary Knudsen number, we have the electrical Knudsen number K_D, Eq. (3.6), and the magnetic Knudsen number K_L, Eq. (3.9). We have briefly discussed the flow regimes according to Knudsen number K_f in chapter III section 2 and according to K_D and K_L in chapter VI section 15. Now we discuss a little more in detail about the flow regimes according to Knudsen number. In this section, we deal with the kinetic theory of a neutral gas without chemical reaction. In section 8, we shall discuss the case with chemical reaction while in section 9, the plasma kinetic theory will be discussed.

In order to give a definite example, we consider the stagnation region of an axisymmetrical blunt body of radius R_b flying at a hypersonic speed. This problem has been discussed thoroughly by Probstein[20] and we summarize some of his results as follows:

There is a detached shock in front of the body with the detached shock distance Δ (Fig. 8.1). The mean free path of the gas behind the shock L_{fs} is much smaller than the mean free path in the free stream $L_{f\infty}$. For a hypersonic flow at a very high temperature, we have approximately

$$\frac{L_{fs}}{L_{f\infty}} \sim \frac{\Delta}{R_b} \sim \epsilon_0 = \frac{\rho_\infty}{\rho_s} \ll 1 \qquad (8.2)$$

The density ratio ϵ_0 across a hypersonic shock is of the order of 0.1 to 0.07 or less. The basic condition for continuous flow may be taken as

$$K_{fs} = \frac{L_{fs}}{\Delta} \ll 1 \qquad (8.3)$$

Even though the flow may be considered as a continuous flow, the actual solution of the problem may be different from the classical results according to the value of K_{fs}. The following subdivision may be taken:

(1) Vorticity interaction regime. In this regime, we have

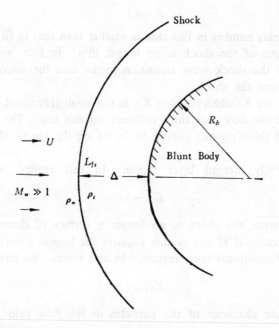

Figure 8.1 Stagnation point flow of a blunt body in a hypersonic flow

$$K_{f_s} \ll \epsilon_0 \qquad (8.4)$$

The thickness of the shock is very thin. Immediately behind the shock, the flow may be considered as inviscid and rotational. There is a very thin boundary layer near the body. This is essentially the classical case as shown in Fig. 7.16.

(2) Viscous layer regime. In this regime, we have

$$K_{f_s} \ll \sqrt{\epsilon_0} \qquad (8.5)$$

The Reynolds number in this case is smaller than that in (i). The thickness of the shock is still thin but the flow immediately behind the shock should be considered as a viscous and heat-conducting flow.

(3) The incipient merged layer regime. In this regime, we have

$$K_{fs} \ll 1 \qquad (8.6)$$

The Reynolds number in this case is smaller than that in (ii). Now the thickness of the shock is no longer thin. In fact, we cannot distinguish the shock wave transition region and the viscous layer region behind the shock.

When the Knudsen number K_{fs} is not negligibly small, we may also define the flow into three different regimes too. The detailed analysis of these regimes should be based on the kinetic theory of gases.

(4) Fully merged layer region. In this regime, we have

$$K_{fs} \sim 1 \qquad (8.7)$$

In this regime, the shock is no longer a surface of discontinuity. A strict treatment of this regime requires the kinetic theory of gas.

(5) Transitional layer regime. In this regime, we have

$$K_{fs} > 1 \qquad (8.8)$$

The discrete character of the particles in the flow field becomes important.

(6) First order collision regime. In this regime, we have the local Knudsen number is large but not large enough to reach the free molecule regime. In the analysis, we need to take into account the collisions between a free stream molecule and a reemitted molecule.

Finally, when the Knudsen number K_{fs} is much larger than unity, the collision between the molecules is negligible and we have the free molecule regime.

When the Knudsen number is of the order of unity or larger, we have to use the kinetic theory of gases to describe the fluid flow. We have discussed briefly the simple kinetic theory of gases in chapter I sections 5 to 9. In this chapter, we discuss the kinetic theory of gases in detail for both its relations with the continuum theory and its use for rarefied gas flow.

In sections 2 to 7, we consider the kinetic theory of a single species of a neutral gas, particularly the relations of the kinetic theory and the continuum theory.[3, 6-9, 20] We shall show that for a first approximation, we may derive the fundamental equations of continuum theory from the Boltzmann equation of the kinetic theory of gases. We shall also show the improvement of the fundamental equations of continuum theory obtained from the kinetic theory. In section 8, we generalize the kinetic theory of a single species to multiple species with chemical reaction and in section 9, we consider the kinetic theory of a plasma.

In sections 10 to 13, we consider some flow problems which cannot be covered by the continuum theory such as free molecule flow (section 10), slip flow (section 11), Couette flow for all Knudsen numbers (section 12), and plasma oscillation (section 13). In section 14, we shall discuss briefly the particle motion of a plasma and in section 15, some dimensional consideration of plasma dynamics will be given based on the Boltzmann equation which is similar to the general discussion in chapter VI section 15. Finally, some experimental results of rarefied gas will be briefly discussed in section 16.

2. Molecular Velocity and its Distribution Function

The modern concept of kinetic theory is beyond the scope of this book. An excellent review of the modern concept of kinetic theory is given by Sandri in reference [21]. In fluid mechanics, we shall use the kinetic theory of gases based on the Boltzmann equation in which the assumption of molecular chaos is introduced. We assume that there is no correlation between the positions and velocities of different particles unless they are within each other's field (see chapter I section 5). Hence we may use the single molecular distribution function $F(r, q_m, t)$ to describe the motion of the molecules. This single molecular distribution function or simply called molecular distribution function represents the expectation of the number of molecules per unit volume at the position r and time t within the molecular velocity range q_m and $q_m + dq_m$. This molecular distribution function $F(r, q_m, t)$ depends only on the position r and velocity q_m of any single particle and is independent of the position and velocity of all the other particles, except due to collision or encounter with other particles. Let the number density of the gas at position r and time t be n, we have

$$dn = F(r, q_m, t)dq_m \qquad (8.9)$$

The average number density of the gas at r and t is then

$$n = \int F(r, q_m, t)dq_m = \int_{-\infty}^{\infty} \int_{-\infty}^{\infty} \int_{-\infty}^{\infty} F(r, q_m, t) \, d\xi d\eta d\zeta \qquad (8.10)$$

where ξ, ζ, and η are respectively the x-, y-, and z-component of the molecular velocity vector q_m.

One of the main purposes of the kinetic theory of gases is to determine the molecular distribution function F. If we have the molecular distribution function F, we may calculate the average of all the physical quantities used in the classical continuum theory. For instance, the average x-component of the velocity of the gas flow is

$$u = \overline{\xi} = \frac{1}{n} \int_{-\infty}^{\infty} \int_{-\infty}^{\infty} \int_{-\infty}^{\infty} F\xi \, d\xi \, d\eta \, d\zeta \qquad (8.11)$$

The velocity u is one of the quantities used in the continuum theory. The average of any physical quantity Q of the gas at a given point r in space and a time t is

$$\overline{Q} = \frac{1}{n} \int_{-\infty}^{\infty} \int_{-\infty}^{\infty} \int_{-\infty}^{\infty} QF d\xi \, d\eta \, d\zeta \qquad (8.12)$$

In section 5, we shall discuss the relations of those average quantities used in the continuum theory with the integral expressions of the kinetic theory (8.12).

If the gas is a mixture of several species, e.g., several kinds of molecules, we should have a molecular distribution function $F_s(r, q_{ms}, t)$ for each (sth species) of the species in the mixture which will be discussed in sections 8 and 9.

3. Boltzmann Equation

The distribution function F is governed by the Boltzmann

equation which is

$$\frac{\partial F}{\partial t}+q_m^i \frac{\partial F}{\partial x^i}+\frac{\phi^i}{m}\ \frac{\partial F}{\partial q_m^i}=\left(\frac{\delta F}{\delta t}\right)_c \qquad (8.13)$$

where q_m^i is the ith component of the molecular velocity q_m of the gas, x^i is the ith component of the spatial vector r, ϕ^i is the ith component of the force acting on a molecule and m is the mass of a molecule of the gas. The summation convention is used here. The left-hand side terms of Eq. (8.13) represent the time rate of change of the distribution function F. The first term is the time rate of change of F, the second term is the change due to the variation of velocity and the third term is the change of F due to the body force ϕ^i which affects the velocity of the molecule. The right-hand side term of Eq. (8.13) is known as the collision term which represents the collision or encounter of the molecule considered with other molecules in the flow field. The exact form of the collision term $(\delta F/\delta t)_c$ depends on the kinetic model used in the kinetic theory. For ordinary gas, we may assume that only binary collision is important. In other words, we need only to consider the collision between two typical particles in the gas. In Fig. 8.2, we consider a particle P associated with the distribution function F and another particle Q associated with the distribution function F'. Without interaction, the particle P is moving in the direction PO and the particle Q in the direction of QM. Because of the interaction between these two particles, the actual path of the particle Q is QS. If we draw a plane ROM through the point O and perpendicular to QM, the distance OM is known as the impact parameter b which is the distance of the closest approach of the two particles P and Q if there were no interaction. The angle between OM and OR is denoted by ϵ. The binary collision term of Eq. (8.13) is then

$$\left(\frac{\delta F}{\delta t}\right)_c=\int\int\int(\overline{FF'}-FF')g_0\ b\ db\ dq'_m\ d\epsilon \qquad (8.14)$$

where the bar refers to the distribution function after the collision and g_0 is the absolute initial relative velocity of the two particles:

$$g_0^2=(\xi-\xi')^2+(\eta-\eta')^2+(\zeta-\zeta')^2 \qquad (8.15)$$

where prime refers to the value of the second particle Q and without prime, those of the particle P. The integral (8.14) should be taken for all values of b from zero to infinity; the angle ϵ is from zero to 2π and the values of ξ', η', and ζ' are from minus infinity to plus infinity. The derivation of Eq. (8.14) may be found in any standard textbook of kinetic theory of gases[6] (see problem 4).

For a plasma, the collision term may include the binary collision term given in Eq. (8.14) and a distance encounter known as the Fokker-Planck mechanism which will be discussed in section 9.

In ordinary kinetic theory of neutral gas, we consider the Boltzmann equation with only binary collision term (8.14) and neglect the

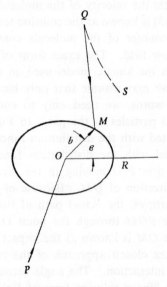

Figure 8.2   Binary collision between gas particles

body force ϕ^i. This simple Boltzmann equation is already a nonlinear differentio-integral equation which is too complicated to be solved for practical flow problems. We can solve only a few very simple flow problems based on this simple Boltzmann equation which will be discussed in sections 10 to 13. However, the Boltzmann equation serves two important aspects in the study of fluid dynamics. In the first place, the fundamental equations of fluid mechanics from the macroscopic point of view may be derived from the Boltzmann equation. For a first approximation, we may obtain the Navier-Stokes equations from the Boltzmann equation. We may also obtain improved fundamental equations of the continuum theory from the Boltzmann equation when the Navier-Stokes equations are no longer a good approximation. Thus we may obtain some information about the validity of the fundamental equations of the macroscopic description, i.e., the theory of a continuum. We shall discuss these points in sections 5 and 6. In the second place, the Boltzmann equation may give us valuable information on the transport coefficients such as coefficient of viscosity, coefficient of heat conductivity, etc. ˙In the continuum theory, these transport coefficients are simply introduced as known functions of physical quantities of the fluid. The relations between the transport coefficients and the physical quantities cannot be determined in the macroscopic analysis. They can be determined from the kinetic theory, as we shall discuss in section 7.

4. Equilibrium Condition of a Gas. Maxwellian Distribution Function[10]

In general it is very difficult to solve Boltzmann equation (8.13) with the collision term (8.14). However, one of the exact solutions of the Boltzmann equation (8.13) has been found and is known as the Maxwellian distribution function [see Eq. (1.26)] which represents a uniform steady state of the gas. It has been extensively used in various flow problems.

In deriving the Maxwellian distribution function, we make the following assumptions:

(i) There is no body force ϕ^i, and

(ii) The state of the gas is uniform so that the distribution function F is independent of the spatial coordinate r, i.e., $F = F(q_m, t)$.

Under these two conditions, the Boltzmann equation (8.13) with binary collision term becomes

$$\frac{\partial F}{\partial t} = 2\pi \int \int (\overline{F}\overline{F}' - FF') g_0 \, b \, db \, dq'_m \qquad (8.16)$$

where we assume that the particle is spherically symmetric and the integration with respect to the angle ϵ can be carried out immediately.

Now we define a *H*-function such that

$$H = \int F \cdot \log F \cdot dq_m \qquad (8.17)$$

From Eqs. (8.16) and (8.17), we have

$$\frac{\partial H}{\partial t} = 2\pi \int \int \int (1 + \log F)(\overline{F}\overline{F}' - FF') g_0 \, b \, db \, dq_m \, dq'_m \qquad (8.18)$$

or

$$\frac{\partial H}{\partial t} = -\frac{1}{2}\pi \int \int . \int \log\left(\frac{\overline{F}\,\overline{F}'}{FF'}\right)(\overline{F}\overline{F}' - FF') g_0 \, b \, db \, dq_m \, dq'_m \qquad (8.19)$$

From Eq. (8.19), we have

$$\frac{\partial H}{\partial t} \leqq 0 \qquad (8.20)$$

because the sign of $\log(\overline{F}\overline{F}'/FF')$ and that of $(\overline{F}\overline{F}' - FF')$ are always the same. Eq. (8.20) is known as the Boltzmann *H*-theorem. The *H*-function is associated with the entropy of the gas.

For a steady state, $\partial H/\partial t = 0$, the integrand of Eq. (8.19) must be zero and we have

$$\overline{F}\overline{F}' = FF' \qquad (8.21)$$

and

$$\log \overline{F} + \log \overline{F}' = \log F + \log F' \qquad (8.22)$$

Since the distribution function F now is a function of q_m only, it can be shown that in order to satisfy the relation (8.22) which is known as the summation invariant, the distribution function F must be of the form:

$$\log F = a_1 m + m a_2 \cdot q_m + \frac{1}{2} a_3 q_m^2 \tag{8.23}$$

where a_1, a_2, and a_3 are parameters independent of r, q_m, and t. Eq. (8.23) may be written in the form:

$$F = a_0 \exp[-b_0(q_m - q)^2] \tag{8.24}$$

where a_0 and b_0 are constants and q is the average value of q_m. The constants a_0 and b_0 may be expressed in terms of the temperature of the gas T and its mass m (see section 5) as follows:

$$F = n \left(\frac{m}{2\pi kT}\right)^{3/2} \exp\left[-\frac{m(q_m - q)^2}{2kT}\right] = F_0 \, (n, \ q, \ T) \tag{8.25}$$

Eq. (8.25) is the well-known Maxwellian distribution function [see Eq. (1.26)] which represents a uniform and steady state of a gas of mean flow velocity q, an absolute temperature T and a number density n.

We may define a random or peculiar velocity c_a with components c_i such that

$$c_i = q_{mi} - q_i \tag{8.26}$$

The number density between the absolute magnitude of random velocity c_a and $c_a + dc_a$ is

$$dn_c = n \left(\frac{1}{\pi c_m^2}\right) \exp\left(-\frac{c_a^2}{c_m^2}\right) dc_a \cdot c_a^2 \cdot 4\pi \tag{8.27}$$

where

$$c_m = (2kT/m)^{1/2} = \text{the most probable speed} \tag{8.28}$$

because at $c_a = c_m$, (dn_c/dc_a) is a maximum. The Maxwellian distribution function F_0 may be written as

$$F_0 = \frac{n}{(\sqrt{\pi} \, c_m)^3} \exp\left(-\frac{c_a^2}{c_m^2}\right) \tag{8.29}$$

Eq. (8.29) has been plotted in Fig. 1.3.

Another important quantity in the kinetic theory of gases is the mean free path which is the average distance travelled by molecules between collisions.   Now we consider the gas at rest under the equilibrium condition of Maxwellian distribution.   For equilibrium condition, the two terms in the right-hand side of Eq. (8.14) must be equal.   Hence we may consider any one of these two terms to determine the average collisions.   The number of collisions between molecules having the velocities in the range c_a and $c_a + dc_a$ and those with velocities in the range of c_1 and $c_1 + dc_1$ per unit volume and per unit time is

$$g_0 F_0(c_a) F_0(c_1) dc_a dc_1 \cdot S_0 \qquad (8.30)$$

where

$$S_0 = 2\pi \int b\, db = \text{collision cross section} \qquad (8.31)$$

The integration of expression (8.30) over all the velocities gives twice the total number of collisions per unit volume per unit time because each collision is counted twice, one as a c-molecule and a c_1-molecule and the other with the roles of these particles reversed. But since each collision terminates two free paths, the integration gives the total number of free path per unit volume and per unit time, i.e.,

$$\int_0^\infty \int_0^\infty g_0 F_0(c_a) F_0(c_1) S_0\, dc_a\, dc_1 = \sqrt{2}\, n^2 S_0 \bar{c}_a = n\, \bar{c}_a / L_f$$

$$(8.32)$$

where $\bar{c}_a = 2c_m / \sqrt{\pi}$ is the average value of c_a.   Since the total distance travelled by all molecules in unit volume per unit time is $n\bar{c}_a$, we have the mean free path L_f by definition as follows:

$$\text{mean free path} = L_f = \frac{n\bar{c}_a}{\sqrt{2}\, n^2 S_0 \bar{c}_a} = \frac{1}{\sqrt{2}\, n S_0} \qquad (8.33)$$

Hence the mean free path L_f is inversely proportional to the collision cross section S_0 and number density n.   The collision cross section may be determined for any given kinetic model which we assume.[6,9]

In most problems of fluid dynamics, we assume that the deviation from the Maxwellian distribution F_0 is small and a perturbation technique may be used such that the molecular distribution function F may be written as

$$F = F_0 (1 + \psi_f) \qquad (8.34)$$

where the absolute value of ψ_f is much smaller than unity and the higher order term of ψ_f is negligible.

5. Relations Between Kinetic Theory and Continuum Theory[3]

Before we discuss the case of flow problems with finite Knudsen number, we are going to define the relations between the distribution function F, the molecular velocity q_m with component q_{mi} and the macroscopic variables such as the velocity component u_i, density $\rho(n)$, pressure p and temperature T of a gas. For engineering problems, our main interests are the values of these macroscopic variables in the whole flow field. We introduce a random or peculiar velocity c_a with components c_i of the molecule such that

$$q_{mi} = u_i + c_i \qquad (8.35)$$

By the definition of the average process (8.12), we have the following relations of the variables used in the continuum theory of macroscopic treatment and the molecular distribution function F:

(1) The density ρ of the gas is

$$\rho = mn = m \int F dc_a \qquad (8.36)$$

where m is the mass of a molecule.

(2) The flow velocity q with component u_i has the relation

$$u_i = \overline{q_{mi}} = \frac{1}{n} \int q_{mi} F dc_a \qquad (8.37)$$

(3) The pressure tensor p_{ij} and the pressure p are respectively

$$p_{ij} = mn \, \overline{c_i c_j} = m \int c_i c_j F dc_a \qquad (8.38)$$

and

$$p = \frac{1}{3}\ (p_{11} + p_{22} + p_{33}) \tag{8.39}$$

The viscous stress tensor τ' with ijth component τ_{ij} is defined as

$$\tau_{ij} = -p_{ij} + p\delta_{ij} \tag{8.40}$$

where $\delta_{ij} = 1$ if $i = j$ and $\delta_{ij} = 0$ if $i \neq j$.

(4) The kinetic temperature T is defined as

$$T = \frac{m}{3k}\ \overline{c_a^2} = \frac{m}{3kn} \int (c_1^2 + c_2^2 + c_3^2)\ F\ dc_a \tag{8.41}$$

(5) The energy flux. The third moment of the peculiar velocity is

$$S_{ijk} = mn\ \overline{c_i c_j c_k} = m \int c_i c_j c_k F dc_a \tag{8.42}$$

The heat flux due to conduction is then

$$Q_{ci} = \frac{1}{2}\ (S_{i11} + S_{i22} + S_{i33}) = \frac{1}{2}\ m \int (c_a^2) c_i\ F\ dc_a \tag{8.43}$$

In the above relations (8.36) to (8.43), we assume that the gas consists of the same kind of molecules. The case of gas consisting of a mixture of several kinds of molecules will be discussed in sections 8 and 9.

6.   Transfer Equations and Thirteen Moment Equations

With the definitions of the macroscopic variables defined in the last section, we may obtain the equations which govern these variables from the Boltzmann equation by taking the proper moments. The resultant equations are known as the transfer equations which are very close to the fundamental equations of continuum theory. For simplicity, let us consider the Boltzmann equation without the body force ϕ^i and with binary collision term (8.14) only. Let Q be any factor which does not involve x_i and t. We multiply Q to the Boltzmann equation (8.13) with $\phi^i = 0$ and integrate the resultant equation with respect to molecular velocity q_m. We have the Maxwell's equation of transfer as follows:

$$\frac{\partial}{\partial t} \int QF dq_m + \frac{\partial}{\partial x_i} \int Q \, q_{mi} \, F \, dq_m$$

$$= \iiiint (\overline{F F'} - F F') \, Q g_0 b \, db \, dt \, dq'_m dq_m$$

$$= \frac{1}{2} \iiiint (\overline{Q'} + \overline{Q} - Q' - Q) F F' \, g_0 \, b \, db \, d\epsilon \, dq'_m dq_m = I \quad (8.44)$$

The integral I vanishes if Q is a summational invariant, i.e., $\overline{Q} + \overline{Q'} = Q + Q'$. Now we may obtain the transfer equations for various values of Q.

(1) Equation of continuity. If we take $Q = m$, Eq. (8.44) gives

$$\frac{\partial \rho}{\partial t} + \frac{\partial \rho u_i}{\partial x_i} = 0 \quad (8.45)$$

where $I = 0$ because $Q = m$ is a summational invariant and the mass is a constant here. We also use the relations of ρ and u_i of last section. Eq. (8.45) is identical to the equation of continuity of continuum theory, i.e., Eq. (3.32).

(2) Equations of motion. If we take $Q = m \, q_{mi}$, Eq. (8.44) gives the ith component of the equations of motion:

$$\frac{\partial}{\partial t} (\rho u_i) + \frac{\partial}{\partial x_j} (\rho u_i u_j + p_{ij}) = 0 \quad (8.46)$$

where I is again vanishes because of the conservation of momentum. Eq. (8.46) becomes the corresponding Navier-Stokes equation, e.g., Eq. (3.97) without body forces if we use the simple relation (3.52) for the pressure tensor p_{ij}. However, in general, the simple relation (3.52) does not hold true, particularly for the case of the large Knudsen number in a rarefied gas. The pressure tensor should be determined by a differential equation as we shall show later. Eq. (3.52) is in fact only a first approximation of this differential equation of pressure tensor when the Knudsen number is very small. We shall show this in next section for which we have the definition of coefficient of viscosity from the kinetic theory of gases point of view.

(3) Equation of energy. If we take $Q = m \, q_m^2$, Eq. (8.44) gives the energy equation of the gas:

$$\frac{\partial}{\partial t}(\rho q^2 + 3p) + \frac{\partial}{\partial x_j}(\rho q^2 u_j + 3pu_j + 2u_i p_{ij} + Q_{cj}) = 0 \qquad (8.47)$$

where

$$q^2 = u_1^2 + u_2^2 + u_3^2 \quad \text{and} \quad q_m^2 = q_{m_1}^2 + q_{m_2}^2 + q_{m_3}^2$$

The integral I in the present case vanishes again because of the conservation of the kinetic energy of the particles during the elastic collision.

By the definitions of ρ, p, and T of Eqs. (8.36), (8.39), and (8.41) we have

$$p = nkT = \rho RT \qquad (8.48)$$

Equation (8.48) is the equation of state of the gas. The translational internal energy per unit mass which is the only internal energy considered in the present case is

$$U_m = c_v T = \frac{3}{2} RT \qquad (8.49)$$

In the present analysis, we consider only the monatomic gas which has only the translational energy given by Eq. (8.49). We shall discuss the cases where other internal energies will be considered in sections 8 and 9. Substituting the relations (8.48) and (8.49) into Eq. (8.47), we have the conventional energy equation (3.102) without heat addition and work done by the electromagnetic force:

$$\rho \frac{DU_m}{Dt} + p \frac{\partial u_i}{\partial x_i} = \tau_{ij} \frac{\partial u_i}{\partial x_j} + \frac{\partial Q_{ci}}{\partial x_i} \qquad (8.50)$$

Of course, Eq. (8.50) becomes the ordinary energy equation in a viscous compressible fluid (3.102) provided that some simple relations of the viscous stresses τ_{ij} and the heat flux due to conduction Q_{ci} are used. In general, both τ_{ij} and Q_{ci} should be determined by differential equations obtained from the Boltzmann equation of the transfer equation (8.44). Only in the case of the very small Knudsen number, we may assume that Q_{ci} is proportional to the temperature gradient, i.e., the Fourier law of Eq. (1.55). We shall show how to obtain this approximation in this chapter. From this approximation, we obtain the definition of the coefficient of thermal conductivity K from the kinetic theory of gases point of view.

Eqs. (8.45), (8.46), and (8.47) show the relations between the kinetic theory of gases and the Navier-Stokes equations of ordinary fluid dynamics. This approach is the well known method of Chapman-Enskog.[6] This method is good only for the very small Knudsen number so that we may use the transport coefficients of viscosity and heat conductivity. This method may be considered as a first approximation of the kinetic theory of gases. A second approximation which was first proposed by Grad is known as the thirteen moment equations[7] in which the equations of the thirteen moments, i.e., n, u_i, T, p_{ij}, and Q_{ci} are derived from the Boltzmann equation. It should be noticed that there is a definite relation between the temperature T and the diagonal elements of p_{ij} and we may consider p_{ij} as five elements together with T. In the thirteen moment equations, the first five equations are still Eqs. (8.45) to (8.47). In addition, we have the following equations:

(4) Equations of pressure stress. If we take $Q = m\,q_{mi}\,q_{mj}$, Eq. (8.44) will give a differentio-integral equation for the pressure stress tensor p_{ij}. In general, this equation of p_{ij} or τ_{ij} contains also the third moments S_{ijk} and the integral I of Eq. (8.44) does not vanish. Hence the system is not closed. Grad introduced the following approximation for the distribution function F. He used the expression (8.34) for the distribution function F with the perturbed distribution function ψ_f as follows:

$$\psi_f = A_i c_i + B \left(c_a^2 - \frac{3}{2} a^2 \right) + B_{ij} \left(c_i c_j - \frac{1}{3} \delta_{ij} c_a^2 \right)$$

$$+ C_i \left(c_a^2 c_i - \frac{5}{2} a^2 c_i \right) \tag{8.51}$$

where

$$a^2 = \frac{2kT}{m} \tag{8.52}$$

The coefficients A_i, etc., are functions of the coordinates and the time to be found from the equations which will be developed. The products and squares of these coefficients are negligible. In this way we may express the third moment S_{ijk} and the integral I in terms of the first thirteen moments, i.e., n, u_i, T, p_{ij}, and Q_{ci}. We will have a closed system of those thirteen moments. The corresponding viscous stress tensor equations are

$$\frac{\partial \tau_{ij}}{\partial t} + \frac{\partial}{\partial x_k}(u_k \tau_{ij}) + \frac{2}{5}\left(\frac{\partial Q_{ci}}{\partial x_j} + \frac{\partial Q_{cj}}{\partial x_i} - \frac{2}{3}\delta_{ij}\frac{\partial Q_{ck}}{\partial x_k}\right)$$

$$+ \tau_{ik}\frac{\partial u_j}{\partial x_k} + \tau_{jk}\frac{\partial u_i}{\partial x_k} - \frac{2}{3}\delta_{ij}\tau_{k1}\frac{\partial u_k}{\partial x_1}$$

$$+ p\left(\frac{\partial u_i}{\partial x_j} + \frac{\partial u_j}{\partial x_i} - \frac{2}{3}\delta_{ij}\frac{\partial u_k}{\partial x_k}\right) = +\frac{p}{\mu}\tau_{ij} \qquad (8.53)$$

The interesting points are that the viscous stress depends on the heat flux and the viscous stress is governed by a very complicated differential equation.

(5) Equation of heat flux. If we take $Q = \frac{1}{2}m\,c_a^2 c_i$ and use the approximation of Eq. (8.51), we have the following equation for the heat flux Q_{ci}.

$$\frac{\partial Q_{ci}}{\partial t} + \frac{\partial}{\partial x_k}(u_k Q_{ci}) + \frac{7}{5}Q_{ck}\frac{\partial u_i}{\partial x_k} + \frac{2}{5}Q_{ck}\frac{\partial u_k}{\partial x_i} + \frac{2}{5}Q_{ci}\frac{\partial u_k}{\partial x_k}$$

$$+ RT\frac{\partial \tau_{ik}}{\partial x_k} + \frac{7}{2}\tau_{ik}\frac{\partial RT}{\partial x_k} + \frac{5}{2}p\frac{\partial RT}{\partial x_i}$$

$$- \frac{\tau_{ik}}{\rho}\left(\frac{\partial \tau_{k1}}{\partial x_1} + \delta_{k1}\frac{\partial p}{\partial x_1}\right) = -\frac{2}{3}\frac{p}{\mu}Q_{ci} \qquad (8.54)$$

The complete set of the thirteen moments equations (8.45), (8.46), (8.47), (8.53), and (8.54) is much more complicated than the Navier-Stokes equations of a compressible viscous fluid, i.e., Eqs. (8.45) to (8.47). Since we can not solve the Navier-Stokes equations for complicated flow problems, our chance to solve the thirteen moment equations for general flow problems is much less than that for the Navier-Stokes equations. Only very simple flow such as a plane Couette flow or shock transition region have been solved based on some simplified version of the thirteen moment equations.

We shall not discuss them here. Those readers who are interested in the solution of the flow problem based on thirteen moment equations may refer to references [7], [14], [15], [24], and [25].

7. Expressions of the Coefficients of Viscosity and Heat Conductivity[6, 9]

One of the major contributions of the Boltzmann equation in the problem of fluid dynamics is that it gives valuable information on the transport coefficients such as coefficient of viscosity μ, coefficient of heat conductivity κ, etc. For a gas consisting of one kind of molecules, we have only two transport coefficients μ and κ. In the macroscopic approach, these transport coefficients μ and κ represent some phenomenological facts. Hence the relations of viscous stresses (3.52) and that of heat flux (1.55) are empirical. We can only determine these transport coefficients experimentally. However, when we derive the Navier-Stokes equations (8.45) to (8.47) from the transfer equation (8.44), we have definite relations between viscous stresses and molecular velocities and a definite relation between heat flux and the molecular velocities. In the kinetic theory, these transport coefficients may be determined from the transfer equations as a first approximation if we assume certain kinetic picture or collision law. There are many collision laws which have been used in the kinetic theory of gases, e.g., we may assume that the molecules are hard elastic spheres or as particles which repel each other with a force dependent on the distance between them. Even though the exact expressions for these coefficients depend on the assumption of the kinetic model of the molecular motion, the essential features of these expressions are quite similar. The exact calculations are complicated. Special treatise such as references [6] and [9] should be referred to for the exact calculations. We shall discuss briefly the general principles of these calculations and some of the well known results only.

From the simple kinetic theory, we have the simple expression for the coefficient of viscosity (1.53) and that for the heat conductivity (1.57). For more accurate results, we have to evaluate certain integrals similar to I in Eq. (8.44) from the definition of viscous stress or heat flux by using different values of Q's. Usually we assume that the actual flow condition deviates slightly from the Maxwellian distribution (8.25) so that the molecular distribution function F may be written in the form of Eq. (8.34). We have to write down the explicit expression for the perturbed distribution function ψ_f in order to evaluate the collision integrals of Eq. (8.44). For instance, the expression (8.51) has been usually used. Substituting Eq. (8.51) into the collision integral (8.44), we may express those coefficients A_i, etc., in terms of the macroscopic variables such as temperature, components of stress tensor, and heat flux. From the collision integral I of Eq. (8.44), the following integral will result:

$$Z_{kj} = \frac{4}{\sqrt{\pi} \, \alpha^{2j+4}} \int_0^\infty \left[g_0^{2j+3} \exp\left(-\frac{g_0^2}{\alpha^2} \right) S_k \right] dg_0 \qquad (8.55)$$

where

$$S_k = 2\pi \int (1 - \cos^k \epsilon) b \, db = \text{collision cross section} \qquad (8.56)$$

and

$$\alpha = (4kT)^{1/2}$$

It should be noticed that the right-hand side terms of Eqs. (8.53) and (8.54) are due to the collision term calculated from Maxwellian distribution function such that some small additional terms are neglected. The left-hand side terms of these equations are due to the convection terms.

(1) Coefficient of viscosity. If we evaluate the collision integral I for the viscous stress and neglect the small terms, we have

$$\tau_{ij} = \frac{5}{2} \frac{(mkT)^{1/2}}{Z_{22}} \left(\frac{\partial u_i}{\partial x_j} + \frac{\partial u_j}{\partial x_i} - \frac{2}{3} \delta_{ij} \frac{\partial u_k}{\partial x_k} \right) = \frac{5}{2} \frac{(mkT)^{1/2}}{Z_{22}} \epsilon_{ij} \qquad (8.57)$$

Hence we have the coefficient of viscosity μ

$$\mu = \frac{5}{2} \frac{(mkT)^{1/2}}{Z_{22}} \qquad (8.58)$$

It is evident that in formula (1.53), the factor $1/Z_{22}$ is replaced by the mean free path. The factor Z_{22} may be evaluated by assuming a given kinetic model. For instance, for a hard sphere of diameter d, we have the coefficient of viscosity μ as follows:

$$\mu = \frac{5}{8} m(\pi RT)^{1/2} (2\pi d^2)^{-1} \qquad (8.59)$$

where $Z_{22} = 8\sqrt{\pi} d^2$.

(2) Coefficient of heat conductivity. Similarly, if we neglect the small terms in the collision integral, we have the coefficient of heat conductivity κ as follows:

$$\kappa = \frac{75k}{8} \left(\frac{kT}{m} \right)^{1/2} \frac{1}{Z_{22}} \qquad (8.60)$$

In deriving the expression of the transport coefficient, we neglect all the small terms in Eqs. (8.53) and (8.54). For a second approximation, Eqs. (8.53) and (8.54) may be written in the following forms respectively:

$$\tau_{ij}^{(2)} = \mu \epsilon_{ij} + S[\tau_{ij}^{(1)}, Q_{ci}^{(1)}] \qquad (8.61)$$

$$Q_{ci}^{(2)} = -\kappa \frac{\partial T}{\partial x_i} + F\left[\tau_{ij}^{(1)}, \ Q_{ci}^{(1)} \right] \qquad (8.62)$$

where S and F are the correction terms calculated from Eqs. (8.53) and (8.54) with the expressions of Navier-Stokes relations for viscous stress (8.57) and Fourier expression for heat conduction flux (1.55). If we substitute the viscous stress and heat conduction flux (8.61) and (8.62) into Eqs. (8.46) and (8.47), we have a second approximation of the fundamental equations of fluid dynamics which are known as the Burnett equations. In Burnett equations, there are temperature gradient terms in the equations of motion and the product of heat conduction flux with velocity gradients in the energy equation. Of course, both the Navier-Stokes equations and the Burnett equations are simplified versions of the thirteen moment equations.

8. Kinetic Theory of Chemically Reacting Gas[16]

In sections 2 to 7, we considered the case where the gas consists of one kind of molecules. For a chemically reacting gas, we have to consider the case where the gas is a mixture of N species. For each species, we have a molecular distribution function. We denote the molecular distribution function of sth species by F_s (r, q_{ms}, t), where q_{ms} is the molecular velocity of sth species particles. We may obtain the number density n_s, the flow velocity q_s, etc., in exactly the same manner as in the case of single kind of particles. In other words:

$$n_s = \int F_s(r, q_{ms}, t) dq_{ms} \tag{8.63}$$

The molecular distribution function F_s is governed by the Boltzmann equation which is of the same form as Eq. (8.13), i.e.,

$$\frac{\partial F_s}{\partial t} + q_{ms}^i \frac{\partial F_s}{\partial x_i} + \frac{\phi_s^i}{m_s} \frac{\partial F_s}{\partial q_{ms}^i} = \left(\frac{\delta F_s}{\delta t} \right)_c \tag{8.64}$$

For a chemically reacting gas, the binary collision is still the most important collision term. Hence we may use a form similar to Eq. (8.14) for the collision term. Since there are N species in the mixture, we have the summation of N collision terms between different species, i.e.,

$$\left(\frac{\delta F_s}{\delta t} \right)_c = \sum_{t=1}^{N} \int \int \int (\bar{F}_s \bar{F}_t - F_s F_t) g_{st} \, b \, db \, d\epsilon \, dq_{mt} = J_s(F_s, F_t) \tag{8.65}$$

In most of the problems of a chemically reacting gas, the masses of various species in the mixture do not differ greatly from one another, the diffusion velocities of various species are usually small. Hence we usually define the temperature of each species with respect to the mean velocity of the mixture as a whole, i.e., q with component u_i, where the velocity of the mixture as a whole is defined by the formula:

$$u_i = \frac{\sum_s n_s m_s u_{si}}{\sum_s n_s m_s} = \frac{\sum_s n_s m_s u_{si}}{\rho} \tag{8.66}$$

where ρ is the density of the mixture as a whole.

We define the peculiar velocity c_{s_a} of the sth species as

$$c_{s_a} = q_{ms} - q \qquad (8.67)$$

We may define the diffusion velocity w_s of sth species as

$$w_{si} = u_{si} - u_i = \frac{1}{n_s} \int F_s c_{si} dc_{s_a} \qquad (8.68)$$

where c_{si} is the ith component of c_{s_a}.

Similarly we may define the pressure tensor p_{sij} as

$$p_{sij} = m_s \int F_s c_{si} c_{sj} dc_{s_a} \qquad (8.69)$$

and the heat conduction flux as

$$Q_{csi} = \frac{1}{2} m_s \int (c_{s_a}^2) c_{si} F_s dc_{s_a} \qquad (8.70)$$

Hence we may generalize our previous analysis for a simple gas in a straight forward manner to the case of chemically reacting gas. However, there are some difference in the resultant transfer equations because of the chemical reaction which is represented by inelastic collision instead of elastic collision and all the Q's are not summation invariants. For instance, we consider the equation of continuity of a species s in the mixture. We multiply Eq. (8.64) by $Q = m_s$ and integrate the resultant equation in the whole velocity domain, we have

$$\frac{\partial \rho_s}{\partial t} + \frac{\partial \rho_s u_{si}}{\partial x_i} = m_s \int \left(\frac{\delta F_s}{\delta t} \right)_c dq_{ms} = \sigma_s \qquad (8.71)$$

The collision integral is no longer zero. We may use the symbol σ_s to represent the value of the collision integral which is exactly the source function of Eq. (3.34). In the macroscopic theory of Eq. (3.34), the source function is not known. In microscopic theory of (8.71), we may calculate the source function if we assume certain kinetic model of the chemical reaction. In reference [16], the evaluation of these collision integrals for simple kinetic model has been discussed.   Similarly, we may find other transport equations of the chemically reacting gas.

If we calculate the equations of motion of sth species and neglect the higher order terms including the thermal diffusion and chemical reaction, we have the following equation for the diffusion velocities:

$$\frac{\rho_s}{\rho}\frac{\partial p}{\partial x_i} - \frac{\partial p_s}{\partial x_i} = \sum_t K_{st}(w_{si} - w_{ti}) \tag{8.72}$$

where

$$K_{st} = K_{ts} = \frac{2\eta\alpha}{3} n_s n_t Z_{st11} = \text{friction coefficient} \tag{8.73}$$

If we compare the friction coefficient K_{st} with the ordinary diffusion coefficient D_{st} as defined in the standard textbook of Chapman and Cowling,[6] we have

$$K_{st} = \frac{n_s n_t p}{n^2 D_{st}} \tag{8.73a}$$

Hence we have the first approximation of the diffusion coefficient:

$$D_{st} = \frac{3}{2}\left(\frac{kT}{2\eta}\right)^{\frac{1}{2}}\frac{1}{Z_{st11}} \tag{8.74}$$

where

$$\eta = \frac{m_s m_t}{m_s + m_t}; \quad \alpha = \left(\frac{2kT}{\eta}\right)^{\frac{1}{2}}$$

and Z_{st11} is given by Eq. (8.55) for $k=1$ and $j=1$ with the proper collision cross section between the species s and t.

The diffusion coefficient D_{st} is another transport coefficient which cannot be determined in the macroscopic theory.

9.   Kinetic Theory of a Plasma[3, 18, 19, 23, 24]

The kinetic theory of a plasma is similar to that of a chemically reacting gas in that we have to consider the molecular distribution functions for each species in a plasma including electrons and ions. In ordinary chemically reacting gas, we may neglect the body force

ϕ_s^i in the Boltzmann equation (8.64) but in plasma, we have to consider the electromagnetic force as the body force on charged particles. Hence we have to solve the Boltzmann equation (8.64) simultaneously with the Maxwell equations of the electromagnetic fields. Secondly, in the kinetic theory of a plasma, the collision terms contain not only the close encounter such as the binary collision $J_s(F_s, F_t)$ of Eq. (8.65) but also the distance encounter, the collision terms known as the Fokker-Planck mechanism. This involves taking account of very small angle scattering and thus of a great number of acts of very small momentum transfer. A certain part of the scattering is, in this way, represented as the cumulative and statistically determined results of a very large number of small-angle deflections by many other particles. The collision term due to this mechanism may be written as

$$\left(\frac{\delta F_s}{\delta t}\right)_2 = \sum_{r=1}^{N}\left[\frac{1}{2}\sum_{i,j}\frac{\partial^2}{\partial v_s^i \partial v_s^j}(F_s<v_{s,r}^i, v_{s,r}^j>)\right.$$

$$\left. -\frac{\partial}{\partial v_s^i}(F_s<v_{s,r}^i>)\right]=K_s(F_s, F_t) \qquad (8.75)$$

where $v_s^i=q_{ms}^i$ and for any quantity x, $<x_r>$ is defined by

$$<x_r>=2\pi\int_0^\infty g_0 F_r\, dq_{mr}\int_{b_c}^{b_m} xb\, db \qquad (8.76)$$

where b_c is the lower limiting value and b_m is the upper cut-off distance of the impact parameter b in a plasma. If one takes b_m to be infinite, the integral (8.76) diverges logarithmically for the case of the Coulomb force. Hence b_m must be taken as a finite value. Usually it may be taken as the Debye length in the plasma. The term $<v_{s,r}^i>$ is the change of the ith component of velocity of a particle of sth species due to the scattering of s and r particles. In fully ionized plasma, the collision terms due to the Fokker-Planck mechanism are more important than those due to binary collisions while in slightly ionized plasma, the binary collision is the more important term between those two collision terms. Since the collision terms determine the electrical conductivity of a plasma, we expect that the behavior of electrical conductivity in a fully ionized plasma is different from that in a slightly ionized plasma, as we have discussed in chapter I, i.e., Eqs. (1.85) and (1.86).

With the above two points in mind, we may treat the kinetic theory of a plasma from the Boltzmann equation (8.64) in exactly the same manner as in the case of a single species or in the case of a mixture of neutral gases. However, there is another point which is worthwhile to be mentioned: In the case of a chemically reacting gas mixture, the masses of various species in the mixture are approximately equal. As a result, the diffusion velocities of various species are small in comparison with the flow velocity of the mixture as a whole. We may use the peculiar velocity c_{s_a} of Eq. (8.67) to define the viscous stresses and heat flux. In a plasma, the mass of an electron is much smaller than those of ions or neutral particles. Hence the diffusion velocity of electrons will be much larger than those of ions and neutral particles. If we use the peculiar velocity defined by Eq. (8.67), the peculiar velocity may not be small and we may not be able to use the perturbation technique as given by Eqs. (8.34) and (8.51). This is particularly true for the case of fully ionized plasma or plasma of high degree of ionization. In these cases, it is better to use the modified peculiar velocity c_s^{*i} which is defined as

$$c_s^{*i} = q_{ms}^i - q_s^i = c_s^i + w_s^i \qquad (8.77)$$

In general, the modified peculiar velocities are small and we may use the perturbation technique in the kinetic theory of a plasma. In other words, we may define the pressure tensor as

$$p_{sij} = m_s \int c_{si}^* c_{sj}^* F_s dc_s^* \qquad (8.78)$$

and the temperature of sth species as

$$T_s = \frac{m_s}{3kn_s} \int (c_{s1}^{*2} + c_{s2}^{*2} + c_{s3}^{*2}) F_s dc_s^* \qquad (8.79)$$

and the energy flux as

$$S_{sijk} = m_s \int c^*_{si} c^*_{sj} c^*_{sk} F_s \, dc^*_s \qquad (8.80)$$

and the heat flux due to conduction as

$$Q_{sci} = \frac{1}{2}(S_{si11} + S_{si22} + S_{si33}) = \frac{1}{2} m_s \int (c^{*2}_s) c^*_{si} F_s dc^*_s$$

$$(8.81)$$

There are definite relations between the pressure tensor, etc., defined with respect to modified peculiar velocity c^*_s and those with respect to peculiar velocity c_{s_a} but the relations are very complicated. When the diffusion velocities are large, it is advisable to use the definitions given here in Eqs. (8.78) to (8.81).

The interaction of the Boltzmann equations of a plasma and the electromagnetic fields is through the electrical current density J which is defined as follows:

$$J = \sum_{s=1}^{N} e_s n_s q_s = \sum_{s=1}^{N} e_s \int q_{ms} F_s dq_{ms} \qquad (8.82)$$

where e_s is the electrical charge of a particle of sth species in a plasma.

We may derive the transport equations in a plasma in exactly the same manner as we discussed before. Of course, in the present case we have additional terms due to the electromagnetic force ϕ^i_s. For instance, the corresponding equation of diffusion velocity (8.71) in a plasma will be

$$\left(\rho_{cs} - \frac{\rho_s \rho_e}{\rho} \right) E_u + \left(J_s - \frac{\rho_s}{\rho} J \right) \times B - \left(\nabla p_s - \frac{\rho_s}{\rho} \nabla p \right)$$

$$= \sum_{t=1}^{N} K_{st}(w_s - w_t) \qquad (8.83)$$

Eq. (8.83) is a generalized Ohm's law of the electric current which contains not only the ordinary generalized Ohm's law (6.11) and the Hall current (6.126) but also some effects of gasdynamics on the electrical current density. We shall discuss this equation in detail in chapter IX section 4. In Eq. (8.83),

$$\rho_{es} = n_s e_s, \quad \rho_s = m_s n_s, \quad \rho_e = \sum_{s=1}^{N} \rho_{es}, \quad J_s = \sum_{s=1}^{N} \rho_{es} q_s$$

and

$$E_u = E + q \times B.$$

The electrical current density may be written as

$$J = \sum_{s=1}^{N} n_s e_s w_s + \rho_e q \qquad (8.84)$$

Hence we have definite relation between J and the diffusion velocity w_s.

10. Free Molecule Flow[1, 22]

In sections 5 to 9, we discussed essentially the relations between the kinetic theory of a gas or a plasma and the corresponding macroscopic theory, particularly the derivations of the fundamental equations of macroscopic theory and the expressions of the transport coefficients from the Boltzmann equations. In the next few sections, we shall discuss a few examples of simple flow problems in which the classical macroscopic theory cannot be used because the mean free path of the gas or the plasma is not small. We have to use equations different greatly from those equations of the type of Navier-Stokes equations as we used in chapters V to VII.

The simplest type of flow of a rarefied gas is the free molecule flow in which the Knudsen number is much larger than unity. In this case, the collision between the molecules is much less than the collision between the molecule and the surface of the body in the flow field. As a result, we may neglect completely the collisions between the particles and consider only the interaction between the gas molecules and the surface of the body in the flow field.

In the free molecule flow, we may divide the gas molecules into two parts: the incident stream in which the flow velocity and temperature are known and the molecular distribution function may be assumed to be the Maxwellian distribution function (8.25) corresponding to the known flow velocity and temperature in the incoming stream ; and the reflected stream which composes of the molecules reflected from the surface of the body in the flow field. The molecular distribution function of the reflected stream could be determined from the known condition of the incident stream and the properties of the surface of the body either theoretically or experimentally. Such a determination of the interaction of the molecules and the surface is rather complicated. However, for engineering problems, our main interests are the heat transfer from the body and the momentum transfer or forces on the body in the flow field. For such engineering problems, it is sufficient to know a thermal accommodation coefficient α and two reflection coefficients σ_1 and σ_2 which are defined as follows:

$$\alpha = \frac{dE_i - dE_r}{dE_i - dE_w} \tag{8.85}$$

$$\sigma_1 = \frac{\tau_i - \tau_r}{\tau_i - \tau_w} (\tau_w = 0) \tag{8.86}$$

$$\sigma_2 = \frac{p_i - p_r}{p_i - p_w} \tag{8.87}$$

where subscript i, r, and w refer to the value of the incident stream, reflected stream, and the wall of the body respectively. dE is the energy flux, τ is the tangential momentum on the surface, and p is the normal momentum on the surface. The values of α, σ_1, and σ_2 depend on the properties of the surface as well as the gas and they give some overall phenomenological average properties. Their values have been determined experimentally for various cases. The value of α are between 0.87 to 0.97 and the values of σ_1 are from 0.79 to 1.00 and no information is available for σ_2.

There are two special cases which are of interest, i.e., the entirely specular reflection and the entirely diffuse reflection. By specular reflection, we mean that the molecule after the collision with the surface has the same tangential velocity before collision but a reversal of its normal velocity component to the surface. By diffuse reflection, we mean that the molecules issue a Maxwellian velocity distribution at a temperature not necessarily equal to that of the surface. For entirely specular reflection with vanishing energy exchange, we have $\alpha = \sigma_1 = \sigma_2 = 0$ while for entirely diffuse reflection which has been completely accommodated to the wall temperature T_w, we have $\alpha = \sigma_1 = \sigma_2 = 1$. Actually the values of α, σ_1, and σ_2 are between 0 and 1 as we have mentioned before.

After the values of α, σ_1, and σ_2 are known, we may calculate the heat transfer and the force on the body from the conditions of the incident stream and that of the surface of the body. In the calculation of the heat transfer, the accommodation coefficient α may be different for different modes of internal energy. For instance, the surface collision will have less effect on the vibrational energy than on the translational or rotational energy. But for a first approximation, we may assume that α is the same for all modes of internal energy.

Let dQ be the heat removed from an elementary area dA of the body per unit time. We have

$$dQ = dE_i - dE_r = \alpha(dE_i - dE_w) \qquad (8.88)$$

where we assume that α is different from zero. The value dE_i may be calculated by assuming that the molecular distribution function of the incident stream is Maxwellian, i.e., F_{0i} with flow velocity q_i and the gas temperature T_i. The value of dE_i is composed of two parts: one is the translational energy dE_{ti} and the other is due to the other internal energy dE_{ji}. These two parts may be written in the following forms:

$$dE_{ti} = \int_0^\infty \int_0^\infty \int_0^\infty \frac{1}{2} m\, q_m^2 q_{m_1} F_{0i}\, dq_{m_1} dq_{m_2} dq_{m_3}\, dA \qquad (8.89)$$

where we assume that the gas consists of one kind of particles of mass m and the coordinate system is so chosen that the direction 1 is normal to the surface dA, and 2 and 3 are the other two orthogonal tangential directions.

The part of internal energy dE_{ji} may be written as

$$dE_{j_i} = \frac{1}{2} j k T n_i = \frac{5-3\gamma}{\gamma-1} \frac{mRT}{2} n_i \qquad (8.90)$$

where n_i is the number of molecules that are incident on dA per unit time. Since it is uncertain for the values of reflection coefficients, from now on, we consider only the technical surface from which the reflection of the molecules is diffuse but the accommodation coefficient of the surface is a constant different or equal to unity. We consider a uniform free stream with a velocity U_∞ and a temperature T_∞ passing over a convex body so that multireflection does not occur. Let us consider an elementary surface on the body dA (Fig. 8.3). We further assume that there is no gain or less of molecules

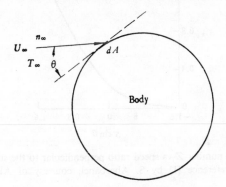

Figure 8.3   A body in a free molecule flow

during the reflection process. The number of the incident molecules crossing area dA in a unit time is

$$n_i = dA \cdot \int_0^\infty \int_0^\infty \int_0^\infty q_{m1} F_{0i} dq_{m_1} dq_{m_2} dq_{m_3}$$

$$= dA \cdot \frac{n_\infty c_{m\infty}}{2\sqrt{\pi}} \Big\{ \exp\left(-S^2 \sin^2 \theta\right)$$

$$+ \sqrt{\pi} \, S \sin \theta \Big[1 + \operatorname{erf}\left(S \sin \theta\right) \Big] \Big\} = dA \cdot n_\infty c_{m\infty} Z \qquad (8.91)$$

where θ is the local angle of attack and subscript ∞ refers to the values in the free stream, c_m is the most probable speed defined in Eq. (8.28) and

$$S = \frac{U_\infty}{c_{m\infty}} = \left(\frac{1}{2}\gamma\right)^{1/2} M_\infty = \text{speed ratio} \qquad (8.92)$$

where M_∞ is the Mach number of the free stream. The parameter Z defined by Eq. (8.91) is a nondimensional number flux which is plotted in Fig. 8.4.

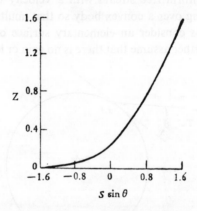

Figure 8.4   Flux number Z vs speed ratio perpendicular to the surface. (Figure 1 of reference [1] by S. Abarbanel, courtesy of AIAA)

The translational energy integral (8.90) with the Maxwellian distribution gives

$$\begin{aligned}
dE_{ti} = dA \cdot \frac{1}{2} \; mn_\infty c_{m\infty}^3 \; \bigg\{ & \left(\frac{2+S^2}{\sqrt{2\,\pi}}\right) \exp\left(-S^2 \sin^2\theta\right) \\
& + \left(\frac{5+2S^2}{4}\right) S \sin\theta \left[1 + \text{erf}\,(S\sin\theta)\right] \bigg\} \\
& = \frac{1}{2} mc_{m\infty}^3 n_i \epsilon\, dA
\end{aligned} \qquad (8.93)$$

Fig. 8.5 shows that variation of the nondimensional translational energy flux ϵ with $\sin\theta$ and S.

The energy flux dE_w is obtained from a Maxwellian distribution with temperature T_w and zero flow velocity, i.e.,

$$dE_w = dA \cdot \left(2 + \frac{1}{2} j \right) kT_w n_i = dA \cdot \frac{\gamma+1}{2(\gamma-1)} kT_w n_i \quad (8.94)$$

From Eq. (8.85), we have the reflected energy flux as

$$\frac{dE_r}{dE_i} = 1 - \alpha \left[1 - \frac{4+j}{2+j} \left(\frac{T_w}{T_\infty} \right) \right] \quad (8.95)$$

Substituting Eqs. (8.89) to (8.94) into Eq. (8.88), we obtain the heat transfer through an elementary area dA. For any given body, we integrate dQ over the whole area of the body and obtain the total heat transfer Q over the body.

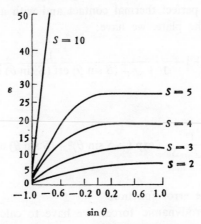

Figure 8.5 Translational energy flux ϵ at various incident angles θ. (Figure 2 of reference [1] by S. Abarbanel, courtesy of AIAA)

After Q and the temperature in the reflected molecules T_r are obtained the heat transfer characteristics of a given body may be expressed in terms of a thermal recovery factor r' and a Stanton number S_t which are defined respectively as follows:

$$r' = \frac{(T_r - T_\infty)(\gamma+1)}{(T_{0\infty} - T_\infty)\gamma} \quad (8.96)$$

and

$$S'_t = \frac{\gamma Q}{A\rho U_\infty c_p (T_r - T_w)(\gamma + 1)} \tag{8.97}$$

where T_∞ is the free stream temperature and $T_{0\infty}$ is the stagnation temperature of the flow such that

$$T_{0\infty} = T_\infty \left(1 + \frac{\gamma - 1}{\gamma} S^2 \right) \tag{8.98}$$

If we neglect the possibility of molecular interreflection such that the body surface is everywhere convex toward the gas stream, the values of r' and S'_t can be easily calculated for simple bodies. For instance, for a flat plate at an angle of attack θ with front and rear surface in perfect thermal contact and with a total area A of both sides of the plate, we have:

$$r' = \frac{1}{S^2}\left[2S^2 + 1 - \frac{1}{1 + \sqrt{\pi}\,(S\sin\theta)\,\mathrm{erf}\,(S\sin\theta)\exp(S\sin\theta)^2} \right] \tag{8.99}$$

and

$$S'_t = \frac{1}{4\sqrt{\pi}\,S}\left[\exp(-S\sin\theta)^2 + (S\sin\theta)\,\mathrm{erf}\,(S\sin\theta) \right] \tag{8.100}$$

where erf is the error function.

For the aerodynamic forces, we have to calculate p_i, τ_i, and p_w as follows:

$$p_i = \int_{-\infty}^{\infty}\int_{-\infty}^{\infty}\int_{0}^{\infty} m q_{m_1}^2 F_{0i}\,dq_{m_1}dq_{m_2}d q_{m_3} \tag{8.101}$$

$$\tau_i = \int_{-\infty}^{\infty}\int_{-\infty}^{\infty}\int_{0}^{\infty} m q_{m_1}q_{m_2} F_{0i}\,dq_{m_1}dq_{m_2}dq_{m_3} \tag{8.102}$$

$$p_w = \frac{1}{2} m (2\pi R T_w)^{\frac{1}{2}} n_i \tag{8.103}$$

The net pressure p and the shearing stress τ on the surface of the body is then

$$p = p_i + p_r = (2 - \sigma_2)p_i + \sigma_2 p_w \qquad (8.104)$$

$$\tau = \tau_i - \tau_r = \sigma_1 \tau_i \qquad (8.105)$$

For simplicity, instead of two reflection coefficients, we may use the Maxwellian method of only one reflection coefficient σ_1 so that a fraction of the incident molecules is reflected diffusely and the remainder $(1 - \sigma_1)$ specularly. Eqs. (8.104) and (8.105) become respectively

$$p = (2 - \sigma_1)p_i + \sigma_1 p_r \qquad (8.106)$$

$$\tau = \sigma_1 \tau_i \qquad (8.107)$$

After p and τ are obtained, we may integrate the forces over the body to obtain the resultant lift L and drag D forces on the body. The drag D and lift L on the body are customarily expressed in terms of the drag coefficient C_D and the lift coefficient C_L such that

$$C_D = \frac{D}{\frac{1}{2}\rho U^2{}_\infty A}; \quad C_L = \frac{L}{\frac{1}{2}\rho U^2{}_\infty A} \qquad (8.108)$$

For simple bodies, the drag and lift coefficients in a free molecule flow can be easily obtained by simple integration. For instance, for a flat plate at an angle of attack θ and an area A of one side of the plate we have for diffuse reflection

$$c_{Dd} = \frac{2}{\sqrt{\pi}\,S}\Big[\exp{(-S \sin \theta)^2}$$

$$+ \sqrt{\pi}\,S \sin \theta \left(1 + \frac{1}{2S^2}\right) \mathrm{erf}\,(S \sin \theta) + \frac{\pi S}{S_w}\sin^2\theta \Big] \qquad (8.109)$$

$$C_{Ld} = \frac{\cos \theta}{S^2}\Big[\mathrm{erf}\,(S \sin \theta) + \sqrt{\pi}\,\frac{S^2 \sin \theta}{S_w} \Big] \qquad (8.110)$$

and for specular reflection:

$$C_{Ds} = \frac{4 \sin \theta}{S^2}\Big\{ (S \sin \theta)\exp[-(S \sin\theta)^2] + \Big[\frac{1}{2} + (S \sin \theta)^2\Big]$$

$$\times \mathrm{erf}(S \sin \theta) \Big\} \qquad (8.111)$$

$$C_{Ls} = C_{Ds} \cdot \cot \theta \tag{8.112}$$

where

$$S_w = U_\infty / (2RT_w)^{1/2} \tag{8.113}$$

The above calculation is based on the assumption that the Knudsen number is very large so that the collision between the molecules can be completely neglected. For the case that the Knudsen number is large but not very large, we should consider some of the collisions between molecules. As a result we may express the molecular distribution function F in terms of power series of $1/K$ such that

$$F = F_0 \left(1 + \frac{1}{K}\phi_1 + \frac{1}{K^2}\phi_2 + \cdots \right) \tag{8.114}$$

In general, the perturbed molecular distribution function ϕ_n should be calculated from the Boltzmann equation. If only the function ϕ_1 is included, the region is known as the first order collision region.

11. Slip Flow and Temperature Jump in a Rarefied Gas[13, 22]

The other limiting case of rarefied gasdynamics is the case in which the Knudsen number is small but not small enough such that the fundamental equations of fluid dynamics are the Navier-Stokes equations with simple expressions for the viscous stresses and heat conduction flux but the no-slip boundary conditions do not hold. There is a slip on the wall of the body and there is also a temperature jump on the wall too. Let us consider a simple case where the fraction of molecules that reflect diffusely is σ_1 and those that reflected specularly is $(1-\sigma_1)$ of the total molecules. Let y be the normal coordinate to the wall and $y=0$ represents the location of the wall. Near the wall, one half of the gas molecules come off the wall and the other half come from a layer of gas a mean free path away. The x-component of the velocity of the gas on the wall $u(0)$ is then

$$u(0) = \frac{1}{2}\left\{ \left[u(0) + L_f\left(\frac{\partial u}{\partial y}\right)_0 \right] + (1-\sigma_1)\left[u(0) + L_f\left(\frac{\partial u}{\partial y}\right) \right] + \sigma_1 \cdot 0 \right\} \tag{8.115}$$

From Eq. (8.115), we have

$$u(0) = \frac{2-\sigma_1}{\sigma_1} L_f \left(\frac{\partial u}{\partial y} \right)_0 \tag{8.116}$$

Eq. (8.116) shows that even $\sigma_1 = 1$, i.e., completely diffuse reflection, there is still a definite slip velocity on the wall which is proportional to the mean free path L_f. However, usually the mean free path L_f is so small that no-slip is a good approximation for Eq. (8.116). In rarefied gas where the mean free path is not negligibly small, the slip velocity is not negligible. For a more accurate calculation, we have

$$u(0) = \frac{2-\sigma_1}{\sigma_1} L_f \left(\frac{\partial u}{\partial y} \right)_0 + \frac{3\mu}{4\rho T} \left(\frac{\partial T}{\partial x} \right)_0 \tag{8.117}$$

and

$$T(0) = \frac{2-\alpha}{\alpha} \frac{2\gamma}{\gamma+1} \frac{L_f}{P_r} \left(\frac{\partial T}{\partial y} \right)_0 + T_w \tag{8.118}$$

For slightly rarefied gas, we may still use the Navier-Stokes equations to describe the flow field but the slip velocity and the temperature jump should be used as the boundary conditions. Such a treatment is referred to as slip flow. Some simplified flow configurations have been analyzed in this manner such as plane Couette flow, flat plate, sphere, cylinder, and cone in a uniform stream.

For a uniform flow over a semi-infinite wedge, the skin friction coefficient in a slip flow is then

$$c_f = c_{f_0} \left\{ 1 - \left(\frac{2-\sigma_1}{\sigma_1} \right) \frac{m(R_e)^{1/2}}{f''(0)} \frac{L_f}{x} + \cdots \right\} \tag{8.119}$$

where $m = \theta/(\pi-\theta)$ and θ is the semi-vertex angle of the wedge. Hence for a flat plate, the first order effect of slip is zero. The skin friction coefficient without slip is c_{f0}, and f is the nondimensional form of the streamfunction, i.e., $\psi = (u_e \nu_g x)^{1/2} f(\eta)$, $\eta = [u_e/(\nu_g x)]^{1/2} y$, $u_e = cx^m$ and $f = d^2 f/dn^2$. The boundary layer approximations are used to calculate the stream function f and the skin friction coefficient c_f but with the slip velocity condition. Eq. (8.119) may be written in the following nondimensional form of Mach number $M = U/a$ and the Reynolds number $R_e = Ux/\nu_g$:

$$c_f = c_{f_0} \left\{ 1 - \frac{2-\sigma_1}{\sigma_1} \frac{1.255 m \sqrt{\gamma}}{f''(0)} \frac{M}{(R_e)^{\frac{1}{2}}} + \cdots \right\} \qquad (8.120)$$

Eq. (8.120) shows that for high Mach number M and low Reynolds number R_e the effect of slip becomes important.

12. Couette Flow [14, 15]

It would be of extreme interest if we had a solution of the flow problem over the whole range of gas density from free molecule flow to the atmospheric condition. In principle if we can find an exact solution of a simple flow configuration from the Boltzmann equation, the result will cover the whole range of gas density from the free molecule flow to the classical Navier-Stokes regime. However, it is extremely difficult to find the exact solution of the Boltzmann equation. In order to get some workable results, some simplifications must be made. One of these simplifications is to consider the transfer equation (8.44) by assuming some reasonable form of the distribution function so that the velocity and the temperature distributions of a simple flow configuration may be obtained, Lees[14] introduced the two-stream Maxwellian distribution functions in the transfer equation and obtained some very interesting results. The assumed distribution function should satisfy the following conditions:

(1) It must have the two-sided character which represents the feature of highly rarefied gas flows.

(2) It must be capable of providing a smooth transition from the rarefied gas flow to the solution of a continuum medium.

(3) It should lead to the simplest possible set of differential equations and boundary conditions consistent with (1) and (2).

In Lees' two-stream distribution functions, he assumed a Maxwellian distribution function F_1 within the cone of influence (Region 1 of Fig. 8.6) and another Maxwellian distribution function F_2 in the region 2 of Fig. 8.6, i.e.,

$$F_1 = \frac{n_1(r,\ t)}{[2\pi RT_1(r,\ t)]^{3/2}}\ \exp\left\{-\frac{[q_m - q_1(r,t)]^2}{2RT_1(r,\ t)}\right\} \quad (8.121)$$

$$F_2 = \frac{n_2(r,\ t)}{[2\pi RT_2(r,\ t)]^{3/2}}\ \exp\left\{-\frac{[q_m - q_2(r,\ t)]^2}{2RT_2(r,\ t)}\right\} \quad (8.122)$$

where n_1, n_2, T_1, T_2, q_1, and q_2 are ten initially determined functions of r and t which may be determined from the Maxwellian moment equations (8.44) and the boundary conditions.

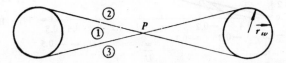

Figure 8.6 Regions of two-stream Maxwellian distribution functions

Lees and Liu[15] applied this method to the plane Couette flow of a gas, i.e., the flow between two parallel plates at different temperatures and the plates moving a relative velocity U. Furthermore, they assumed that the boundary conditions correspond to completely diffuse reemission. The velocity and temperature distributions for various values of Reynolds number R_e and Mach number $M = U/a_{\mathrm{II}}$ are shown in Figs. 8.7 and 8.8 respectively. The interesting results are as follows:

(1) The ratio of the shear stress to the product of ordinary viscosity and velocity gradient, which is unity for a Newtownian fluid, depends on the gas density, the plate temperatures, and the plate spacing. This ratio decreases rapidly with increasing plate Mach number when the plate temperatures are fixed. At a fixed Mach number based on the temperature of one plate, this ratio approaches to unity as the temperature of the other plate increases.

(2) The ratio of the heat flux to the product of the ordinary heat conductivity coefficient and the temperature gradient, which is unity in ordinary gasdynamics, depends also on the gas density, the plate temperatures, and the plate spacing.

(3) In the Navier-Stokes regime, most of the gas follows the hot plate, because the gas viscosity is large there. As the gas density

decreases, the situation reverses, because the velocity slip is larger at the hot plate than at the cold plate. In the limiting case of a highly rarefied gas most of the gas follows the cold plate.

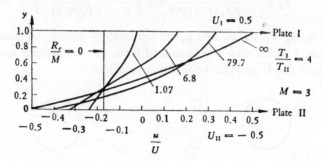

Figure 8.7   Velocity distribution for a plane Couette flow over a large range of R_e/M. (Figure 8 or reference [15], courtesy of the authors)

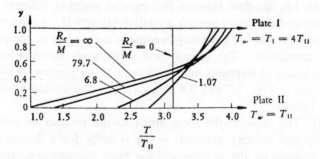

Figure 8.8   Temperature distributions for a plane Couette flow over a large range of R_e/M. (Figure 9 of reference [15], courtesy of the authors)

13. Plasma Oscillation[18, 23]

In our previous discussion, we did not try to solve the Boltzmann equation for the flow problem. In this section, we use the Boltzmann equation to describe a simple flow problem. We consider the electron oscillations in a homogeneous electrically neutral plasma in the absence of external electric and magnetic fields. We also neglect the collision terms. This is the simplest case of a wave motion in an ideal plasma which will be discussed from the continuum theory point of view in chapter IX section 4 too. The result is the well known phenomenon of plasma oscillation. The Boltzmann equation for the one-dimensional motion of electrons is then:

$$\frac{\partial F}{\partial t} + v\,\frac{\partial F}{\partial x} - \frac{eE}{m}\,\frac{\partial F}{\partial v} = 0 \qquad (8.123)$$

where $F(x, v, t)$ is the distribution function of the electrons of mass m and electric charge $-e$. The average electric field E satisfies the Poisson's equation:

$$\frac{\partial E}{\partial x} = e\left(n_0 - \int F dv\right) \qquad (8.124)$$

where n_0 is the number density of the electrons in the undisturbed state.

For a small disturbance, we write

$$F = F_0 + F_1 \qquad (8.125)$$

where $F_0 = F_0(v)$ is the undisturbed distribution function of the electrons and $F_1 = F_1(x, v, t)$ is the perturbed distribution function which is small in comparison with F_0. The resultant linearized equations for F_1 and E are respectively

$$\frac{\partial F_1}{\partial t} + v\,\frac{\partial F_1}{\partial x} - \frac{eE}{m}\,\frac{\partial F_0}{\partial v} = 0 \qquad (8.126)$$

and

$$\frac{\partial E}{\partial x} = -e \int F_1\,dv \qquad (8.127)$$

We assume that at $t=0$, an initial disturbance $F_1(x, v, 0)$ is applied to the equilibrium condition specified by $F_0(v)$. We solve Eqs. (8 126) and (8.127) from the initial condition by the method of Fourier transform.

We define the following Fourier transforms:

$$F_1(\omega, \ v, \ k) = \int_0^\infty dt \int_{-\infty}^\infty dx \exp\left[-i(kx - \omega t)\right] F_1(x, \ v, \ t)$$

(8.128)

$$E(\omega, k) = \int_0^\infty dt \ \int_{-\infty}^\infty dx \exp\left[-i(kx - \omega t)\right] E(x, \ t)$$

(8.129)

and the reciprocal relations

$$F_1(x,v,t) = \frac{1}{(2\pi)^2} \int_W d\omega \int_{-\infty}^\infty dk \cdot \exp\left[\ i(kx - \omega t)\right] \cdot F_1(\omega, \ v, \ k)$$

(8.130)

$$E(x, \ t) = \frac{1}{(2\pi)^2} \int_W d\omega \int_{-\infty}^\infty dk \cdot \exp\left[i(kx - \omega t)\right] \cdot E(\omega, \ k)$$

(8.131)

where the contour W is chosen in the upper plane parallel to the real ω-axis, above any poles in $F_1(\omega, \ v, \ k)$ or $E(\omega, \ k)$.

Multiplying Eq. (8.126) by $\exp[-i(kx - \omega t)]$ and integrating the resultant equations over the space and time and using the relations (8.128) and (8.129), we have

$$F_1(\omega, \ v, \ k) = \frac{1}{i(kv - \omega)}\left[\phi(v, \ k) + \frac{e}{m} \ \frac{\partial F_0}{\partial v} E(v, \ k)\right] \quad (8.132)$$

where the initial value term $\phi(v, \ k)$ is defined as

$$\phi(v, \ k) = \int_{-\infty}^\infty dx \cdot \exp(-ikx) \cdot F_1(x, \ v, \ 0) \qquad (8.133)$$

Similarly from Eq. (8.127), we have

$$E(\omega, \ k) = \frac{e}{\omega_p^2 H\left(\dfrac{\omega}{k}, \ k\right)} \int \frac{\phi(v, \ k)}{v - \dfrac{\omega}{k}} dv \qquad (8.134)$$

where

$$H\left(k, \frac{\omega}{k}\right) = \frac{k^2}{\omega_p^2} - \int \frac{G(v)}{v - \frac{\omega}{k}} dv \qquad (8.135)$$

$$\omega_p = e(n_0/m)^{\frac{1}{2}} = \text{frequency of plasma oscillation} \qquad (8.136)$$

$$G(v) = \frac{1}{n_0} \frac{\partial F_0}{\partial v} \qquad (8.137)$$

By the relations (8.130) and (8.131), we obtain $F_1(x, v, t)$ and $E(x, t)$ from Eqs. (8.132) and (8.134) respectively. Let us consider the electric field $E(x, t)$:

$$E(x, t) = \frac{e}{4\pi^2 \omega_p^2} \int_{-\infty}^{\infty} dk \int_W d\omega \cdot \exp(ikx - i\omega t) \frac{\int \frac{\phi(v, k) \, dv}{\left(v - \frac{\omega}{k}\right)}}{H\left(k, \frac{\omega}{k}\right)}$$

$$(8.138)$$

With a reasonable initial disturbance $\phi(v, k)$, the integral will not have poles in the complex plane and then the characteristic frequencies of the plasma oscillation are determined by the dispersion equation:

$$H\left(k, \frac{\omega}{k}\right) = 0 \qquad (8.139)$$

For a Maxwellian distribution and for the case that ω/k is large in comparison with v in $G(v)$, we may expand the denominator in the integral of H and obtain the following results:

$$\frac{k^2}{\omega_p^2} = -\frac{k}{\omega} \int G(v) \left(1 + \frac{kv}{\omega} + \frac{k^2 v^2}{\omega^2} + \cdots\right) dv \qquad (8.140)$$

From Eq. (8.140), we have

$$\omega^2 = \omega_p^2 \left(1 + \frac{3k^2 \overline{v^2}}{\omega_p^2} + \cdots\right) \qquad (8.141)$$

where $\overline{v^2} = kT_0/m = a_T^2 = $ mean square of the velocity of the distribution $F_0(v)$. Eq. (8.141) with only the first two terms gives the simple result of plasma oscillation.

With the first two terms of Eq. (8.141) for ω^2, it is easy to show that the imaginary part of ω is

$$\text{Im}(\omega) = -\frac{\pi}{8}\omega_p\left(\frac{\omega_p}{ka_T}\right)^3 \exp\left(-\frac{\omega^2}{2k^2a_T^2}\right) \qquad (8.142)$$

The expression (8.142) shows that the plasma oscillation will be damped out, which is known as Landau damping of a plasma oscillation. The plasma oscillations are damped rapidly for the wave number k of the order of Debye wave length $k_D = \omega_p/a_T$. For large wave lengths, the damping constant vanishes exponentially so that there is no damping for such long wave of plasma oscillation. In chapter IX section 4, we consider the plasma oscillation from the macroscopic point of view, i.e., only the long wave lengths will be considered. As a result, we shall not have the Landau damping there. It shows that the study based on Boltzmann equation will give much more information than those based on the transfer equations in the continuum theory. In general, the solution of the Boltzmann equation is much more difficult to obtain. At the present time, it is not possible to obtain the solution of the Boltzmann equation for complicated engineering flow problems.

14. Particle Motion of a Plasma[4, 5]

Even though the Boltzmann equation may describe the motion of a highly rarefied gas, sometimes it is convenient to describe the motion of a highly rarefied gas by considering the motion of individual particles, particularly for the charged particles whose motion is largely dominated by the electromagnetic forces. Such an analysis has been widely used for highly rarefied plasma. For a charged particle, ion or electron, the equation of motion under the electromagnetic force is

$$m_s\frac{dq_s}{dt} = e_s(E + q_s \times B) \qquad (8.143)$$

where m_s is the mass of a particle of the sth species in a plasma, q_s is its velocity vector and e_s, its electrical charge. Eq. (8.143) should be solved with Maxwell's equations for the electromagnetic fields. Without going into detailed analysis, Eq. (8.143) gives many interesting results. We shall consider a few simple cases from Eq. (8.143).

(1) Constant electric field $E=$ constant and no magnetic field $B=0$. Eq. (8.143) under these conditions becomes

$$\frac{dq_s}{dt} = \frac{e_s E}{m_s} \tag{8.144}$$

The particle has a constant acceleration in the direction of the electric field E.

(2) No electric field, $E=0$ and constant magnetic field $B=$ constant. For this case, it is convenient to use cylindrical coordinates r, θ, z with z in the direction of B. The final equilibrium condition will be the case when the particle has no radial velocity component q_r but has only the axial and tangential velocity components q_z and q_θ. The axial velocity component q_z is determined by the initial condition of the particle and is unaffected by the magnetic field. The tangential velocity component q_θ is determined by the equilibrium condition between the centrifugal force and the electromagnetic force, i.e.,

$$m_s \frac{q_\theta^2}{r_1} = e_s q_\theta B \tag{8.145}$$

where r_1 is the radius of the circle where the particle moves in the equilibrium condition. Eq. (8.145) may be written as

$$\omega_c = \frac{q_\theta}{r_1} = \frac{e_s B}{m_s} \tag{8.146}$$

where ω_c is known as the cyclotron frequency or Larmor frequency of the charged particle m_s. If q_z is zero, the particle moves on a circle of radius r_1; while if q_z is different from zero, the particle will move in a helical path with a radius r_1 of the helix. The radius r_1 is sometimes called the Larmor radius. If we assume q_θ to be the mean molecular velocity at a given temperature T, i.e.,

$$q_\theta = \left(\frac{8kT}{\pi m_s}\right)^{\frac{1}{2}} \tag{8.147}$$

then the Larmor radius of this particle m_s is

$$r_1 = \left(\frac{8kTm_s}{\pi}\right)^{1/2} \frac{1}{e_sB} \qquad (8.148)$$

(3) Constant electric field and constant magnetic field and E is perpendicular to B. In this case, it is convenient to define a new velocity of the particle v_s such that

$$v_s = q_s - \frac{E \times B}{B^2} = q_s - q_d \qquad (8.149)$$

where

$$q_d = \frac{E \times B}{B^2} = i_d \frac{E}{B} = \text{drift velocity} \qquad (8.150)$$

where i_d is the unit vector in the direction of $E \times B$.

The velocity v_s satisfies the following equation:

$$m_s \frac{dv_s}{dt} = e_s v_s \times B \qquad (8.151)$$

The velocity v_s is independent of the electric field E.

If the electric field E is not perpendicular to B, we may resolve the electric field into two components: one is perpendicular to B, i.e., $E_\perp$, and the other is parallel to B, i.e., $E_\parallel$. We use the component $E_\perp$ in Eq. (8.150) to calculate the drift velocity q_d and the component $E_\parallel$ gives an axial acceleration of the particle along the direction of B.

(4) If both the electric field E and the magnetic induction B are not constant, the motion of the particle is very complicated. However, if the variations of the electric and the magnetic fields are small, the motion of the particle will be qualitatively the same as we have described above. Some of the interesting problems of plasma may be brought out by studying the particle motion in a slowly varying field such as stability of the plasma by magnetic fields.

In Eq. (8.143), we assume that the particle is subjected to an electromagnetic force only. We may generalize the motion of the particle by considering other forces, such as gravitational force. In general, the gravitational force has similar effect as the electric field on a charged particle.

15. Some Dimensional Considerations of Plasma Dynamics[11]

In chapter VI section 15, we discussed various flow regimes of plasma dynamics. Such an analysis will be more clear if we apply the dimensional consideration to the Boltzmann equation of a plasma. In ordinary gasdynamics, we have only one mean free path which is associated with the collision process and which characterizes the rarefication of the gas. But in plasma dynamics, we have three characteristic lengths: the mean free path L_f, the Larmor radius L_L, and the Debye length L_D, which all have some influence on the rarefication of the plasma. Hence the classification of the flow regime in a plasma is much more complicated than that of ordinary gas dynamics discussed in section 1. We shall discuss the case of a plasma as follows:

A plasma may be considered as a mixture of N species, charged and neutral particles. For each species, we may define a molecular distribution function F_s. Let F_s be the molecular distribution function of the sth species. The Boltzmann equation for the sth species is then

$$\frac{\partial F_s}{\partial t} + q_{ms}^i \frac{\partial F_s}{\partial x^i} + \frac{e_s}{m_s} \left[E^i + (q_{ms} \times B)^i \right] \frac{\partial F_s}{\partial q_{ms}^i} = \frac{\delta F_s}{\delta t} \quad (8.152)$$

where q_{ms} is the molecular velocity vector of the sth species with component q_{ms}^i, e_s is the electric charge on a particle of the sth species and m_s is the mass of a particle of sth species.

The most difficult part in this analysis of the Boltzmann equation is the evaluation of the collision term. If we are interested in only the order of magnitude of various terms, we may introduce a relaxation time t_r such that[11]

$$\frac{\delta F_s}{\delta t} = \frac{F_{s_0} - F_s}{t_r} \quad (8.153)$$

where F_{s_0} is the Maxwellian distribution function toward which the distribution function F_s will relax. The relaxation time may be written roughly as

$$t_r = \frac{L_f}{U} \tag{8.154}$$

where L_f is the mean free path of the plasma and U is a typical velocity. Now we write

$$q_{ms} = q_s + q_{sc} \tag{8.155}$$

where q_s is the flow velocity of the sth species. Furthermore, we introduce the following nondimensional quantities:

$$t^* = \frac{Ut}{L}, \ x^{*i} = \frac{x^i}{L}, \ q_s^{*i} = \frac{q_s^i}{U}, \ q_{sc}^{*i} = \frac{q_{sc}^i}{U}, \ B^* = \frac{B}{B} \tag{8.156}$$

where L, U, and B are respectively the reference values of length, velocity, and magnetic induction.

Substituting Eqs. (8.153) and (8.156) into Eq. (8.152), we have the nondimensional form of the Boltzmann equation as follows:

$$\frac{\partial \ln F_s}{\partial t} + q_{ms}^{*i} \frac{\partial \ln F_s}{\partial x^{*i}} + \frac{e_s L}{m_s U^2} E_u^i \frac{\partial \ln F_s}{\partial q_{ms}^{*i}}$$
$$+ \frac{L}{L_{Ls}} (q_{sc}^* \times B^*)^i \frac{\partial \ln F_s}{\partial q_{ms}^{*i}} = \frac{L}{L_f} \left(\frac{F_{s0}}{F_s} - 1 \right) \tag{8.157}$$

where $E_u^i = E^i + (q_s \times B)^i$ is the electric field measured in a system moving with a velocity q_s

$$L_{Ls} = \frac{m_s U}{e_s B} = \text{Larmor radius of } s\text{th species} \tag{8.158}$$

For neutral particles, $e_s = 0$, Eq. (8.157) becomes

$$\frac{\partial \ln F_s}{\partial t} + q_{ms}^{*i} \frac{\partial \ln F_s}{\partial x^{*i}} = \frac{L}{L_f} \left(\frac{F_{s0}}{F_s} - 1 \right) \tag{8.159}$$

Eq. (8.159) is the Boltzmann equation for ordinary gasdynamics in which the main parameter is the Knudsen number (L_f/L). If the Knudsen number $K_f = L_f/L$ is very small, the gas may be considered as a continuum and the Navier-Stokes equations may be used to describe the flow field. If $K_f \ll 1$, we have

$$\left(\frac{F_{s_0}}{F_s} - i \right) \ll 1 \qquad (8.160)$$

because the terms on the left-hand side of Eq. (8.159) are of the order of unity. Eq. (8.160) shows that for very small Knudsen number, the deviation of the distribution function F_s from the Maxwellian distribution function F_{s_0} is small so that the perturbation technique (8.34) may be used to find F_s. This is, of course, also true for a single gas consisting of the same type of molecules as we discussed in sections 2 to 7.

If the Knudsen number is very large, the collision term may be neglected. When we have the free molecule flow, the distribution function F_s is again approximately the Maxwellian distribution function because the Maxwellian distribution function is a solution of Eq. (8.159) with the right-hand side term of Eq. (8.159) equal to zero.

For the finite Knudsen number, the collision term is rather complicated. We have the slip flow and the transition region which have been discussed before.

For charged particles, we should consider Eq. (8.157). In addition to the mean free path L_f or the ordinary Knudsen number K_f, we have a new important parameter, the Larmor radius L_{Ls} or the magnetic Knudsen number:

$$K_m = \frac{L_{Ls}}{L} \qquad (8.161)$$

Furthermore, we also have a term due to the electric field, the third term on the left-hand side of Eq. (8.157). For the time being, we shall assume that this term is of the same order of magnitude as the first two terms. This means that the force due to the electric field is of the same order of magnitude as the inertial force, i.e.,

$$e_s E_u = O\left(\frac{m_s U^2}{L} \right) \qquad (8.162)$$

where E_u is a typical magnitude of the electric field E_u. If we vary the value arbitrarily, we may have some new flow regimes corresponding to the electrogasdynamics discussed in chapter VI. We shall discuss here only the relative magnitude of the mean free path L_f and the Larmor radius L_{Ls} in a plasma according to the suggestion of Kantrowitz and Petschek:[11]

We may divide the plasma dynamics into five domains according to the values of the Knudsen number K_f and the magnetic Knudsen number K_m as follows:

(1) *S*-domain. In this domain, we have $K_f \ll 1$ and $K_m \gg 1$. The continuum theory may be used and the electrical conductivity of the plasma may be considered as a scalar quantity because $L_f / L_{Ls} \ll 1$.

(2) *T*-domain. In this domain, the electron Larmor radius L_{Le} becomes smaller than the mean free path and K_f is still much smaller than unity. In this domain, the Hall current becomes important and the electrical conductivity of the plasma should be considered as a tensor quantity. The magnetic term, i.e., the fourth term in Eq. (8.157) for the electron $s = e$ becomes the most important term. Since the ion Larmor radius L_{Li} is still larger than the mean free path L_f, the collision term is still dominant in the ion Boltzmann equation. The transport coefficient such as the coefficients of viscosity and heat conductivity will be approximately the same as those in the *S*-domain.

(3) *M*-domain. In this domain, the ion Larmor radius L_{Li} becomes smaller than the mean free path of L_f. Now the magnetic term becomes the most important term in the Boltzmann equation. The Larmor radius is more important than the mean free path in the determination of the transfer equation. It is expected that the equation of transfer for this case should be different from those cases where the mean free path is the dominant factor, i.e., the Navier-Stokes equations. However, no simple equations of transfer have been obtained for this case yet. This is a rather important domain because it is the region where the fusion reactor will operate and it is also a region where some astronomical phenomena occur.

(4) *EM*-domain. In this domain, $K_{mi} = L_{Li}/L \ll 1$ and $L_{Li} < L_f$. In this region, the ions will move as free particles subject

to the electromagnetic force. Both the electric and magnetic fields are important.

(5) *E*-domain. In this domain, $K_{me} = L_{Le}/L \ll 1$ and $L_{Le} < L_f$. Now the main force in the Boltzmann equation is the electric force. It is a region of very low density in which many glow discharges operate.

Besides the five domains, if the Debye length is larger than *L*, there are no interaction between gas particles. One should treat the problem by considering individual particles instead of treating it as a gas or a plasma as a whole as we discussed in section 14.

16. Some Experimental Results of Rarefied Gas

The free molecule flow has been first studied by Knudsen[12] in 1909. Knudsen studied essentially the effusion of gases through a small orifice with a large mean free path. It is important in connection with vacuum installation. Recently Liepmann[17] repeated and improved Knudsen's experiments on effusive flow. This new experiment of Liepmann is of great interest because it gives an example showing the transition from gasdynamic region to the free molecule region. Fig. 8.9 shows some of Liepmann's results in which Γ_k is the free molecule flow value of the reduced mass flow

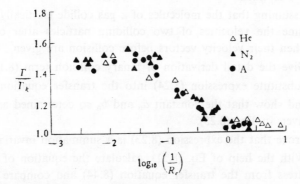

Figure 8.9   Reduced mass flow as a function of the Reynolds number (Figure 3 of reference [17] by H. W. Liepmann, courtesy of Pergamon Press and the author)

parameter.   As the Reynolds number increases, the ratio of reduced mass flow parameter Γ to the corresponding value for free molecule flow Γ_k increases, from the free molecule flow value of unity to a continuous flow value through a transition region.

Experimental results on simple bodies such as spheres, cylinders, flat plates, cones, have been obtained in various low density wind tunnels for the drag coefficients and Nusselt numbers and thermal recovery factors.[22]   The experimental results verify the general trend predicated by the simple theory.   However, the experimental results are usually scattered.

PROBLEMS

1.   Plot Mach number against the standard atmospheric altitude for an aircraft of a typical length of 5 meters for the boundarys of various flow regimes: free molecule flow, transition flow, slip flow, and continuum flow (use NACA Tech.   Note 1200, 1947 for the standard atmosphere).

2.   For a satellite flying at a speed of 10,000 m/sec, estimate the corresponding flow regimes of rarefied gasdynamics at the following altitudes: (i) 10 miles, (ii) 20 miles, (iii) 50 miles, and (iv) 80 miles.

3.   Assuming that the molecules of a gas collide elastically, determine the velocities of two colliding particles after collision when their velocity vectors before collision are given.

4.   Give the detail derivation of binary collision term (8.14).

5.   Substitute expression (8.24) into the transfer equation (8.44), and show that the constant a_0 and b_0 so determined are those given in Eq. (8.25).

6.   Prove that the expression (8.23) is a summation invariant.

7.   With the help of Eq. (8.51), calculate the equation of viscous stress from the transfer equation (8.44) and compare it with Eq. (8.53).

8.   With the help of Eq. (8.51), calculate the equation of heat flux from the transfer equation (8.44) and compare it with Eq.

(8.54).

9. Derive the expression of the coefficient of viscosity (8.57) from the equation of viscous stresses obtained in problem 7.

10. Derive the expression of the coefficient of heat conductivity from the equation of heat flux obtained in problem 8.

11. With the help of Eq. (8.51), derive the diffusion equation (8.72) from the transport equation (8.44), and then derive the coefficient of diffusion for a binary mixture.

12. Give the detailed derivation of the Fokker-Planck collision expression (8.75).

13. From the transport equation (8.44) and the expression (8.51), derive the generalized Ohm's law (8.83).

14. Give the detailed derivations of the formulas (8.99) and (8.100).

15. Calculate the temperature distribution of a thermally insulated flat plate in a free molecule flow when the thermal radiative flux is not negligible.

16. Find the drag coefficient in a free molecule flow for a right cone with its axis normal to the free stream flow for (i) diffuse reflection and (ii) specular reflection.

17. Find the drag of a sphere in a free molecule flow for (i) diffuse reflection and (ii) specular reflection.

18. Write down the Burnett equations of rarefied gasdynamics and discuss their properties for the case of a plane Couette flow.

19. Calculate the drag of a flat plate in a nearly free molecule flow where the first-order collision term should be included.

20. Calculate the plane Couette flow by using the Navier-Stokes equations and the slip boundary conditions.

21. Calculate the drag coefficient of a sphere by using the Navier-Stokes equations of an incompressible fluid and the slip boundary conditions.

22. Give the detailed derivation of the formula of skin friction (8.119) and extend it to the second-order term of (L_f/L).

23. Study the Rayleigh problem in which an infinite flat plate is moving impulsively to a uniform speed U in a rarefied gas by the Grad thirteen moment equations.

24. Discuss the solution of the Boltzmann equation of a plasma

without the collision term for the case of two uniform streams of electrons moving in opposite directions.

25. Discuss the motion of a charged particle under the influence of a homogeneous stationary magnetic field where are present uniform and stationary gravitational and electric forces.

26. Calculate the value of Larmor radius of electron as well as that of a hydrogen ion under a magnetic induction of 10,000.00 gausess. Compare these Larmor radii with the mean free path of air in the standard atmosphere (problem 1) at the altitudes (i) sea level, (ii) 30 miles, and (iii) 100 miles.

27. Calculate the Debye lengths at altitudes (i) sea level, (ii) 30 miles, and (iii) 100 miles in standard atmosphere if we assume the air to be fully ionized plasma with the density and temperatures of the values at these standard atmospheres.

REFERENCES

[1]` Abarbanel, S. "Radiative heat transfer in free molecule flow", *Jour. Aero. Sci.*, Vol. 28, No. 4, 1951, pp. 299—308.

[2]   Bhatnagar, P. L., Gross, E. P., and Krook, M. "A model of collision processes in gases. I. Small amplitude processes in charged and neutral one-component systems", *Phys. Rev.*, Vol. 94, No. 3, 1954, pp. 511—525.

[3]   Burgers, J. M. "The bridge between particle mechanics and continuum mechanics", in Proc. of Plasma Dynamics, F. Clauser (ed.). Addison Wesley Publishing Co., 1960, pp. 119—186.

[4]   Burgers, J. M. "Aspects of particle motion and gas flow in magnetic fields". Notes of Grad. School of Aerospace Eng., Cornell Univ., 1965.

[5]   Chandrasekhar, S. Plasma Physics. University of Chicago Press, 1960.

[6]   Chapman, S. and Cowling, T. G. The Mathematical Theory of Non-Uniform Gases. Cambridge University Press, 1939.

[7]   Grad, H. "On the kinetic theory of rarefied gases". *Comm. on Pure and Applied Math.*, Vol. 2, No. 4, 1949, pp. 331—407.

[8]   Grad, H. "Principles of the kinetic theory of gases". Handbuch der Physik, Vol. 12, 1958.

[9]   Hirschfelder, J. O., Curtis, C. F., and Bird, R. B. Molecular Theory of Gases and Liquid. John Wiley, 1954.

[10] Jeans, J. H. The Dynamical Theory of Gases. Dover Pub., 1954.

[11] Kantrowitz, A. R. and Petschek, H. E. "An introductory discussion of magnetohydrodynamics", in A Symposium on Magnetohydrodynamics. R. K. M. Landshoff (ed.). Stanford Univ. Press, 1957, pp. 3—15.

[12] Knudsen, M. "Die Gesetze der Molekularstroemung und der inneren Reibungstroemung der Gase durch Roehren". *A. Physik*, Vol. 28, 1909, pp. 75—130.

[13] Laurmann, J. A. "Slip flow over a short flat plate", in Rarefied Gas Dynamics. Pergamon Press, 1960, pp. 293—316.

[14] Lees, L. "A kinetic theory of description of rarefied gas flows". Mem. No. 51, Hypersonic Research Proj., Gug. Aero. Lab., Calif. Inst. of Tech., 1959.

[15] Lees, L. and Liu, C. Y. "Kinetic theory description of plane compressible Couette flow". Mem. No. 58, Hypersonic Res. Proj., Gug. Aero. Lab., Calif. Inst. of Tech., 1960.

[16] Li, T. Y. "Gas kinetic equations and their applications to non-equilibrium flows with diffusion and chemical reaction", in Dynamics of Fluids and Plasmas. S. I. Pai (ed.). Academic Press, 1966, pp. 155—178.

[17] Liepmann, H. W. "A study of effusive flow", in Aeronautics and Astronautics. Pergamon Press, 1960, pp. 153—160.

[18] Pai, S.-I. Magnetogasdynamics and Plasma Dynamics. Springer-Verlag, 1962.

[19] Patterson, G. N. "Mechanics of rarefied gases and plasmas". UTIAS Rev. No. 18, Inst. for Aerospace Studies, Univ. of Toronto, 1964.

[20] Probstein, R. F. "Continuum theory and rarefied hypersonic aerodynamics", in Rarefied Gas Dynamics. Pergamon Press, 1960, pp. 416—431.

[21] Sandri, G. "The physical foundations of modern kinetic theory", in Dynamics of Fluids & Plasmas. S. I. Pai (ed.). Academic Press, 1966, pp. 341—398.

[22] Schaaf, S. A. and Chambre, P. L. "Flow of rarefied gases", Sec. H of Fundamentals of Gas Dynamics. E. W. Emmons (ed.). Vol. III of High Speed Aerodynamics and Jet Propulsion. Princeton Univ. Press, 1958.

[23] Spitzer, L. Jr. Physics of Fully Ionized Gases. Interscience Publ., 1956.

[24] Wu, T. Y. "Kinetic theory of gases and plasmas". Lecture notes of Polytechnic Institute of Brooklyn and Academic Sinica, 1964.

[25] Yang, H. T. and Lees, L. "Rayleigh's problem at low Reynolds number according to the kinetic theory of gases", in Rarefied Gas Dynamics. Pergamon Press, 1960, pp. 201—238.

CHAPTER

IX

SPECIAL TOPICS OF
MODERN FLUID MECHANICS

1. Introduction

In our discussions of the fluid flow problems in the last eight chapters, three basic assumptions have been implicitly used: the first one is that the properties of the fluid are different entirely from those of a solid; the second one is that the fluid considered is more or less a homogeneous medium whose local properties can be simply described; and the third one is that the fluid flow behaves according to the principles of classical physics. There are many flow problems of special fluids in which one or more of these assumptions are not valid. Hence special attention must be made to treat these flow problems. Peculiar phenomena may occur in these cases. In this chapter, we discuss briefly the essential points of these special topics.

The fluids which we usually consider in classical fluid mechanics, as well as those in major portion of the last eight chapters, are the so-called Newtonian fluids which satisfy the Navier-Stokes relations (3.52). However, there are many fluids which do not satisfy these simple relations (3.52). These fluids may be called non-Newtonian fluids. The rarefied gas discussed in chapter VIII may be considered as a non-Newtonian fluid. There are many other non-Newtonian fluids. Some of the fluids behave partly as a solid and partly as a fluid. A special science known as Rheology treats the flow of non-Newtonian fluids. We shall discuss some basic facts of non-Newtonian fluid flow in section 2.

In all the previous discussions, we consider the fluid more or less as a homogeneous medium. The fluid is either in a liquid state or in a gaseous state. We did not discuss flow problems of a fluid which is a mixture of materials in more than one state. There are many practical problems of flow field in which the medium exists in more than one state, particularly the two-phase flow problems. Hence we have two phase flows which may be a mixture of solid and liquid, a mixture of solid and gas, or a mixture of liquid and gas. Let us briefly consider the case of a mixture of solid and gas. In the exhaust plume of a solid fuel rocket, there are many solid particles in the jet plume from the rocket. Since the properties of these solid particles differ greatly from those of the gas, the resultant flow field with solid particles differs considerably from that without solid particles. The behavior of the solid particles in the flow field and their interaction with the gas flow need special treatment. For the mixture of liquid and gas, the evaporation and condensation of the gas phase to the liquid phase need special treatment. Hence there are many interesting and difficult problems in these two-phase flow problems which will be discussed in section 3.

In our treatment of the flow of a plasma, we consider only the simple case that the temperatures of all species in a plasma are the same. Since the electrons are much lighter than those ions or neutral particles, its temperature may differ considerably from those of heavy particles in the flow field. Hence when the degree of ionization is high, it might not be a good approximation to use one temperature for all species at each point in space for a plasma. Furthermore, in the mixture of neutral gases, the diffusion velocities are usually small and we may use the simple expression of diffusion coefficient to express the diffusion velocities in the treatment of a mixture of gases, such as we did in chapter V. For a plasma, the diffusion velocity of electrons must be very large. It might not be a good approximation, if we use some simple expression to deal with the diffusion velocity. We have mentioned briefly that the generalized Ohm's law (6.11) is only a first approximation. In order to improve the treatment of plasma flow, it is better to use the multifluid theory. We may derive the transfer equations for each of the species and solve the transfer equations of all species simultaneously. In section 4, we shall discuss the fundamental equations of multifluid theory of a plasma and give some simple results of the multifluid theory. We shall show that the results of the multifluid theory give us much more information about the flow field which cannot be obtained in the classical treatment of single fluid theory. We shall also discuss the relations between the classical fluid theory and the multifluid theory.

At very low temperature, near to the absolute zero degree of temperature, the quantum effects begin to be of importance in the properties of fluids. Such a subject may be called quantum fluid mechanics. The fluid under such a condition is the liquid helium. The most important property of such a fluid is its superfluidity, i.e., its viscosity and electrical resistivity tend to be zero. We call such fluids as superfluid. There are many peculiar properties of such fluids. We shall discuss them in section 5.

In the previous chapters, we consider that the velocity of the fluid, both the macroscopic motion and the microscopic motion, is much smaller than the velocity of light; the relativistic effects are negligibly small. When the velocity of the fluid, whether it is the microscopic motion of the fluid particles or the macroscopic motion, is not negligibly small in comparison with the velocity of light, we have to consider the relativistic effects. The study of the fluid mechanics with relativistic effects may be called relativistic fluid mechanics which will be discussed in section 6. In radiation gasdynamics, the thermal radiation may be considered as a stream of photons which move at the speed of light. Hence we should use the relativistic Boltzmann equation to study the motion of photons, which will also be discussed in section 6.

What we have studied are the interactions of various physical phenomena. Recently, with the expansion of the biological sciences, there is a trend such that many biological problems are studied quantitatively by the help of the principles of physics and chemistry. In fact, there has been a recent announcement from Stanford University that life has been produced in the laboratory. The study of the interrelation between biology and physics and chemistry is at the frontier of science nowadays. It is natural that the science of biomechanics or biofluid mechanics has been extensively studied recently. In section 7, we shall give a few examples of the problems of biofluid mechanics. This field is too broad to give a comprehensive survey, and some of the basic concepts are still in the developmental stage. For instance, the distinction of life and death requires notions which do not fit into the terminology of physics. The inclusion of this section of biofluid mechanics is simply to inform the readers that such a new field of fluid mechanics is included in the research of current fluid mechanics, even though many of these problems have been studied for a long time (see section 7). For instance, Poiseuille, who was a physician and who was interested in the blood flow, studied the flow in pipes, which study is well known in fluid dynamics as the famous Poiseuille flow. The modern research in blood flow is much more advanced than the results of the classical Poiseuille flow, which will be discussed in section 7.

2. Rheology. The Non-Newtonian Fluids

Fluids may be divided into two classes: the Newtonian fluids which satisfy the relation (3.52) in which the coefficient of viscosity is a function of the state of the fluid, and the non-Newtonian fluids. Ordinary liquids and gases such as water and air under normal sea level conditions are Newtonian fluids. However, rarefied gas is an example of a non-Newtonian fluid. Since we have already discussed rarefied gasdynamics in chapter VIII, we shall not discuss it here. Many macromolecular materials such as colloids, plastics, high polymers, and so on, are non-Newtonian fluids. Important technological applications of these non-Newtonian fluids have been found in recent years. Since the kinetic theory of these non-Newtonian fluids has not been developed, we have to study them by macroscopic theory by formulating some stress-strain relations. Thus we may divide these non-Newtonian fluids into the following three classes:

(1) Visco-inelastic fluids. These fluids known as Stokesian fluids which are characterized by the fact that they are isotropic and homogeneous when at rest, and the resultant stress depends only on the rate of change of strain when in motion. The Newtonian fluid is a special case of this class. Fig. 1.9 of chapter I shows several cases of this class of fluids. This kind of fluid may be subdivided into the following types:

(a) Dilatant fluids. For a two-dimensional shear motion, the shearing stress of this type satisfies the relation:

$$\tau_{xy} = \mu^{1/n} \left(\frac{\partial u}{\partial y} \right)^{1/n} \qquad (9.1)$$

where n is less than unity. Eq. (9.1) is known as Ostwald's law. If $n = 1$, we have the Newtonian fluid. The coefficient of viscosity is μ.

(b) Pseudoplastic. This fluid satisfies also the relation (9.1) but the value of n is larger than unity.

(c) Bingham plastics. This fluid behaves like the Newtonian fluid when the flow starts but it is able to sustain a certain finite stress, called the yield-stress before the flow begins.

(d) Reiner-Rivlin fluid. This fluid is characterized by a non-linear relation between viscous stress τ_v and the rate of strain A, and there are more than one coefficient of viscosity in the relation between the stress and the rate of strain. For instance, the following relation has been used by Reiner and Rivlin for certain fluids:[36]

$$\tau_v = -p \, I + a_1 \, A + a_3 \, A^2 \qquad (9.2)$$

where I is a unit tensor, a_1 may be regarded as a coefficient of viscosity and a_3 is called the coefficient of cross viscosity. Because of the existence of the cross viscosity, some new phenomena, known as Weissenberg's effect[45] and Merrington's effect,[20] occur. The Weissenberg's effect shows the climbing of the non-Newtonian fluid along the inner fixed cylinder against the centrifugal force when the fluid is sheared between two coaxial cylinders by rotating the outer cylinder. The Merrington's effect shows the swelling of the stream of the non-Newtonian fluid at the exit of the circular pipe through which it is flowing.

(2) Time dependent fluids. Time dependent fluids, when subjected to a steady rate of shear under isothermal condition, show either increase or decrease in the viscosity as time passes. The fluid is called Rheopectic fluid of which the viscosity increases with time; the fluid is called Thixotropic fluid of which the viscosity decreases with time. This behavior of a time dependent fluid is due to the whole sequence of structural changes that takes place

following an impressed disturbance. Hence the mechanism of breaking and of reformation of the molecular chains determines the relation of the stress and strain. At the present time, our knowledge in the mechanism of this fluid property is very limited; no definite constitutive equation has been established yet.

(3) Visco-elastic fluids. These fluids behave partly like ordinary fluid and partly like elastic solid. Hence the stresses in these fluids depend on both the strain and the rate of strain. In other words, a certain amount of energy of the flow is stored up in the material as strain energy in addition to viscous dissipation. These materials strains are determined by the stress history of the fluid and cannot be specified kinematically in terms of the large overall movements of the fluid. During the flow, the natural state of the fluid changes constantly and tries to attain the instantaneous state of the deformed state, but it does not succeed completely. This lag is a measure of the elasticity or the so-called memory of the material. We may introduce some relaxation time to express the effect of elasticity. It is similar to the relaxation phenomena of the internal degree of freedom of gas particles as we discussed in chapter V. Another approach is due to Rivlin and Ericksen who assume that the material is to be isotropic and homogeneous when at rest and the stress components at any instant can be expressed as a polynomial in tensors formed by various order material derivatives of velocity. The coefficients of various terms have to be functions of invariants of tensors in addition to the material properties in order to preserve their forms under translation and rotation. It is clear that in this approach we assume the end result, namely the constitutive equation, and then work backwards to trace the effect of the previous history of the fluid. This approach has an advantage in that the theory of visco-inelastic fluids can be deduced as a particular case of the general theory.[33]

The fundamental equations of the flow of a non-Newtonian fluid are still the same as those of the flow of a Newtonian fluid, discussed in chapter III, and the only difference is the constitutive equation, i.e., the stress and strain relation (3.52) is replaced by the proper relation such as Eq. (9.2). Once the constitutive equation is given, the fundamental equations, equations of motion, equation of continuity, etc., can be solved for given boundary conditions. In fact, we may reexamine all the flow problems in Newtonian fluids, i.e., ordinary fluid dynamical problems, for the non-Newtonian fluid cases. Because of the complicated constitutive equations, only flows with simple configurations have been investigated. In references [9], [21], [30], [33], and [36], many problems of non-Newtonian fluid flows have been discussed. As a simple example, we consider the power law fluid in the fully developed flow between two parallel plates (similar to the problem of chapter VI section 7): For an incompressible fluid, the fundamental equation in this problem is

$$\frac{d\tau_{xy}}{dy} = \frac{dp}{dx} = \text{constant} \tag{9.3}$$

For a power law fluid, we write

$$\tau_{xy} = K\left(\frac{du}{dy}\right)^n \tag{9.4}$$

where K is a constant and n is also a constant for a given power law fluid. Substituting Eq. (9.4) into Eq. (9.3), we have the velocity distribution:

$$u - u_0 = \left\{ \frac{n}{(n+1)A} (Ay+C)^{\frac{n+1}{n}} - (Ay_0+C)^{\frac{n+1}{n}} \right\} \tag{9.5}$$

where $u = u_0$ when $y = y_0$, $A = (1/K) \, dp/dx$ and C is a constant which is determined by the boundary condition. It is interesting to know that for Plane Poiseuille flow, the velocity distribution depends on the power factor n but for plane Couette flow, the velocity distribution is always linear for all n's.

3. Two-Phase Flows[46]

There are many practical problems of flow in which a mixture of substances of more than one state of matter exists. In the problems of cavitation and atomization, we consider the mixture of gas and liquid. In the case of sandstorm and transport of gravel and suspended solids in river, we consider the mixture of solid and liquid. Finally in the propulsion system of solid fuel and in the problem of interior ballistics, etc., we have to consider the mixture of solid and gas. If the temperature of the mixture is very high, we shall consider the mixture of solid and plasma. Since the properties of solid, liquid, gas, and plasma are not the same, we have to treat the above various kinds of two-phase flow separately. We are going to discuss some essential points of these two-phase flows as follows:

(1) Liquid-gas flows. When we study the mixture of liquid and gas, the resultant flow field depends on the relative amount of liquid and gas. The phenomenon of a small amount of gas in a large amount of liquid is different from the case in which a small amount of liquid is in a large amount of gas.

(a) Cavitation. Cavitation represents the phenomenon of a small amount of gas in a large amount of liquid. In a liquid whose stagnation or undisturbed pressure p_0 is larger than its vapor pressure p_v, the motion of this liquid may reduce the pressure of the liquid p to a value less than its vapor pressure p_v. Vapor is then suddenly formed and the pressure of the liquid rises to p_v. Because of the formation of the vapor in the liquid, the streamlines of the flow field will be different from the case without such cavitation. When the pressure of the liquid rises above the vapor pressure, the cavitation may collapse suddenly with a large sound. If there is no gas in the liquid, a cavity may not be formed even if the liquid pressure is below the vapor pressure. However, if there are some gases in the liquid cavitation begins when the vapor pressure is reached. Cavitation frequently begins when the flow of the liquid separates from the wall. Such phenomenon can easily be observed in the divergent channel where the flow separates.[10]

Cavitation is very important in the high speed flow of water such as high speed water turbines, ship propellers, hydrofoils, etc.[10] In these cases one should try to avoid the cavitation. The nondimensional parameter which characterizes the cavitation flow is the cavitation parameter R_{ca}:

$$R_{c_a} = \frac{p_0 - p_v}{\frac{1}{2}\rho U^2} \qquad (9.6)$$

The lift and drag coefficient of high speed hydrofoils have been correlated with this cavitation parameter.

Another similar phenomenon to ordinary cavitation is the boiling effect of superspeed hydrodynamics,[42] which may become important in the near future. We consider a very slender body or a flat plate moving at a very high speed in the ocean. If the speed of the body is high enough, the temperature on the surface of the body due to skin friction will be so high that the water begins to boil at the surface of the body. If the speed of the body is very high, we expect that there would be a water vapor boundary layer surrounding this thin body which would affect significantly the skin friction and heat transfer of the body. The critical speed above which the boiling phenomenon on the surface occurs depends on the depth of operation in the ocean, which determines the water pressure. At the surface of the ocean, this critical speed is about 1700 knots (or 3120 km/hr) while at a depth of 2260 meters in the ocean, the critical speed will be a little larger than the sound speed of sea water which is 3000 knots (or 5520 km/hr). In the future, if the speed of seagoing craft reaches supersonic speeds, we should consider such a vaporization phenomenon in determining the motion of such a craft. It should be noticed that this superspeed hydrodynamics and the related shock wave phenomena are also important in underwater explosions.[6]

(b) Condensation of water vapor. The other extreme case of liquid-gas flow is a small amount of liquid in a large amount of gas. The most common example of this case is the rain, i.e., water drops in the air. If the size of the water drop is very small, the shape of the drop is practically spherical. We may consider the water drops as small rigid spheres in the air. The Stokes formula of very slow motion may be used to determine the drag force of these spheres. For large drops, the drag force will be proportional to the square of the velocity of the drops. Because of the pressure force on the drop of large size, it will break into small drops.

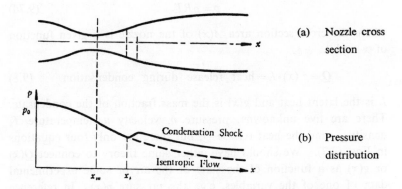

(a) Nozzle cross section

(b) Pressure distribution

Figure 9.1 Sketch of the condensation process of water in a supersonic nozzle

Another interesting phenomenon of the present problem is the condensation of water vapor or the evaporation of water drop in the gas flow in which heat addition or absorption would affect the flow field. Let us consider the steady flow of moist air in a supersonic nozzle of different pressure gradient (Fig. 9.1a). If there is no moisture in the nozzle, the one-dimensional flow solution of ordinary gasdynamics is well known. If the nozzle is so designed, we have isentropic flow in the nozzle. The pressure distribution along the nozzle is shown in the dotted curve of Fig. 9.1b. We assume that the amount of water vapor is small, it will practically not affect the pressure distribution if the water vapor does not condense. Our problems are: (i) when the condensation process of water vapor starts and (ii) what is the effect of the condensation of the water vapor on the flow. We may use the quasi-one-dimensional analysis to this problem, and the fundamental equations are:

$$\frac{dp}{\rho} + \frac{dA}{A} + \frac{du}{u} = 0 \tag{9.7a}$$

$$dp + \rho u \, du = 0 \tag{9.7b}$$

$$c_p \, dT - \frac{1}{\rho} \, dp - dQ = 0 \tag{9.7c}$$

$$p = \rho RT \qquad (9.7d)$$

where the cross section area $A(x)$ of the nozzle is a given function of x and

$$Q = g(x) \cdot L = \text{heat release during condensation} \qquad (9.8)$$

L is the latent heat and $g(x)$ is the mass fraction of the condensate. There are five unknowns: pressure p, velocity u, temperature T, density ρ, and the heat release Q, but there are only four equations in Eqs. (9.7). We should have either some theory to connect $Q(x)$ or $g(x)$ as a function of the state of the air or some experimental data of one of the variables, e.g., the pressure $p(x)$. In reference [44], the nucleation theory of condensation has been proposed. The theoretical results have been checked with experimental results by measuring the pressure distribution $p(x)$ along the nozzle. If $p(x)$ is known, we may numerically integrate Eq. (9.7). The main effect on the flow field due to condensation is the well known condensation shock which was first shown by Prandtl in 1935.[29] As the water vapor condenses, the pressure of the air stream increases above the isentropic value as shown in Fig. 9.1b. The deviation of pressure from isentropic process starts at $x = x_s$ where the condensation of water vapor begins.

(c) Atomization of liquid. When a jet of liquid issues into a medium of gas, the slender jet will break up into drops due to the instability of the jet. The instability of the jet depends on the critical Reynolds number which is usually very small for a free jet and is of the order of 10. For a light liquid such as water, the jet will easily break up into drops. For a very viscous liquid such as syrup, the jet may not break up for a long distance from the exit of a nozzle.

The break-up of a jet is important in many practical problems such as in the carburator of an internal combustion engine. It is desirable to break up the fuel into fine particles and to mix it thoroughly with the air. The atomization of the fuel is improved if there is a high speed gas surrounding the liquid.

(d) Ablation. One of the most important problems during the reentry of a space vehicle is to protect the surface of the space vehicle from overheating. One of the effective methods of heat protection for a blunt body is ablation which is a process of absorbing heat energy by removal of surface material, either by melting (possibly accompanied by vaporization of the molten material), or by sublimation. Let us consider a blunt nose body of revolution in a hypersonic flow (Fig. 9.2). There is a bow shock wave in front of the body. Behind the bow shock, the gas in the shock layer is very hot and is partly or fully dissociated. On the surface of the body, we have the boundary layer of the gas. In ablation, the surface of the body is allowed to melt or vaporize. Hence there will be a molten liquid layer between the gas boundary layer and the solid body. The melting rate, which is the rate at which the surface of the body receded, depends on the external conditions and the boundary layer flow of the liquid layer. The problem is complicated by the fact that the interface temperature between the gaseous boundary layer and the liquid layer is not known *a priori*. We have to match the solutions

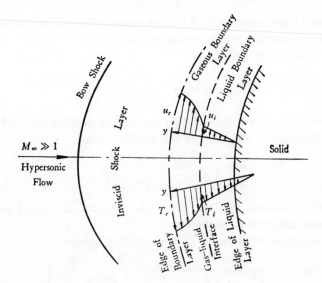

Figure 9.2 Two-phase boundary layer over melting blunt body in a hypersonic flow

of these two boundary layers. Hence our problem consists of three parts: the first is the gaseous boundary layer problem; the second is the liquid boundary layer problem; and the third is the interaction between the gas phase with the liquid phase.

In the gaseous phase laminar boundary layer, we assume that the gas is a mixture consisting of air atoms, air molecules, and vaporizing molecules. It is assumed that the flow is quasisteady and vaporizing material will not affect the thermodynamic properties of the air appreciably and has the same properties as the air molecules. Hence the gaseous boundary layer may be treated by the method of dissociated gas boundary layer of chapter V section 15. The unknowns are velocity components u and v, the temperature T, density ρ, mass concentration of atom c_A and mass concentration of vaporizing material c_k. The fundamental equations for these unknowns are:

Equation of state

$$p = \rho RT \qquad (9.9a)$$

where the gas constant R of the gas mixture depends on the concentration of various species in the mixture.

Equation of continuity

$$\frac{\partial}{\partial x}(\rho u r_0) + \frac{\partial}{\partial y}(\rho v r_0) = 0 \qquad (9.9b)$$

where $r_0(x)$ is the radius of the cross section of the blunt body.

Equation of diffusion of the atomic species

$$\rho u \frac{\partial c_A}{\partial x} + \rho v \frac{\partial c_A}{\partial y} = \frac{\partial}{\partial y}\left(\rho D_{Aj} \frac{\partial c_A}{\partial y} \right) \qquad (9.9c)$$

where we assume that the gas phase reactions are frozen and the thermal diffusion is negligible.

Equation of diffusion of the vaporing species:

$$\rho u \frac{\partial c_k}{\partial x} + \rho v \frac{\partial c_k}{\partial y} = \frac{\partial}{\partial y}\left[\rho D_{Ak} \frac{\partial c_k}{\partial y} \right] \qquad (9.9d)$$

Equation of motion:

$$\rho u \frac{\partial u}{\partial x} + \rho v \frac{\partial u}{\partial y} = -\frac{\partial p}{\partial x} + \frac{\partial}{\partial y}\left(\mu \frac{\partial u}{\partial y}\right) \tag{9.9e}$$

Equation of energy of the mixture:

$$c_p \, \rho \left(u \frac{\partial T}{\partial x} + v \frac{\partial T}{\partial y} \right) = u \frac{\partial p}{\partial x} + \mu \left(\frac{\partial u}{\partial y} \right)^2 + \frac{\partial}{\partial y}\left(\kappa \frac{\partial T}{\partial y} \right)$$

$$+ \sum_n \left(c_{pn} D_{nj} \frac{\partial c_A}{\partial y} \right) \frac{\partial T}{\partial y} \tag{9.9f}$$

We are going to study the ablation near the stagnation point based on the above equations and the following boundary conditions:

$$\left. \begin{array}{l} y=0: \ u=u_i, \ T=T_i, \ v=v_i, \ c_A=c_{Ai}, c_k=c_{ki} \\ y\to\infty: \ u\to u_e = xu_{e0}, \ T\to T_e, \ c_A\to c_{Ae}, c_k\to 0 \end{array} \right\} \tag{9.10}$$

It should be noticed that the values at the interface, i.e., T_i, u_i, etc., are not known *a priori*. The values at the outer boundary of the gaseous boundary layer are given by the solution of the inviscid shock layer.

In the liquid phase laminar boundary layer, the fluid is an incompressible liquid and we assume that the surface is melting at a steady rate such that the derivative with respect to time is much smaller than the spatial derivatives. The aerodynamical pressure gradient and shear stress cause the molten material to flow. The fundamental equations are then:

Equation of continuity:

$$\frac{\partial}{\partial x}(u r_0) + \frac{\partial}{\partial y}(v r_0) = 0 \tag{9.11a}$$

Equation of motion:

$$\rho u \frac{\partial u}{\partial x} + \rho v \frac{\partial u}{\partial y} = -\frac{\partial p}{\partial x} + \frac{\partial}{\partial y}\left(\mu \frac{\partial u}{\partial y}\right) \tag{9.11b}$$

Equation of energy:

$$\rho c_p \left(u \frac{\partial T}{\partial x} + v \frac{\partial T}{\partial y} \right) = u \frac{\partial p}{\partial x} + \frac{\partial}{\partial y}\left(\kappa \frac{\partial T}{\partial y} \right) + \mu \left(\frac{\partial u}{\partial y} \right)^2$$

$$\tag{9.11c}$$

We shall take the coordinate system such that the y-coordinate is zero at the interface. Hence the boundary conditions for the liquid layer at $y = 0$ are the same as those given in Eq. (9.10). If we take y as positive toward the solid body, the boundary condition at infinity are

$$y \to \infty : \quad u \to 0, \quad T \to T_b \tag{9.12}$$

where T_b is the temperature of the solid body.

The next step is to match the solutions of the liquid phase and gas phase. A unique solution is obtained when the interface mass transfer $\dot{m}_i$, the interface velocity u_i, the interface shear stress τ_i, the interface temperature T_i, the interface energy transfer Q_i, and the mass friction of vaporizing species c_{ki} are the same for both the gas and the liquid phase.

It has been found that u_i / u_e is of the order of 10^{-4}. Hence as far as the velocity distribution in the gaseous boundary layer is concerned, we may consider $u = 0$ at $y = 0$. The solution to the liquid phase equation at the proper match point will yield the value u_i / u_e a posteriori. The quantities to be matched are as follows:

In the gaseous boundary layer, we have

$$\dot{m}_i = \dot{m}_i \, (c_{ki}, \; T_i) \tag{9.13a}$$

$$\tau_i = \tau_i (\dot{m}_i, \; T_i) \tag{9.13b}$$

$$Q_i = Q_i \, (\dot{m}_i, \; T_i) \tag{9.13c}$$

Eq. (9.13a) means that the mass transfer is a function of the gaseous mass concentration of evaporated species at the interface and the interface temperature. This relation may be obtained by solving the gaseous boundary layer equations for various values of c_{ki} and T_i. Similarly, we may obtain the relations (9.13b) and (9.13c).

The solution to the liquid phase boundary layer equations gives

$$Q_i = Q_i (\tau_i, \; \dot{m}_i, \; T_i) \tag{9.13d}$$

Finally from the assumed dependence of vapor pressure on temperature, we obtain the relation:

$$c_{ki} = c_{ki}(T_i) \tag{9.13e}$$

From the five relations of Eqs. (9.13), we may solve for the five unknowns: $\dot{m}_i$, τ_i, Q_i, T_i, and c_{ki}. For a given condition of the environment, we may obtain the solution of this two-phase boundary layer problem. Reference [16] gives an example of this problem.

(2) Solid-liquid flows. The solid-liquid flow may also be divided into two classes: one is the motion of solid particles in a liquid which is the problem of sedimentation and the other is the motion of liquid in a solid which is the flow problem through porous media. We are going to discuss some basic concepts of these problems as follows:

(a) Sedimentation. The motion of sand, gravel, stone, and other solid particles in the river belongs to this case of two-phase flow problems. It is very important for hydraulic engineering.

For the motion of any simple stone in the river, we may calculate the flow field by the equation of motion of water over a given body. Actually, it is very difficult to carry out this calculation because the forces on the solid may be mainly due to turbulent stresses, and our knowledge of the turbulent stresses is still meager. The number of the solid particles in the river is very large. Even if we know the exact form of the forces on a particle, it is still difficult to perform the total calculations. Some statistical mechanics should be used. We shall discuss this point further in the next section when we study the solid-gas flow. Hence in hydraulic engineering, empirical formulas are usually used (see Reference [2]). Of course, these formulas hold true only for special cases. For every particular case, a special formula should be used.

For sand in a river, because the weight of the sand is about the same as that of water, the diffusion velocity of the sand in the water is small. For a first approximation, we may assume that the velocity of the sand in the stream is about the same as that of water with a small diffusion velocity which may be expressed in terms of a turbulent diffusion coefficient. Hence, the turbulent diffusion theory may be used to analyze the concentration of the sand in the river. The theory of sand in a laminar flow of water is a special case for the motion of small amount of solid particles in a large amount of gas which, will be discussed later.

(b) Flow through porous media. This represents another extreme case in which the amount of solid is large but the fluid is small. Strictly speaking, we might not call it two phase flow because in this case, the solid will not move. For instance, if we let the water flow from the bottom of a sand bed, the sand will not be disturbed if the flow of the water is small. This phenomenon may be considered as a special case of flow through a porous medium. A porous medium is simply a solid with holes in it. However, in many practical problems, the shapes, sizes, and interconnections of these holes are not known. We have to use some average properties of such a medium such as the averaging resistance to the flow of fluid to characterize the flow. Hence the treatment of flow through a porous medium is essentially empirical. The most well known empirical law for flow through porous media is Darcy's law:

$$q_f = -\frac{k_p A}{\mu}\left(\frac{dp}{dL} + \rho g \sin Q \right) \qquad (9.14)$$

where q_f is the flow rate at a given point along the porous medium at a given time. The direction of the flow is in the direction of L and is at an angle Q with the horizontal plane and Q is positive when the flow is upward. A is the cross-sectional area of the porous medium in the direction perpendicular to L. The coefficient of viscosity is μ and k_p is the coefficient of permeability of the medium. The most important thing in the study of the flow through a porous medium is to determine the value of permeability k_p.

In derivation of Eq. (9.14), it is assumed that the flow is laminar and in one direction. For a three-dimensional case, if we assume that the gravitational acceleration g is in the z-direction, the flow rate per unit area is then:

$$q_f = -\frac{k_p}{\mu}(\nabla p + i_z \rho g) \qquad (9.15)$$

where i_z is the unit vector in the z-direction. The velocity vector q_f has the component u^i. The equation of continuity (3.32) becomes

$$\frac{\partial\left(\frac{k_p}{\mu}\frac{\partial p}{\partial x}\right)}{\partial x}+\frac{\partial\left(\frac{k_p}{\mu}\frac{\partial p}{\partial y}\right)}{\partial y}+\frac{\partial\left[\frac{k_p}{\mu}\left(\frac{\partial p}{\partial z}+\rho g\right)\right]}{\partial z}=-\phi\frac{\partial\rho}{\partial t} \quad (9.16)$$

where ϕ is the porosity. Eq. (9.16) has been used to solve the general flow problem of flow through a porous medium.[31] For the isothermal steady state flow of an incompressible fluid, Eq. (9.16) reduces simply to the Lapace equation of the pressure p which is easily to be solved.

(3) Solid-gas flows. This is a well known problem in internal ballistics, i.e., the flow of gases behind a projectile being expelled from a gun powder. Recently, this problem has been extensively. studied because it connects with the solid fuel problem in a large rocket and it also connects with the tektite problem of lunar ash flows. This problem may be divided into several stages in which the flow behaves differently in different stages as follows:

For simplicity, let us consider well packed gun powder. In the first stage, when the gun powder just starts to burn, there is a very small amount of flow of gas. The gun powder is practically undisturbed. The flow is similar to the case of gas flowing through a porous medium. This stage of flow may be called the fixed-bed flow. As the combustion continues, the amount of gas flow increases. As the gas flux reaches a critical value, called the flow for fluidization, at which the character of the solid powder changes abruptly to a pseudofluid, waves can be set up in the gun powder. The pseudofluid tends to form a level surface. The flow field at this stage may be called dense phase of the fluidized bed. Ordinarily the overall density of the mixture of solid and gas has decreased only fractionally — say 10 to 50% — as compared with the fixed bed case.

If the flow rate is further increased, the flow is seen to become irregular; bubbles of gas rise through the packed gun powder and burst. This process is known as slugging. Further increase of the flow rate will cause a disturbed and irregular regime in which the flow becomes so rapid that it will push the bullet and the gas will carry some gun powder with it. Beyond this point, the region is called the dilute phase in which the solid matter occupies less than 5 % of the total volume of the mixture and mixes with the gas more or less uniformly in the flow field.

In the study of the jet plume from a solid fuel rocket exhaust, we deal with the dilute phase of the mixture of solid and gas. Recently, considerable progress in the treatment of the gas-solid mixture flow has been made.[18, 19, 22, 23] We are going to outline the analysis of this problem below:

We consider a gas flow in which a certain amount of solid particles exist. We assume that the size of the solid particles is small enough so that we may assume that the average properties of these particles are the same and that the number of the particles is large enough so that the continuum theory may be used for the particle motion as well as that of the gas motion. Under such conditions, the unknowns in the flow field of the gas-solid mixture are:

The velocity vector of the gas q with components u_i,
The density of the gas ρ,
The pressure of the gas p,
The temperature of the gas T,
The velocity vector of the solid particles q_p with components v_i,
The temperature of the solid particles T_p, and
The number density of the solid particles n_p.

There are eleven unknowns in the present problem, so we have to find eleven fundamental equations for this problem.[46]

(1)   The equation of state of the gas

$$p = \rho R T \tag{9.17a}$$

(2)   Equation of continuity of the gas

$$\frac{\partial \rho}{\partial t} + \nabla \cdot (\rho q) = 0 \tag{9.17b}$$

(3)   Equation of motion of the gas

$$\rho \frac{Dq}{Dt} = -\nabla p + \nabla \cdot \tau_v + F_p \tag{9.17c}$$

where F_p is the force exerted upon the gas by the solid particles.

One of the difficult points of the theory of the dynamics of gas-particle mixture is to determine this force. The drag of the particles will act as an additional force to the gas. The resultant force of all the particles in the gas gives the force F_p. Since the particles of the solid are of very small size, we may use the Stokes formula for the drag force of these particles as a first approximation. For a high speed flow, the validity of the Stokes formula is questionable. Furthermore, the Stokes formula is for the use of a uniform free stream. In the present case, the particles either accelerate or decelerate. Again the validity of the Stokes formula is questionable. At the present stage, we do not have any better formula than the Stokes formula; the following formula is used to F_p:

$$F_p = n_p \, 6\pi \, d\mu \, (q_p - q) \qquad (9.18)$$

where d is the average particle radius. We assume that the particles are all spheres of the same radius d. The coefficient of viscosity of the gas is μ.

(4) Equation of energy of the gas:

$$\rho \frac{DU_m}{Dt} + p \, (\nabla \cdot q) = \Phi + \nabla \cdot (\kappa \nabla T) + \Phi_p + Q_v \qquad (9.17d)$$

where U_m is the internal energy of the gas, Φ is the viscous dissipation of the gas, κ is the coefficient of heat conductivity of the gas, Φ_p is the work done on the gas by the solid particles which may be written as

$$\Phi_p = (q_p - q) \cdot F_p \qquad (9.19)$$

and Q_p is the total heat transfer of the solid particles. The other difficult problem of the present analysis is to determine this heat transfer term Q_p. In the first place, we do not know exactly the heat transfer for each particle, and thus we know even less for the resultant effect of the heat transfer of all the particles. The heat transfer of the particles consists of both the heat transfer by conduction and by thermal radiation. For a first approximation, we may write:

$$Q_p = Q_{pc} + Q_{pR} \qquad (9.20)$$

To the same approximation as Stokes formula, the heat transfer by conduction may be written as:

$$Q_{pc} = n_p(\kappa/d)\, 4\pi d^2 (T_p - T) \tag{9.21}$$

and the heat transfer by thermal radiation may be written as

$$Q_{PR} = n_p \cdot 4\, D_R\, a_R\, T_p^3 \left(\frac{T_p - T}{d} \right) \tag{9.22}$$

where the Rosseland approximation is used.

(5) Equation of continuity of the solid particles:

$$\frac{\partial n_p}{\partial t} + \nabla \cdot (n_p q_p) = 0 \tag{9.17e}$$

(6) Equation of motion of the solid particles:

$$m_p\, n_p \left(\frac{\partial}{\partial t} + q_p \cdot \nabla \right) q_p = -F_p \tag{9.17f}$$

For more accurate analysis, we may add the stress tensor due to the momentum transport between the particle and gas motion in Eq. (9.17f). We neglect this term here because of the uncertainty of the collision terms between the particles and the gas molecules. Since F_p is not exact, we may consider the forces due to the stress tensor to be a part of F_p.

(7) Equation of energy of the solid particles:

$$m_p n_p c_s \left(\frac{\partial T_p}{\partial t} + q_p \cdot \nabla T_p \right) = - Q_p \tag{9.17g}$$

For a more accurate analysis, we may introduce the heat flux due to the energy transfer between the particles and the molecules of the gas. We neglect it here because of some uncertainty about Q_p. We may consider that Q_p includes such heat flux term.

Eqs. (9.17) are the fundamental equations for the dynamics of the gas-solid particle mixture. Since these equations are similar to those of ordinary gasdynamics, we may reexamine all the gas-dynamic problems by including the effects of solid particles in the gas flow. Since the solid particles introduce four physical quantities

—— the mass of a particle m_p, the number density of particles n_p, the radius of a particle d, and the specific heat of solid particles c_s —— we have four new nondimensional parameters to characterize the influences of these solid particles. Various forms of these new parameters may be introduced. According to Marble,[15] the following parameters may be used:

(1) Velocity equilibration parameter

$$R_1 = \frac{\lambda_v}{L} = \frac{2}{9} \frac{m_p n_p}{\rho} \frac{d^2 u_0}{L \nu_g} \qquad (9.23a)$$

(2) Thermal equilibration parameter

$$R_2 = \frac{\lambda_T}{L} = \frac{1}{3} P_r \left(\frac{c_s}{c_p} \right) \frac{m_p n_p}{\rho} \frac{d^2 u_0}{L \nu_g} \qquad (9.23b)$$

(3) Momentum interaction parameter

$$R_3 = \frac{m_p n_p}{\rho} \qquad (9.23c)$$

(4) Thermal interaction parameter

$$R_4 = \frac{m_p n_p}{\rho} \frac{c_s}{c_p} \qquad (9.23d)$$

where L and u_0 are respectively reference length and reference velocity. λ_v is a characteristic length required for particle velocity to reach that of the gas and λ_T is a characteristic length required for the temperature of particle to reach that of the gas.

In reference [18], several simple flow problems such as one-dimensional nozzle flow, normal shock, Prandtl-Meyer expansion, laminar boundary layer on a flat plate have been studied. For instance, the solution of a normal shock is sketched in Fig. 9.3. The shock structure has a relaxation zone in the tail. In general the solution is complicated and numerical integration may be required. But if the ratio R_1 is much smaller than unity or much larger than unity, the small perturbation technique may be used by expanding the solution in a power series of R_1 or $(1/R_1)$ whichever is a small quantity.

In actual flow, the velocity of the solid particles q_p is different from that of the gas q and their temperatures are also different.

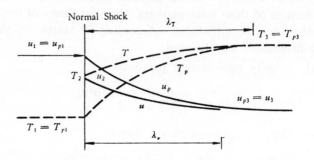

Figure 9.3   Normal shock transition region of a gas-solid particle mixture

One of the ideal cases, known as equilibrium flow, is that $q = q_p$ and $T_p = T$. In a uniform flow, we may assume that the equilibrium flow condition exists. In a nonuniform flow, a velocity lag and a temperature lag may occur, as shown in the normal shock case of Fig. 9.3.

Another ideal case, known as the frozen flow, is the case where the temperature of solid particles retains its original value. In this case, the velocity of the particles q_p differs from that of the gas q. A simple analysis of such a flow shows that the gas equations are hyperbolic in nature while that of the solid particles are parabolic if we neglect the viscous and heat conduction terms. Numerical calculation may be used to solve any initial value problem.

Because of the existence of the solid particles in the mixture, the effective sound speed of the mixture, i.e., the speed of propagation of infinitesimal disturbance, is different from that of the gas alone. In general, the effective sound speed is smaller than the sound speed of the gas at the same temperature.

If the mixture is seeding so that the solid particles are electrified, the flow phenomenon is different from that without electrification. For the two-phase flow in a pipe, the solid particles tend to stay in the center of a pipe when they are electrified.

4. Multifluid Theory of a Plasma

In chapter VI, we considered mainly the classical magnetofluid dynamics of single fluid theory in which we assume that (i) the fluid has a scalar electrical conductivity, (ii) a simple generalized Ohm's law may be used for the electrical current density equation, and (iii) the temperatures of all species in a plasma are the same. Such a system of equations gives good results for an electrically conducting liquid as well as an ionized gas if the strength of the magnetic field is not too large and the density of the gas is not too small. However, in current practice, the strength of the magnetic field gradually increases and the density of the gas decreases. As a result, the Hall current and ion slip will be important and we should not assume that the electric conductivity of the gas is a scalar quantity. When the degree of ionization is large and the density of the gas is low, gasdynamic forces will affect the electric current density distribution and the temperature of electrons may be different considerably from that of the heavy particles. We have to improve our fundamental equations of classical magnetogasdynamics of chapter VI so that these two new phenomena may be taken into account. One way to improve the classical magnetogasdynamic equations is to use a complicated generalized Ohm's law including the Hall current, ion slip, and simple gasdynamic effects, as shown in chapter VI section 13. Such improvements may still be insufficient for many other effects, such as different temperatures between species, large diffusion velocity, etc. A better and more logical approach is the multifluid theory of plasma which will be discussed in this section.

We assume that the plasma consists of N species, which consist of ions, electrons, and neutral particles. From macroscopic point of view, the following variables should be used to describe the flow field of a plasma:

The temperature of sth species T_s,

The pressure of sth species p_s,

The density of sth species $\rho_s = m_s n_s$,

The velocity vector of sth species q_s with components u_s^i,

The electric field strength E with component E^i, and

The magnetic field strength H with component H^i,

where $s = 1, 2, \cdots,$ or N, $i = 1, 2,$ or 3 which represents one of the three spatial directions. The number density of sth species is n_s, the mass of a particle of sth species is m_s, and the electric charge on a particle in sth species is e_s. There are $6N+6$ variables in this multifluid theory instead of 16 variables in the classical theory of electromagnetogasdynamics, which was discussed in chapter VI.

The variables T_s, etc., are known as partial variables in the multifluid theory. The relation between these partial variables and the gross variables of the mixture as a whole are as follows:

$$\text{pressure of the mixture} = p = \sum_{s=1}^{N} p_s \qquad (9.24)$$

$$\text{number density of the mixture} = n = \sum_{s=1}^{N} n_s \qquad (9.25)$$

$$\text{density of the mixture} = \rho = mn = \sum_{s=1}^{N} \rho_s = \sum_{s=1}^{N} m_s n_s \qquad (9.26)$$

where m is the mean mass of a particle in the mixture which is a function of the composition of the mixture.

$$\text{Temperature of the mixture} = T = \frac{1}{n} \sum_{s=1}^{N} n_s T_s \qquad (9.27)$$

$$i\text{th velocity component of the mixture} = u^i = \frac{1}{\rho} \sum_{s=1}^{N} \rho_s u_s^i \qquad (9.28)$$

$$i\text{th component of diffusion velocity of } s\text{th species} = w_s^i = u_s^i - u^i \qquad (9.29)$$

From the definition of the flow velocity u^i and the diffusion velocity

w_s^i we have

$$\sum_{s=1}^{N} \rho_s w_s^i = 0 \qquad (9.30)$$

$$\text{Excess electric charge} = \rho_e = \sum_{s=1}^{N} \rho_{es} = \sum_{s=1}^{N} e_s n_s \qquad (9.31)$$

ith component of electric current density $= j^i = \sum_{s=1}^{N} J_s^i$

$$= \sum_{s=1}^{N} \rho_{es} u_s^i = \sum_{s=1}^{N} \rho_{es} u_s^i + u^i \sum_{s=1}^{N} \rho_{es} = I^i + \rho_e u^i \qquad (9.32)$$

where I^i is the ith component of the electric conduction current density and $\rho_e u^i$ is the ith component of the electric convection current density.

In the classical theory of magnetogasdynamics of chapter VI, we use the gross variables: p, ρ, T, u^i, ρ_e, J^i, E^i, and H^i. Except E^i and H^i where are the same in both the multifluid theory and single fluid theory, the fundamental equations of the classical single fluid theory of magnetogasdynamics may be derived from the fundamental equations of the multifluid theory as follows:

(1) Equation of state. The ideal gas law may be used. For sth species, we have

$$p_s = k n_s T_s \qquad (9.33)$$

The sum of N equations of the type of Eq. (9.33) gives the equation of state of the mixture

$$p = k n T = \rho R T \qquad (9.34)$$

where $R = k/m$ is the gas constant of the mixture which is a function of the composition of the mixture and which is identical to Eq. (6.1).

(2) Equation of continuity. The conservation of mass of each species gives

$$\frac{\partial \rho_s}{\partial t} + \nabla \cdot (\rho_s \, q_s) = \sigma_s \tag{9.35}$$

where σ_s is the mass source per unit volume of sth species. The sum of the source function must be zero because of the conservation of total mass of the mixture, i.e.,

$$\sum_{s=1}^{N} \sigma_s = 0 \tag{9.36}$$

Eq. (9.35) may be derived from the Boltzmann equation of sth species [see Eq. (8.70)].

The sum of N equations of the type of Eq. (9.35) with the help of Eq. (9.36) gives

$$\frac{\partial \rho}{\partial t} + \nabla \cdot (\rho \, q) = 0 \tag{9.37}$$

It is interesting to notice that Eq. (9.35) is the diffusion equation used in ordinary gasdynamics (3.34), but usually some simplified assumptions on the diffusion velocities are made so that we need not solve simultaneously the equations of diffusion velocities. In the multifluid theory, we do not make simple assumptions on the diffusion velocities (3.36). Eq. (9.37) is the same as that of the single fluid theory (6.2).

If we multiply $\gamma_s = e_s / m_s$ to Eq. (9.35), we have

$$\frac{\partial \rho_{es}}{\partial t} + \nabla \cdot (\rho_{es} \, q_s) = \gamma_s \, \sigma_s \tag{9.38}$$

On principle, Eq. (9.38) is the same as Eq. (9.35). By conservation of total electric charge, we have

$$\sum_{s=1}^{N} \gamma_s \, \sigma_s = 0 \tag{9.39}$$

The sum of N equations of the type of Eq. (9.39) with the help of Eq. (9.39) gives

$$\frac{\partial \rho_e}{\partial t} + \nabla \cdot J = 0 \tag{9.40}$$

Eq. (9.40) is the equation of conservation of electric charge (6.13) which is one of the fundamental equations of the single fluid theory of electromagnetogasdynamics but in the multifluid theory, it is simply another form of the equation of continuity and may be used to replace one of the equations of continuity (9.35).

(3) Equation of motion. The conservation of momentum of sth species gives the equation of motion of sth species as follows:

$$\frac{\partial \rho_s u_s^i}{\partial t} + \frac{\partial}{\partial x^i}(\rho_s u_s^i u_s^j - \tau_s^{i\,j}) = X_s^i + \sigma_s Z_s^i \qquad (9.41)$$

where the summation convention is used for the repeated tensorial indices i but not for the indices distinguish the species s. Eq. (9.41) may be derived from the transfer equation of Boltzmann equation of sth species. The term $\sigma_s Z_s^i$ is the ith component of the momentum source of sth species. In general, we have

$$\sum_{s=1}^{N} \sigma_s Z_s^i = 0 \qquad (9.42)$$

The term τ_s^{ij} is the ijth component of the stress tensor of sth species defined by Eq. (8.78). For a small mean free path, we may use the Navier-Stokes relations for τ_s^{ij} in which the velocity gradient of u_s^i will be used. In the most general case, a differential equation for τ_s^{ij} will be used, as we have discussed in chapter VIII. In this section, we use the simple expressions of Navier-Stokes relations for τ_s^{ij}. The coefficient of viscosity μ_s for each species may be different from that of the other species.

The body force X_s^i consists of the electromagnetic force F_{es}^i, a nonelectric force such as gravitational force F_{gs}^i, and the interaction force between species F_{os}^i which are given below:

$$F_{es}^i = \rho_{es}[E^i + (q_s \times B)^i] \qquad (9.43)$$

$$F_{gs}^i = \rho_s g^i \qquad (9.44)$$

$$F_{os}^i = \sum_{t=1}^{N} K_{st}(u_t^i - u_s^i) \qquad (9.45)$$

where K_{st} is known as the friction coefficient between species t and s and

$$\sum_{s=1}^{N} F_{os}^{i} = 0 \tag{9.46}$$

The ith component of gravitational acceleration is g^i.

If we add all the N equations of the type of Eq. (9.41), we have the equation of motion of the mixture as a whole:

$$\frac{\partial \rho u^i}{\partial t} + \frac{\partial \rho u^i u^j}{\partial x^j} = \rho \left(\frac{\partial u^i}{\partial t} + u^j \frac{\partial u^i}{\partial x^j} \right) = \rho \frac{Du^i}{Dt}$$

$$= -\frac{\partial p}{\partial x^i} + \frac{\partial \tau_v^{ij}}{\partial x^j} + F_e^i + F_g^i \tag{9.47}$$

The form of Eq. (9.47) is exactly the same as that of the single fluid theory (6.3) but it is interesting to find out the difference by examining the definition of various terms, particularly the stress tensor.

The nonelectric body force such as gravitational force is simple and is

$$F_g^i = \sum_{s=1}^{N} F_{gs}^i = \rho g^i \tag{9.48}$$

The electromagnetic force is

$$F_e^i = \sum_{s=1}^{N} F_{es}^i = \rho_e E^i + (J \times B)^i \tag{9.49}$$

The stress tensor of sth species may be written as

$$\tau_s^{ij} = -p_s \delta^{ij} + \tau_{os}^{ij} \tag{9.50}$$

and

$$\tau_{os}^{ij} = \mu_s \left(\frac{\partial u_s^i}{\partial x^j} + \frac{\partial u_s^j}{\partial x^i} \right) - \frac{2}{3} \mu_s \frac{\partial u_s^k}{\partial x^k} \delta^{ij} \tag{9.51}$$

The viscous stress tensor of the plasma as a whole is

$$\tau_v^{ij} = \sum_{s=1}^{N} \tau_{os}^{ij} - \sum_{s=1}^{N} \rho_s \, w_s^i \, w_s^j \tag{9.52}$$

The interesting point is that the viscous stress τ_v^{ij} depends on the viscous stress of each species and the diffusion velocities between the species. Only when the diffusion velocities are small and negligible, we may use the simple expression such as the Navier-Stokes relation for the viscous stress of the mixture as a whole as we did in chapter VI, and in that case, the coefficient of viscosity is a function of the composition of the mixture as well as the state of the mixture. In a more general case, we should solve the equation of diffusion velocity simultaneously with other fundamental equations of the plasma dynamics. The difference of Eqs. (9.41) and (9.47) gives the equation of diffusion velocity $w_s^i = u_s^i - u^i$ as follows:

$$\frac{\partial w_s^i}{\partial t} + u_s^j \frac{\partial u_s^i}{\partial x^j} - u^j \frac{\partial u^i}{\partial x^j} = \frac{1}{\rho} \frac{\partial p}{\partial x^i} + \frac{1}{\rho_s} \frac{\partial \tau_s^{ij}}{\partial x^j} - \frac{1}{\rho} \frac{\partial \tau_v^{ij}}{\partial x^j}$$

$$+ \frac{X_s^i}{\rho_s} - \frac{F_e^i + F_g^i}{\rho} + \frac{\sigma_s}{\rho_s}(Z_s^i - u_s^i) \tag{9.53}$$

We shall show later that the generalized Ohm's law (6.11) and (6.124) may be derived from Eq. (9.53) under a number of assumptions. Of course, it is better to use Eq. (9.53) to replace the generalized Ohm's law in the analysis of the dynamics of a plasma.

(4) Equation of energy. The conservation of energy of sth species gives the energy equation for the sth species:

$$\frac{\partial \bar{e}_s}{\partial t} + \frac{\partial}{\partial x^j}(\bar{e}_s u_s^i - u_s^i \tau_s^{ij} - Q_s^j) = \bar{\mathscr{E}}_s. \tag{9.54}$$

where $\bar{e}_s = \rho_s \bar{e}_{ms}$ is the total energy of sth species of the mixture per unit volume which consists of the internal energy U_{ms} of sth species, kinetic energy $\frac{1}{2} u_s^i u_s^i$ of sth species, and potential energy of sth species. The heat conduction flux of sth species is Q_s^j. The energy source $\bar{\mathscr{E}}_s$ consists of the terms due to electromagnetic field $\bar{\mathscr{E}}_{es}$, that due to chemical reaction $\bar{\mathscr{E}}_{cs}$, and that due to elastic collision between species $\bar{\mathscr{E}}_{os}$.

Now if we add all the N equations of the type of Eq. (9.54), we have the energy equation of the mixture as a whole:

$$\frac{\partial \rho \overline{e}_m}{\partial t} + \frac{\partial \rho u^j \overline{e}_m}{\partial x^j} = -\frac{\partial p u^j}{\partial x^j} + \frac{\partial u^i \tau_v^{ij}}{\partial x^j} - \frac{\partial Q^j}{\partial x^j} + \overline{\mathscr{E}}_T \qquad (9.55)$$

The form of Eq. (9.55) is the same as that of the single fluid theory, Eq. (6.5), but the meaning of various terms is different.

The total energy $\overline{e}_m$ consists of the internal energy, kinetic energy, and potential energy of the mixture as a whole but the definition of the internal energy per unit mass of the mixture U_m consists of the diffusion kinetic energy as well as the sum of ordinary molecular internal energy of all the species, i.e.,

$$U_m = \frac{1}{\rho} \sum_{s=1}^{N} \left(\rho_s \, U_{ms} + \frac{1}{2} \rho_s \, w_s^2 \right) \qquad (9.56)$$

The total energy is the sum of the internal energy of all species and the diffusion energy of all species. If the diffusion velocities are not small, we have to solve the energy equation with the equations of diffusion velocities. Since the internal energy of each species depends on its partial temperature T_s, we have to solve the equations for T_s with Eq. (9.55) if the temperatures of all species are not equal. In chapter VI, we assume that the diffusion velocities are all small and the diffusion energy is negligible and that all the species have the same temperature T, the internal energy can be expressed in terms of a specific heat at constant volume $c_v(T)$ which is a function of temperature T only, or simply a constant. The heat flux Q^j consists of the sum of heat conduction fluxes of all species as well as terms due to diffusion velocities. In chapter VI, we again neglect the part due to diffusion velocities and then the heat flux of the mixture as a whole may be expressed in terms of the coefficient of heat conductivity of the mixture as a whole which is a function of the temperature as well as the composition of the mixture.

The energy source $\overline{\mathscr{E}}_T$ consists of the resultant energy source due to chemical reaction $\overline{\mathscr{E}}_c$ and that due to electromagnetic field $\overline{\mathscr{E}}_e = E^i J^i$.

The electromagnetic field equations are the same as those in the single fluid theory, i.e., Eqs. (6.6) and (6.7).

There are two ways to describe the flow of a plasma by the multifluid theory:

(a) We may use the partial variables $(T_s, p_s, n_s, u_s^i, E^i, H^i)$ to describe the flow field of a plasma where $s = 1, 2, \cdots, N$. The fundamental equations are Eqs. (9.33), (9.35), (9.41), (9.54), (6.6), and (6.7); and

(b) We may use the gross variable (T, p, n, u^i, E^i, H^i) and some of the partial variables (T_r, p_r, n_r, u_r^i) where $r = 1, 2, \cdots, (N-1)$. The fundamental equations are Eqs. (9.34), (9.37), (9.47), (9.55), (6.6), (6.7) together with the corresponding equations of the partial variables (9.33), (9.35), (9.41), and (9.54). We may also replace the partial velocity u_r^i by the diffusion velocity w_r^i and the partial density n_r or ρ_r by the concentration $c_r = n_r/n$ or by the mass concentration $k_r = \rho_r/\rho$.

In principle, the above two approaches are identical. But from the easy application of the multifluid theory to the flow problem point of view, approach (a) is preferable, because we do not need to consider the complicated function U_m, etc., which includes many diffusion phenomena. We shall use approach (a) to discuss some simple flow problems later.

It seems to be easier to find the relation between the multifluid theory and the classical single fluid theory from approach (b). To show this relation, we consider a case of partially ionized monatomic gas which consists of atoms, singly charged ions, and electrons, i.e., $N = 3$. According to the approach (b), we should use the following 24 variables to describe the flow field of such a plasma:

$$T, p, \rho, u^i, E^i, H^i, \rho_e, J^i, T_1, p_1, \rho_1, w_2^i, T_2, p_2$$

where subscript 1 refers to the value of electrons and subscript 2 refers to the value of ions.

For the classical single fluid theory of a plasma, discussed in chapter VI, we use only the following 16 variables:

$$T, p, \rho, u^i, E^i, H^i, \rho_e, J^i$$

The reduction of the 24 variables to the 16 variables is based on the following approximations:

(i) We assume that the temperature of all species are equal, i.e.,

$$T = T_1 = T_2 \qquad (9.57)$$

Hence we do not have to consider the variables T_1 and T_2. This assumption is reasonably good if the masses of all species in the mixture of gases are approximately equal. Thus it is a good approximation for neutral air, as we used in chapter V for the problems of flow with chemical reaction. But for a partially ionized air, the electrons are much lighter than ions and neutral particles. The temperature of electrons T_1 may be different considerably from that of ions T_2 or that of atoms T_a. At high temperature range, it is advisable to consider at least two temperatures in the flow field of a plasma: one is the electron temperature T_1, and the other is the temperature of heavy particles $T_2 = T_a$.

(ii) Usually we are not interested in the partial pressure p_1 and p_2 in most engineering problems. Furthermore, if we know the temperature and density of a species, we can easily calculate the partial pressure from the equation of state (9.33). Hence we do not need to analyze the partial pressure in our calculation of the flow variables.

(iii) If the masses of all species in the mixture are approximately equal, the diffusion velocities w_r^i are usually small. We may either neglect the diffusion velocities completely or use some simple formulae for the diffusion velocities, such as Eq. (3.36). If the diffusion velocities may be completely neglected, the concentration of all species in a mixture of gases are constant. As a result, we need to consider the gross variables p, ρ, T, and u^i only. This is the reason why the single fluid theory gives good results for a mixture of neutral gases such as ordinary air. However, in a plasma, the electrons are much lighter than all the other heavy particles and we can not neglect the diffusion velocity of electrons w_1^i. If the diffusion velocity of the ions w_2^i is still negligible, the electric current density J^i is proportional to the electron diffusion velocity w_1^i. The diffusion phenomena of electrons may be expressed in terms of the two variables J^i (w_1^i) and $\rho_e (\rho_1)$.

With the assumption (i) and (iii), the mixture of neutral particles, ions, and electrons can be treated by the single fluid theory of classical electromagnetogasdynamics with the variables p, ρ, T, u^i, E^i, H^i, J^i, and ρ_e, as we did in chapter VI. Strictly speaking, J^i and/or w_1^i should be governed by a differential equation such as Eq. (9.53) or similar equation. Since Eq. (9.53) is too complicated to be useful in the single fluid theory, we use some simple formula instead of Eq. (9.53). We shall discuss these simple formulas later.

When the degree of ionization of the plasma and the temperature range of the flow field is large, the assumptions (i) and (iii) may not be good. The temperature of electrons may differ significantly from that of the heavy particles and the gasdynamic forces may have large influence on the electric current density. Thus the classical theory of single fluid will not be sufficient to predict the flow field of a plasma. It is then advisable to use the multifluid theory to analyze the flow problem of a plasma, particularly by approach (a). We shall give an example later.

(1) Generalized Ohm's law. Hall current and ion slip.

Let us examine the equation of the electrical current density from the multifluid theory point of view. The electric current density J^i consists of an electric conduction current density I^i and an electric convection current density $\rho_e u^i$. Since the convection current $\rho_e u^i$ is already expressed in terms of ρ_e and u^i, we need to find an equation for the electric conduction current density I^i only. Eq. (9.32) shows that the electric conduction current I^i depends on the diffusion velocities w_r^i. Hence we may find a formula for I^i by studying the diffusion velocities w_r^i.

We may calculate the diffusion velocity w_r^i from Eq. (9.53). The generalized Ohm's law (6.11) or (6.124) may be derived from Eq. (9.53) under the following conditions:

(a) The equation is explicitly independent of time and spatial coordinates.

(b) The electromagnetic force is the only dominant force, and

(c) There is no source term, i.e., $\sigma_r = 0$.

Under these three conditions, Eq. (9.53) reduces to the form:

$$\frac{F_{er}^i + F_{or}^i}{\rho_r} = \frac{F_e^i}{\rho} \tag{9.58}$$

or

$$(\rho\rho_{er} - \rho_r\,\rho_e)E_u + (\rho\rho_{er}\,w_r - \rho_r I) \times B = \rho \sum_{s=1}^{N} K_{sr}(w_r - w_s)$$

$$\tag{9.59}$$

where $E_u = E + (q \times B)$. If we consider that ρ_r, ρ_{er}, E_u, B, and K_{sr} are given, Eq. (9.59) gives a set of linear algebraic equations for w_r because I is a linear function of w_r. We may solve for w_r from Eq. (9.59).

To illustrate the solution of Eq. (9.59), we consider the case of a partially ionized monatomic gas which consists of (i) electrons (subscript e), (ii) singly charged ions (subscript i), and (iii) atoms (subscript a). The number densities of electrons, ions, and atoms are respectively n_e, n_i, and n_a. We consider a slightly ionized gas so that $n_e \cong n_i \ll n_a$. The excess electric charge of the gas is then $\rho_e = -en_e + en_i = 0$. The density of the gas as a whole is

$$\rho = m_e n_e + m_i n_i + m_a n_a \cong m_a(n_e + n_a) \tag{9.60}$$

where $m_e \ll m_i \cong m_a$.

From the relation (9.30), we have

$$m_e n_e w_e^i + m_i n_i w_i^i + m_a n_a w_a^i = 0 \tag{9.61}$$

Since $m_e \ll m_i \cong m_a$ and $n_e \cong n_i \ll n_a$ and the three terms of Eq. (9.61) should be of the same order of magnitude, we conclude that

$$w_e^i \gg w_i^i \gg w_a^i \tag{9.62}$$

From Eq. (9.59), we have

   (a)   For electrons

$$-en_e E_u - en_e w_e \times B = K_{ie}(w_e - w_i) + K_{ea}(w_e - w_a) \tag{9.63}$$

   (b)   For ions

$$en_e E_u + en_e w_i \times B = K_{ie}(w_i - w_e) + K_{ia}(w_i - w_a) \tag{9.64}$$

where we neglect the term of the order of m_e/m_a. Eqs. (9.63) and (9.64) have three unknowns w_e, w_i, and w_a. By the relation (9.62), we see that (i) for first approximation, we may neglect both w_i and w_a, and Eq. (9.63) gives w_e and then the conduction current I; (ii) for the second approximation, we may neglect w_a and solve Eqs. (9.63) and (9.64) for w_e and w_i and then I; and (iii) for the third approximation, we may estimate w_a from Eq. (9.61), i.e.,

$$w_a = -\frac{n_e}{n_a} w_i \qquad (9.65)$$

and solve Eqs. (9.63) to (9.65) for w_e and w_i and then I.

After w_e and w_i are obtained, the total electric conduction current density is

$$I = en_e(w_i - w_e) = I_i + I_e \qquad (9.66)$$

and the electron current density is

$$I_e = -en_e w_e \qquad (9.67)$$

From Eqs. (9.63) to (9.65), we find the relation between I and I_e as follows:

$$I_e = \frac{1}{1 + \epsilon_0}\left(I + \beta_i \frac{B}{B} \times I \right) \qquad (9.68)$$

where

$$\epsilon_0 = \frac{K_{ea}}{\dfrac{n_e}{n_a} K_{ea} + \left(1 + \dfrac{n_e}{n_a} \right) K_{ai}} \qquad (9.69)$$

$$\beta_i = \frac{en_e B}{\dfrac{n_e}{n_a} K_{ea} + \left(1 + \dfrac{n_e}{n_a} \right) K_{ai}} = \text{ion slip factor} \qquad (9.70)$$

From Eqs. (9.63) to (9.68), we obtain the generalized Ohm's law as follows:

$$\sigma_e E_u = A_1 I + A_2 I \times B + A_3 (I \times B) \times B \qquad (9.71)$$

where

$$A_1 = 1 + \beta_a \tag{9.72a}$$

$$A_2 = \frac{\sigma_e}{en_e(1+\epsilon_0)}\left[1 - \epsilon_0\left(1 + \frac{n_e}{n_a}\right)\right] \tag{9.72b}$$

$$A_3 = - \frac{1}{1+\epsilon_0}\frac{\beta_i\beta_e}{B^2} \tag{9.72c}$$

$$\beta = \frac{K_{ea}}{K_{ei}}\frac{1 - \dfrac{n_e}{n_a}\epsilon_0}{1+\epsilon_0} = \text{atom collision factor} \tag{9.72d}$$

$$\beta_e = \frac{en_e\,B}{K_{ei}} = \frac{\omega_c}{f} = \text{Hall current factor} \tag{9.72e}$$

$$\sigma_e = \frac{e^2 n_e^2}{K_{ei}} = \text{scalar electric conductivity} \tag{9.72f}$$

Eq. (9.71) reduces to Eq. (9.124) if we neglect the collision between atoms and electrons, i.e., $K_{ea}=0$. We have also discussed the reduction of Eq. (6.124) to Eq. (6.11) in chapter VI. We do not repeat it here.

In the derivation of Eq. (9.71), we neglect completely the gas-dynamic effects. In fact, the diffusion velocities of the species in a plasma do depend on the gasdynamic forces as shown in Eq. (9.53). A simple expression including some of the gasdynamic forces can be obtained from Eq. (9.53), especially when the pressure gradient of electrons is considered. For instance, if we neglect ion slip but include the pressure gradient of electrons, Eq. (9.63) becomes

$$I + \frac{\sigma_e}{en_e}I \times B = \sigma_e\left(E_u + \frac{1}{en_e}\nabla p_e\right) = \sigma_e E_{uT} \tag{9.73}$$

The pressure gradient of electrons introduces an additional effective electric field. Eq. (9.73) corresponds to Eq. (6.126). We may easily obtain an equation similar to Eq. (6.124) by including the pressure gradient of electrons. Since such piecewise improvement of the electric current density equation will not include all the gas-dynamic effects, the author prefers to use the multifluid theory, when the gasdynamic effects on the electric current density are studied.

(2)   Wave motions in a plasma based on multifluid theory.

Now we use the multifluid theory to study the wave of infinitesimal amplitude in a plasma similar to the case of chapter VI section 8.   For simplicity, we consider the case of a fully ionized plasma in which $N=2$.   We consider the plasma as a mixture of singly charge ions (subscript 2) and electrons (subscript 1).   Originally, the plasma is at rest under an external magnetic uniform field H_0, i.e.,

$$H_0 = iH_x + jH_y + k \cdot 0 \tag{9.74}$$

where H_x and H_y are constants which may or may not be zero. The unit vectors in the x-, y-, and z-directions are respectively i, j, and k.   In the undisturbed state of the field, there is neither electric current, nor excess electric charge, nor electric field.   In the disturbed state, we have:

$$\left.\begin{aligned} &q_s = iu_s(x,\ t) + jv_s(x,\ t) + kw_s(x,\ t) \\ &p_s = p_0 + p'_s(x,\ t),\ n_s = n_0 + n'_s(x,\ t),\ T_s = T_0 + T'_s(x,\ t) \\ &E = iE_x(x,\ t) + jE_y(x,\ t) + kE_z(x,\ t) \\ &H = H_0 + h(x,\ t) = H_0 + ih_x(x,\ t) + jh_y(x,\ t) + kh_z(x,\ t) \end{aligned}\right\} \tag{9.75}$$

where $s=1$ or 2.   All the perturbed quantities are function of x and t and subscript 0 refers to the value in the undisturbed state.

We make the following assumptions in our fundamental equations:

(i)   Both ions and electrons may be considered as an inviscid and nonheat-conducting perfect gas.

(ii)   The interaction forces between ions and electrons are proportional to the difference between their mean velocities:

$$F_{12} = K_{ie}(q_1 - q_2) = -F_{21} \tag{9.76}$$

The linearized fundamental equations for the two-fluid theory of a plasma under the above two assumptions are as follows:

(a)   Maxwell's equations for the electromagnetic field:

$$\frac{\partial \mu_e h_x}{\partial t} = 0 \tag{9.77a}$$

$$\frac{\partial \mu_e h_y}{\partial t} = \frac{\partial E_z}{\partial x} \qquad (9.77b)$$

$$\frac{\partial \mu_e h_z}{\partial t} = -\frac{\partial E_y}{\partial x} \qquad (9.77c)$$

$$\frac{\partial \epsilon E_x}{\partial t} + e n_0 (u_2 - u_1) = 0 \qquad (9.77d)$$

$$\frac{\partial \epsilon E_y}{\partial t} + e n_0 (v_2 - v_1) = -\frac{\partial h_z}{\partial x} \qquad (9.77e)$$

$$\frac{\partial \epsilon E_z}{\partial t} + e n_0 (w_2 - w_1) = -\frac{\partial h_y}{\partial x} \qquad (9.77f)$$

(b) Equation of state for sth species:

$$\frac{p'_s}{p_0} = \frac{n'_s}{n_0} + \frac{T'_s}{T_0} \qquad (9.77g)$$

(c) Equation of continuity for sth species:

$$\frac{\partial n'_s}{\partial t} + n_0 \frac{\partial u_s}{\partial x} = 0 \qquad (9.77h)$$

(d) Equations of motion of sth species:

$$m_s n_0 \frac{\partial u_s}{\partial t} = -\frac{\partial p'_s}{\partial x} + e_s n_0 E_x - e_s n_0 B_y w_s + K_{st}(u_s - u_t) \quad (9.77i)$$

$$m_s n_0 \frac{\partial v_s}{\partial t} = e_s n_0 E_y + e_s n_0 B_x w_s + K_{st}(v_s - v_t) \qquad (9.77j)$$

$$m_s n_0 \frac{\partial w_s}{\partial t} = e_s n_0 E_z + e_s n_0 (u_s B_y - v_s B_x) + K_{st}(w_s - w_t) \quad (9.77k)$$

where

$$e_1 = -e, \ e_2 = e, \ B_0 = \mu_e H_0, \ s, \ t = 1 \text{ or } 2$$

(e) Energy equation for sth species:

$$m_s n_0 c_{ps} \frac{\partial T'_s}{\partial t} = \frac{\partial p'_s}{\partial t} \qquad (9.77l)$$

where

$$m_s\, c_{ps} = 5k/2.$$

Among the 18 perturbed quantities, only h_x is independent of all the other 17 quantities, and is given by Eq. (9.77a) alone. This is similar to the result of the single fluid theory and we may take $h_x = 0$. In contrast to the single fluid theory, all the other 17 perturbed quantities are interrelated by Eqs. (9.77b) to (9.77l). However, we may consider these 17 perturbed quantities as four basic modes of waves which are interacted by the external magnetic field. Without the external magnetic field, the 17 perturbed quantities may be divided into three independent groups:

(i)   The first group consists of ν_1, ν_2, E_y, and h_z, which represents a transverse wave.

(ii)   The second group consists of w_1, w_2, E_z, and h_y, which represents another transverse wave.

(iii)   The third group consists of u_1, u_2, p'_1, p'_2, n'_1, n'_2, T'_1, T'_2, and E_x, which represents two longitudinal (sound) waves.

Now we are looking for the periodic solution similar to the single fluid theory by using the form of Eq. (6.75) and we obtain the dispersion relations for these basic waves.

(a) Transverse wave.   The dispersion relations for the two basic transverse waves are the same, because, without the external magnetic field, we cannot distinguish y and z. The dispersion relation for the basic transverse wave is then:

$$c^2\lambda^2 = \omega^2 - \omega_e^2\left[(1 - iK_{ie}^*)/(1 + K_{ie}^{*2})\right] \tag{9.78}$$

where c is the velocity of light, $i = (-1)^{\frac{1}{2}}$,

$$\omega_e = e\left(\frac{n_0}{\epsilon m_1}\right)^{\frac{1}{2}} = \text{electron plasma frequency} \tag{9.79}$$

and

$$K_{ie}^* = \frac{K_{ie}}{m_1 \omega n_0} \tag{9.80}$$

For an ideal plasma, $K_{ie} = 0$, Eq. (9.80) gives the speed of propagation of this transverse wave as

$$V = \frac{\omega}{\lambda} = \frac{c}{\left(1 - \frac{\omega_e^2}{\omega^2}\right)^{1/2}} \qquad (9.81)$$

Eq. (9.81) is the well known results of elementary theory of plasma oscillation. It shows that there is a large damping for the low frequencies ($\omega < \omega_e$) which is the essential mechanism of blackout of radio communication during the reentry of a space vehicle. This result cannot be obtained in the single fluid theory of chapter VI, but we obtain it in the microscopic theory of the Boltzmann equation in chapter VIII section 13.

In vacuum, $n_0 = 0$ and $\omega_e = 0$, Eq. (9.81) shows that $V = c$ and that the electromagnetic wave is propagated in vacuum at a speed of light c.

The effect of electrical conductivity σ_e or K_{ie} is to introduce a damping on this wave.

(b) Longitudinal waves. The dispersion relation of the longitudinal wave gives two roots of λ^2 for the longitudinal wave of group (c) as follows:

$$\lambda^2 = \frac{\omega^2}{2a_i^2}\left\{\left[\left(1 - 2\frac{\omega_i^2}{\omega^2}\right) + 2iK_{ie}^* \frac{m_e}{m_i}\right] \pm \left\{\left[\left(1 - 2\frac{\omega_i^2}{\omega^2}\right) + 2iK_{ie}^* \frac{m_e}{m_1}\right]^2 \right.\right.$$
$$\left.\left. - 4\frac{a_i^2}{a_e^2}\left(1 - \frac{\omega_e^2}{\omega^2}\right) + iK_{ie}^* \right\}^{1/2}\right\} \qquad (9.82)$$

where

$$a_s = [\gamma p_0/(m_s n_0)]^{1/2} = \text{sound speed of } s\text{th species} \qquad (9.83)$$

$$\omega_s = e[n_0/(m_s \epsilon)]^{1/2} = \text{plasma frequency of } s\text{th species} \qquad (9.84)$$

There are two basic longitudinal waves from Eq. (9.82). One is the ion sound speed (corresponding to $+$ sign and shown in Fig. 9.4) which is always an undamped wave. As the frequency ω of this wave approaches to zero, the velocity of propagation of this wave is $V_i = \sqrt{2}\,a_i = a_p$ which is the sound speed of the plasma as a whole. The value of V_i decreases continuously as ω increases and as $\omega \to \infty$, $V_i = a_i$, the sound speed of ions alone. Hence this mode is closely associated with ions and we may call it the ion sound wave. The other longitudinal wave is the electron sound wave (corresponding to the − sign and shown in Fig. 9.5). This wave is a damped wave when $\omega < \omega_e$ and an undamped wave when $\omega > \omega_e$. As $\omega \to \infty$, the speed of propagation of this sound wave is $V_e = a_e$, the speed of sound wave for electrons alone. Hence we call this mode the electron sound wave. The single-fluid theory of chapter VI gives the results corresponding to $\omega \to 0$ only.

(c) Waves under longitudinal external magnetic field. If $H_x \neq 0$ and $H_y = 0$, the basic longitudinal waves are not affected by the external magnetic field and the two basic transverse waves are interacted. The resultant dispersion relation for these interacted transverse waves is

$$A_0 \lambda^4 + 2A_2 \lambda^2 + A_4 = 0 \qquad (9.85)$$

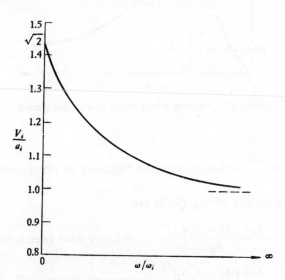

Figure 9.4   Ion sound speed in an ideal plasma

The coefficients A_0, A_2, and A_4 are functions of plasma frequencies, cyclotron frequencies of the species, frequency ω, and friction coefficient K_{ie}^*. For an ideal plasma, we have the following expressions for these coefficients:

$$A_0 = \left(1 - \frac{\omega_{xi}^2}{\omega^2}\right)\left(1 - \frac{\omega_{xe}^2}{\omega^2}\right)$$

$$A_2 = \frac{\omega_e^2}{c^2}\left[\left(1 - \frac{\omega^2}{\omega_e^2}\right) - \frac{\omega_{xi}\omega_{xe}}{\omega^2}\left(1 - \frac{\omega^2}{\omega_i^2} + \frac{\omega_{xi}\omega_{xe}}{\omega_e^2}\right)\right]$$

$$A_4 = \frac{\omega_e^4}{c^4}\left[\left(1 - \frac{\omega^2}{\omega_e^2} + \frac{\omega_{xi}\omega_{xe}}{\omega^2}\right)^2 - \frac{\omega_{xi}\omega_{xe}\omega^2}{\omega_i^2\omega_e^2}\right]$$

$$\left.\vphantom{\begin{array}{c}1\\1\\1\\1\\1\\1\end{array}}\right\} \quad (9.86)$$

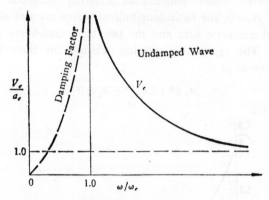

Figure 9.5   Electron sound speed in an ideal plasma

and

$$\omega_{xs} = \frac{eB_x}{m_s} = x\text{-wise cyclotron frequency of }s\text{th species} \quad (9.87)$$

The two solutions of Eq. (9.85) are:

$$\lambda_1^2 = \frac{-A_2 - (A_2^2 - A_0A_4)^{\frac{1}{2}}}{A_0} = \text{ordinary wave (electrons)} \quad (9.88)$$

$$\lambda_2^2 = \frac{-A_2 + (A_2^2 - A_0A_4)^{\frac{1}{2}}}{A_0} = \text{extraordinary wave (ions)} \quad (9.89)$$

The variations of λ_1^2 and λ_2^2 with frequency ω are shown in Fig. 9.6. One interesting result is that as $\omega \to 0$, the speed of wave propagation of these two interacted transverse waves, is given by the same formula:

$$V = \frac{\omega}{\lambda} = \frac{V_x}{\left(1 + \dfrac{V_x^2}{c^2}\right)^{1/2}} \qquad (9.90)$$

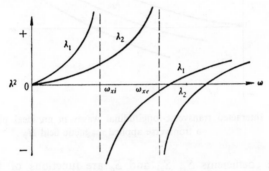

Figure 9.6 Interacted transverse wave in an ideal plasma under longitudinal applied magnetic field H_x

where $V_x = H_x(\mu_e/\rho_0)^{1/2}$ is the speed of Alfven's wave based on H_x. As $V_x \ll c$, Eq. (9.90) reduces to Eq. (6.79). Here again, the single fluid theory gives only the results at $\omega \to 0$. With the applied longitudinal magnetic field, we have undamped transverse waves at low frequencies, instead of damped waves for the case without applied magnetic field. Thus the applied magnetic field improves the transverse wave propagation at low frequencies. In the intermediate frequencies, one or both of the transverse waves may change into damped waves ($\lambda^2 < 0$), but at high frequencies, they become undamped waves again. As $\omega \to \infty$, these two basic waves are not interacted.

(d) Waves under transverse magnetic field. If $H_x = 0$ and $H_y \neq 0$, the first basic transverse wave (v_1, v_2, E_y, and h_z) is not affected by the applied magnetic field and the second transverse wave and the two longitudinal waves are interacted and three new transverse-longitudinal waves result. The dispersion relation of these interacted waves is

$$\lambda^6 + S_4\lambda^4 + S_2\lambda^2 + S_0 = 0 \qquad (9.91)$$

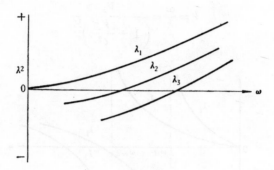

Figure 9.7   Interacted transverse-longitudinal waves in an ideal plasma under
a transverse applied magnetic field H_y

where the coefficients S_4, S_2, and S_0 are functions of the sound speeds, plasma frequencies, cyclotron frequencies of the two species, and the friction coefficients. For an ideal plasma, the solution of Eq. (9.91) are shown in Fig. 9.7. There are three interacted waves. The first wave λ_1^2 represents the interaction between the ion sound wave and the magnetic field H_y, and it is always an undamped wave. As $\omega \to 0$, the velocity of propagation of this wave when $V_y = H_y(\mu_e/\rho_0)^{1/2} \ll c$, is

$$V_1 = (a_p^2 + V_y^2)^{1/2} \qquad (9.92)$$

Eq. (9.92) is the same as Eq. (6.51). Hence the single fluid theory gives again the results of multifluid theory at $\omega \to 0$.

   At low frequencies, the other two interacted waves are damped waves. At high frequencies, these two damped waves turn into undamped waves. The points at which these damped waves turn into undamped waves depend on the relative magnitude of the plasma frequencies of the species to the cyclotron frequencies of the species. At very high frequencies, all three basic waves do not interact.[25]

   (e) Waves under arbitrarily oriented magnetic field. If both H_x and H_y are different from zero, all the four basic waves interact and we have four new modes of waves in a plasma whose dispersion relation is

$$C_0\lambda^8 + C_1\lambda^6 + C_2\lambda^4 + C_3\lambda^2 + C_4 = 0 \qquad (9.93)$$

The coefficients C_0, C_1, etc., are functions of plasma frequencies, cyclotron frequencies, sound speeds of these two species, and friction coefficient. In Fig. 9.8, we sketch the solutions of Eq. (9.93) for the case of an ideal plasma, $K_{ie} = 0$. At low frequencies, there are three undamped waves and one damped wave. These three undamped waves are the fast wave, slow wave, and transverse wave, as those given in the single fluid theory of chapter VI section 8 when $\omega \to 0$. Hence the single fluid theory gives only the results at very low frequencies. As the frequency of the wave increases, only one of the undamped waves remains undamped all the time; the other two waves may change first into damped waves at the ion and electron x-wise cyclotron frequencies respectively. But at high frequencies, they will return to the state of undamped wave. The fourth wave which is associated with the electron sound wave becomes undamped wave when $\omega > \omega_e$.

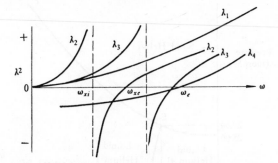

Figure 9.8 Waves in an ideal plasma under applied magnetic field of arbitrary orientation H_x and H_y

From the above results, we see that the multifluid theory gives us much more information than that from the classical single-fluid theory of chapter VI. If the plasma is a partially ionized plasma consisting of three species, electrons, ions and one kind of neutral particles, there will be a sound wave for the neutral particles in addition to all the other waves discussed in the two-fluid theory.

We may reexamine all the problems in the single-fluid theory of chapter VI by the multifluid theory. We will have much more information by using the multifluid theory, but the computation from the multifluid theory is much more complicated. The most important effects of the multifluid theory are (i) the difference of temperature of electrons and that of heavy particles, (ii) the non-equilibrium in degree of ionization as compared with the results of the Saha relation, and (iii) the effects of large diffusion velocities. Some of these results are given in references [26] and [27]. In general, there are only a very few examples of flow problems based on the multifluid theory reported in the literature.[47]

5.   Superfluids

At very low temperature near the absolute zero degree, quantum effects are of importance in the properties of the fluids even in the

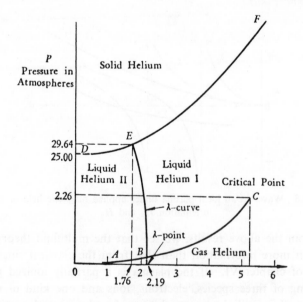

Figure 9.9   A sketch of the phase diagram of helium near the temperature of absolute zero

macroscopic point of view. Since near the absolute zero temperature, helium is the only matter that exists in the liquid state, superfluid is essentially concerned with liquid helium II. The first successful liquefaction of helium was accomplished by Kamerlingh Onnes[14] in 1908. In Fig. 9.9, the phase diagram of helium is sketched. The critical temperature below which the helium may be liquefied is 5.20°K and the boiling point is 4.21°K. Liquid helium near the critical point is known as liquid helium I, which is an ordinary liquid. At the boiling point, liquid helium I boils vigorously. At a temperature of 2.19°K, known as λ-point, because the curves *ABC-BM* looks like an inverted λ, a second order phase transition occurs. Below the λ-point, liquid helium I will change into liquid helium II which is an abnormal liquid. Liquid helium II appears as a transparent, quiescent liquid which refuses to boil. It has a very high thermal conductivity which may be 1000 times greater than that of pure copper at room temperature. The transition from liquid helium I to liquid helium II involves no latent heat and no discontinuous change in volume. Hence it is not a first order transition but a second order transition. The behavior of liquid helium II is so remarkable that quantum effects at this low temperature are predominant. Hence we may call it a quantum liquid. Special treatizes should be referred to for the quantum theory of superfluids (see references [4], [8] and [17]).

In the macroscopic theory of the flow of liquid helium II, we may consider the liquid helium II as a mixture of two fluids: one is the superfluid which has no viscous force, nor heat transfer and which flow is always a potential flow; and the other part is a normal viscous liquid. It should be noted that actually the liquid helium II cannot be separated into such two parts. Separation into two parts is useful in the analysis of the dynamics of liquid helium II. Under this macroscopic theory, we may apply the two-fluid theory as discussed in the last section to analyze the flow of liquid helium II as follows:

Let $\rho_n =$ the density of the normal part of the fluid

$\rho_s =$ the density of the superfluid part

$\rho =$ the total density of the fluid

then

$$\rho = \rho_n + \rho_s \qquad (9.94)$$

The velocity vector of the normal fluid q_n is different from that of the superfluid part q_s. We have to consider two velocity vectors q_n and q_s in the flow field of liquid helium II. The total mass flux J_m is then

$$J_m = \rho q = \rho_n q_n + \rho_s q_s \qquad (9.95)$$

The densities ρ_n and ρ_s depend in general on the velocities as well as the temperature of the fluid. In liquid helium II, we may assume that the temperatures of these two parts are the same. Hence we need only one temperature T in the present analysis. At high velocities, liquid helium actually ceases to be superfluid. At low velocities, we may regard the both ρ_n and ρ_s depend on temperature T only.

From the above assumptions, the unknowns in the dynamics of liquid helium II are the velocity vectors q_n and q_s, the pressure of the fluid p, and the temperature of the fluid T. The four equations which govern these unknowns are:

Equation of motion of the superfluid part is

$$\rho_s \left(\frac{\partial}{\partial t} + q_s \cdot \nabla \right) q_s = -\frac{\rho_s}{\rho} \nabla p + \rho_s S \nabla T \qquad (9.96)$$

where the interaction force between these two parts is zero.

Equation of motion of the normal fluid part is

$$\rho_n \left(\frac{\partial}{\partial t} + q_n \cdot \nabla \right) q_n = -\frac{\rho_n}{\rho} \nabla p - \rho_s S \nabla T$$
$$+ \mu_n \left(\nabla^2 q_n + \frac{1}{3} \nabla (\nabla \cdot q_n) \right) \qquad (9.97)$$

where μ_n is the coefficient of viscosity of the normal fluid.

Equation of continuity is

$$\frac{\partial \rho}{\partial t} + \nabla \cdot J_m = 0 \qquad (9.98)$$

and the equation of energy is

$$\frac{\partial \rho \overline{e}_m}{\partial t} + \nabla \cdot Q_f = 0 \tag{9.99}$$

where $\overline{e}_m$ is the total energy per unit mass of the fluid and Q_f is the energy flux density. The total energy $\overline{e}_m$ contains the internal energy, kinetic energy, and potential energy, and the energy flux density Q_f contains the energy flux by convection, heat conduction, and viscous dissipation.

In the case when the dissipative processes are neglected, Eq. (9.99) reduces to the equation of conservation of entropy S, i.e.,

$$\frac{\partial \rho S}{\partial t} + \nabla \cdot (\rho \, S q_n) = 0 \tag{9.100}$$

because the entropy change is associated with the normal fluid part only.

In the equations of motion, there are two terms which depend on the gradient of temperature. This expresses the fact that the superfluid part tends to move towards regions of high temperature and the normal fluid part tends to move towards lower temperature. The densities ρ_n and ρ_s vary according to the temperature. At absolute zero temperature $\rho_n = 0$ and liquid helium II becomes completely superfluid; while on the λ-curve, $\rho_s = 0$, the liquid helium II becomes completely normal fluid.

We may use Eqs. (9.96) to (9.100) to investigate the flow problems of liquid helium II for various boundary conditions. The boundary conditions for the normal fluid are $q_n = 0$ on the solid wall, i.e., the no-slip boundary conditions may be used and q_s is unrestricted. The last assumption has not been verified experimentally.

There are many abnormal phenomena of liquid helium II in comparison with those of normal fluid. The following are some of the well known results of liquid helium II flow:

(1) The liquid helium II will flow through the small gap of two plates with very little viscosity. The coefficient of viscosity determined by the flow rate of liquid helium II is less than 10^{-11}

poise while that of liquid helium I is 2×10^{-5} poise. The reason is that the superfluid part of the liquid is capable of flowing through the gap at very high speed so that the total quantity flowing increases and the effective coefficient of viscosity is reduced. The flow is almost independent of the pressure head and varies in a complicated manner with the width of the gap.

(2) We may determine the coefficient of viscosity by observing the damping of the torsional oscillation of the disk immersed in liquid helium II. If the peripheral velocity cf the disk is not too great, the value of the coefficient of viscosity so obtained is of the order of magnitude as that of liquid helium I, i.e., of the order of 10^{-5} poise. The reason for the high viscosity in this test is that the normal fluid part of the liquid is responsible for the damping of the motion of an oscillating disk. Andronikashvilli[3] tested the torsional oscillation of a pile of disks in liquid helium II. The gap between the disks was very small so that the normal fluid part of the liquid was dragged round with them and increases the moment of inertia of the system. He found that the period of oscillation continuously decreased as the temperature was lowered below the λ-point. From this experiment, Andronikashvilli showed that the fraction of the superfluid part increases from zero at the λ-point to unity at absolute zero temperature.

(3) Any surface dipping in liquid helium II is covered by a film which is about 3×11^{-6} cm thick at a height of 1 cm above the surface of the bulk liquid. The superfluid part moves freely through this thin film. If one dips an empty vessel in a path of the liquid helium II, it gradually fills with the liquid through the film until the levels of the liquid are equal inside and outside. There is a maximum flow rate of the superfluid in the film. The film is rapidly accelerated up to a critical velocity which is determined by the critical rate of flow, and the velocity remains unchanged thereafter. The critical rate of flow is a function of temperature and independent of the pressure head. The concept of critical rate of flow holds also for the flow in a small gap, i.e., case (1).

(4) If we empty the liquid helium II through some very small orifice from a given vessel, the temperature of the liquid in the vessel

will rise. The reason is that only the superfluid part will be able to flow through these small orifices and the liquid flowing out left its entropy behind to warm up the remaining liquid in the vessel.

(5) Fountain effect. Allen and Jones[1] used an apparatus shown in Fig. 9.10 in which a vessel U is dipped in a bath of liquid helium II. The liquid can flow into the vessel through the packed emery powder. When a light is shone on the emery powder whose temperature rises, the liquid will rush in with such a force that it forms a fountain of height as much as 30 cm. The explanation is that the superfluid part has a tendency to move towards regions of high-temperature. Because only the superfluid part can flow freely through the emery powder gaps, when the temperature increases, it will rush into the vessel to form the fountain.

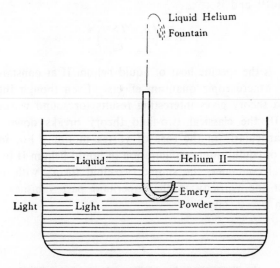

Figure 9.10   Fountain effect of liquid helium II

(6) Second sound wave. In the last section, we showed that the mixture of two fluids usually has two different sound speeds which represent two different modes of propagation of small disturbances: the ion sound wave, and the electron sound wave. This phenomenon is true for all mixtures of two different species. Since

liquid helium II may be regarded as a mixture of two fluids, super-fluid and normal fluid, one would expect that there are two sound waves in it. Indeed it has been found both experimentally and theoretically that there are two sound waves in liquid helium II. The speeds of these two sound waves can be easily obtained from the linearized equations of the dynamics of liquid helium II. If one neglects the viscous dissipation, the sound speed of the first sound wave is the same as that of ordinary sound speed,[4] i.e., the speed of first sound wave in liquid helium II is

$$a_1 = \left(\frac{\partial p}{\partial \rho} \right)_s^{\frac{1}{2}} \qquad (9.101)$$

The speed of the second sound wave depends on the density of the superfluid[4] and is

$$a_2 = \left(\frac{\rho_s}{\rho_n} \frac{TS^2}{c_v} \right)^{\frac{1}{2}} \qquad (9.102)$$

where c_v is the specific heat of liquid helium II at constant volume.

(7) Macroscopic quantum effects. Even though the ordinary two fluid theory gives interesting results for sound waves in liquid helium II, the classical two-fluid theory breaks down in certain cases, particularly in the case of vortex motion. For instance, in the case of equilibrium configuration of liquid helium II in a rotating cylindrical container, the simple two-fluid theory with the assumption that the vorticity in the superfluid part vanishes $\nabla \times q_s = 0$ and then $q_s = 0$, gives

$$z = \frac{\rho_n}{\rho} \frac{\omega^2 r^2}{2g} \qquad (9.103)$$

where z is the height of the surface along the axis of rotation and ω is the angular velocity of the container, r is the perpendicular distance from the axis of rotation, and g is the gravitational accelera-tion. Eq. (9.103) depends on the temperature of liquid helium II because of the factor ρ_n/ρ. However, from careful experimental results, one finds that z is independent of the temperature and is given by the classical formula:

$$z = \frac{\omega^2 r^2}{2g} \qquad (9.104)$$

We have to modify the classical two fluid theory in order that the theoretical result agrees with the experimental observation. This modification is to quantize the vortex. For simplicity, we assume that a vortex filament in superfluid is represented by the velocity distribution:

(i) $\qquad\qquad 0 \leq r \leq r_0: \ q_s = \omega_s r$

and

(ii) $\qquad\qquad r \geq r_0 : q_s = \frac{A}{r} i_\theta$

where A is a constant. We determine the constant A by quantization, i.e.,

$$\oint q_s \, r \, d\theta = L\Gamma, \ L = 0 \ ,1, \ 2, \ \cdots \qquad (9.105)$$

and

$$A = \frac{L\Gamma}{2\pi} \qquad (9.105a)$$

where Γ is the strength of singly quantized vortex. Hence the properties of the quantized vortices are still the same as classical vortices but their strength is the integral multiple of Γ.

If we assume a uniform distribution of vortices in the cylindrical container considered above, we have approximately the classical result of Eq. (9.104) if the density of the vortex filaments were $n_0 = 2\omega/\Gamma$. Since Γ is equal to 9.97×10^{-4} cm²/sec, n_0 is 2000 vortices lines per square centimeter.

The quantized vortex rings had been observed experimentally.[8]

There are two isotopes of helium, i.e., the helium four He^4 and the helium three He^3. The abnormalities of liquid helium belong to He^4 only. The helium three He^3 does not have the property of superfluidity at temperature as low as $0.1\,°K$. The explanation can be obtained only from quantum theory. The He^4 contains an even number of fundamental particles, i.e., two protons, two neutrons, and two electrons, and must obey Bose-Einstein statistics. The abnormalities of liquid helium have been ascribed to the peculiarities of these statistics.   The He^3 atom contains an odd number of fundamental particles, i.e., two protons, one neutron, and two electrons, and must obey Fermi-Dirac statistics.   Hence no superfluidity will result.   A study of the mixture of He^3 and He^4 would be interesting in the superfluidity of liquid helium II.

6.   Relativistic Fluid Mechanics

When the velocity of fluid particles, of which photons are considered as a special kind, is not small in comparison with the velocity of light, the relativistic effect should be considered in the analysis of the fluid motion.   We are going to discuss briefly the relativistic analysis of the equations of motion of fluid flow.   Of course, these equations will reduce to ordinary gasdynamical equations when the velocity of the fluid q is much smaller than the velocity of light c; in other words, the relativistic parameter $R_r = q/c$ of Eq. (4.74) or its square is much smaller than unity.

We shall derive the equations of relativistic fluid mechanics based on the special theory of relativity.   The basic postulates of relativity are:

(i)   It is impossible to measure or detect the unaccelerated translatory motion of a system through free space, and

(ii)   The velocity of light in free space is the same for all observers independent of the relative velocity of the source of light and the observer.

Because of these two postulates, we have a relativistic conception of space and time.   In classical mechanics, which we used exclusively before, we consider only the three-dimensional space. The location at a point in the space may be represented by a vector r which has three spatial components x_i, $i = 1$, 2, or 3.   The time t is considered as a scalar.   However, in the theory of relativity, we have to use four dimensional space to describe physical quantities. The coordinates of a certain reference system x_α has four components, i.e., $\alpha = 1$, 2, 3, and 4.   The first three components refer to the ordinary spatial coordinates and the fourth one $x_4 = ct$ where t is the time. Hence in the theory of relativity, we consider time as the fourth dimension. Let us consider two Cartesian coordinate systems (x, y, z, t) and (x', y', z', t') with a relative velocity V in the direction of the x-axis between the two systems. If at time $t = 0$, the origins of the two systems coincide, Lorentz found that the relations between these two systems are:

$$x' = \frac{x - Vt}{(1 - V^2/c^2)^{1/2}}, \ y' = y, \ z' = z, \ t' = \frac{t - xV/c^2}{(1 - V^2/c^2)^{1/2}} \quad (9.106)$$

Eq. (9.106) is known as Lorentz transformation. From Eq. (9.106), we have the Lorentz invariant of the elementary length in the four-dimensional space ds given by the relation:

$$-ds^2 = dx^2 + dy^2 + dz^2 - c^2 dt^2 = dx'^2 + dy'^2 + dz'^2 - c^2 dt'^2 \quad (9.107)$$

In classical mechanics, the term V^2/c^2 is negligible and the Lorentz transformation (9.106) will reduce to the Galilean transformation and the time will be unchanged.

In relativistic mechanics, we consider four-dimensional space and all vectors have four components. We have to generalize all definitions of the vectors in the three-dimensional space of classical mechanics to the four-dimensional space of relativistic mechanics. We would like to know, especially, what is the meaning of the fourth component of a vector in terms of the quantities in classical mechanics. Before we discuss some important 4-vectors in relativistic mechanics, we would like to discuss a little about the properties of the 4-vector in general. From now on, we use the Greek letters α, β, etc., to denote the components in the four-dimensional space, i.e., $\alpha = 1, 2, 3,$ or 4 and the English letter i, j, etc., to denote the ordinary components in the three-dimensional space, i.e., $i = 1, 2,$ or 3.

The Lorentz transformation (9.106) is a special case of a general linear transformation:

$$x'_\alpha = A_{\alpha\beta} x_\beta \quad (9.108)$$

where the summation convention is used, i.e.,

$$x'_\alpha = A_{\alpha\beta}x_\beta = A_{\alpha1}x_1 + A_{\alpha2}x_2 + A_{\alpha3}x_3 + A_{\alpha4}x_4 \qquad (9.108a)$$

where x_β is the coordinate in the β-axis of a rectangular system. If the unit vector along the β-axis is i_β, the 4-vector of position x is then

$$x = i_\beta x_\beta \qquad (9.109)$$

Since the coordinate system is assumed to be orthogonal, the scalar product of the two-unit vector gives

$$i_\alpha \cdot i_\beta = \delta_{\alpha\beta} \qquad (9.110)$$

where $\delta_{\alpha\beta} = 0$ if $\alpha \neq \beta$ and $\delta_{\alpha\beta} = 1$ if $\alpha = \beta$.

Eq. (9.108) gives the relation of coordinates for a given vector x in the four-dimensional space in a system X' with components x'_α to those in a system X with component x_β. Now we can easily find the matrix $A_{\alpha\beta}$ from the definition of these components, i.e.,

$$x'_\alpha = i'_\alpha \cdot x = i'_\alpha \cdot (i_\beta x_\beta) = A_{\alpha\beta}x_\beta \qquad (9.111)$$

where

$$A_{\alpha\beta} = i'_\alpha \cdot i_\beta \qquad (9.112)$$

Comparing Eq. (9.111) with the Lorentz transformation (9.106), we have

$$x_1 = x, \quad x_2 = y, \quad x_3 = z, \quad x_4 = ict \qquad (9.113)$$

and

$$A = \begin{vmatrix} \dfrac{1}{(1-V^2/c^2)^{1/2}} & 0 & 0 & \dfrac{iV/c}{(1-V^2/c^2)^{1/2}} \\ 0 & 1 & 0 & 0 \\ 0 & 0 & 1 & 0 \\ \dfrac{-iV/c}{(1-V^2/c^2)^{1/2}} & 0 & 0 & \dfrac{1}{(1-V^2/c^2)^{1/2}} \end{vmatrix} \qquad (9.114)$$

where $i = \sqrt{-1}$. Since the magnitude of the vector x is an invariant under the rotation of the coordinate system, we have immediately Eq. (9.107) in the following form:

$$-ds^2 = x_\alpha x_\alpha = x'_\alpha x'_\alpha \tag{9.115}$$

where ds is the elementary length in the four-dimensional space.

Now we would like to know what are the components of a velocity vector in the four-dimensional space. In the three-dimensional space, the velocity component q_i is defined as $q_i = dx_i/dt$. Since dt is not an invariant in the theory of relativity, we should define the velocity vector slightly different in the theory of relativity from that in the classical mechanics. The velocity vector in the four-dimensional space is defined in terms of the invariant element ds defined in Eq. (9.115) as follows:

$$q_\alpha = \frac{dx_\alpha}{ds} \tag{9.116}$$

From Eqs. (9.107) and (9.115), we have

$$ds = c\,dt \left(1 - \frac{dx_i dx_i}{c^2\,dt^2} \right)^{1/2} = c\,dt \left(1 - \frac{q^2}{c^2} \right)^{1/2} \tag{9.117}$$

where $q^2 = q_i q_i$. From Eqs. (9.116) and (9.117), we see that the velocity vector in the four-dimensional space is a nondimensional quantity with components:

$$q_1 = \frac{u}{cb_0}, \quad q_2 = \frac{v}{cb_0}, \quad q_3 = \frac{w}{cb_0}, \quad q_4 = \frac{i}{b_0} \tag{9.118}$$

where $b_0^2 = 1 - (u^2 + v^2 + w^2)/c^2$ and u, v, and w are respectively the velocity components in the sense of classical mechanics along the x-, y-, and z-axis. Similarly, we define the acceleration of a particle in the four-dimensional space as dq_α/ds.

The conservation of number of fluid particles, the equation of continuity, may be expressed in terms of the particle flux n_α which has four components and which may be expressed in terms of the velocity vector q_α as follows:

$$n_\alpha = nq_\alpha \tag{9.119}$$

where n is a scalar which is the number density of the particles. The equation of continuity is then

$$\frac{\partial}{\partial x_\alpha}(nq_\alpha)=0 \qquad (9.120)$$

Eq. (9.120) reduces to the ordinary equation of continuity of fluid mechanics (9.37) if the relativistic effect is neglected, i.e., $b_0=1$ and $m=$ constant.

One of the main features of relativistic mechanics is that the mass of a particle is not a constant but increases with velocity of the particle q according to the Einstein formula:

$$m=\frac{m_0}{(1-q^2/c^2)^{\frac{1}{2}}} \qquad (9.121)$$

where m_0 is the rest mass of a particle when it is not moving and m is the mass of the particle when it moves with a velocity q. The velocity q is referred to the velocity in the three-dimensional space, i.e., $q^2=q_iq_i=u^2+v^2+w^2$. It should be noted that the velocity vector in the four-dimensional space q_α is essentially a unit vector, i.e.,

$$q_\alpha^2=q_\alpha q_\alpha=\frac{dx_\alpha dx_\alpha}{ds^2}=-1 \qquad (9.122)$$

Newton's second law of motion in the three-dimensional space is

$$F_i=\frac{d(mq_i)}{dt}=\frac{dp_i}{dt} \qquad (9.123)$$

where F_i is the ith component of the force acting on a particle of a mass m and velocity q_i and momentum $p_i=mq_i$. In classical mechanics, $R_s=q^2/c^2\ll1$ and the mass m may be assumed to be a constant. Hence we may write

$$F_i=m\frac{dq_i}{dt} \qquad (9.124)$$

However in relativistic mechanics, the mass m is no longer a constant. Thus it is convenient to use the momentum vector p_i instead of the velocity vector q_i. In the four-dimensional space, the momentum vector p_α is defined by the following expression:

$$p_\alpha = m_0 c^2 q_\alpha \qquad (9.125)$$

Now if we use $p_i = mq_i$ as the momentum in the three-dimensional space of ordinary classical mechanics, the first three components of the 4-momentum vector p_α are simply cp_i while the 4th component $p_4 = imc^2 = iE_m$ where E_m is the total energy of the particle at velocity q. Hence the 4-momentum vector is sometimes referred to as the momentum-energy vector.

The equation of motion in relativistic fluid mechanics may be expressed in terms of a momentum-energy tensor $T_{\alpha\beta}$. In the proper system in which the fluid element is at rest, the momentum-energy tensor $T_{\alpha\beta}$ has the following components:

$$T_{\alpha\beta} = \begin{vmatrix} p & 0 & 0 & 0 \\ 0 & p & 0 & 0 \\ 0 & 0 & p & 0 \\ 0 & 0 & 0 & U_m \end{vmatrix} \qquad (9.126)$$

where p is the pressure of the fluid and U_m is its internal energy.

In other coordinate system, $T_{\alpha\beta}$ may be expressed in terms of the velocity vector and dissipation terms. If we neglect the dissipative processes which correspond to the Euler's equations in classical fluid mechanics, we have

$$T_{\alpha\beta} = A_m q_\alpha q_\beta + pg_{\alpha\beta} \qquad (9.127)$$

where $A_m = U_m + p$ and the value of $g_{\alpha\beta}$ are $g_{11} = g_{22} = g_{33} = 1$, $g_{44} = -1$, $g_{\alpha\beta} = 0$ if $\alpha \neq \beta$.

In the three-dimensional space, we have

$$\left. \begin{array}{l} T_{ij} = \dfrac{A_m q_i q_j}{c^2(1-q^2/c^2)} + p, \quad T_{i4} = \dfrac{iA_m q_i}{c(1-q^2/c^2)} \\[4mm] T_{44} = \dfrac{A_m}{1-q^2/c^2} - p = \dfrac{U_m + pq^2/c^2}{1-q^2/c^2} \end{array} \right\} \qquad (9.128)$$

The equation of motion in relativistic fluid mechanics is given by the divergence of the momentum-energy tensor, i.e.,

$$\frac{\partial T_\alpha^\beta}{\partial x^\beta} = 0 \tag{9.129}$$

Substituting Eq. (9.127) into Eq. (9.129), we have

$$\frac{\partial T_\alpha^\beta}{\partial x^\alpha} = q_\alpha \frac{\partial A_m q^\beta}{\partial x^\beta} + A_m q^\beta \frac{\partial q_\alpha}{\partial x^\beta} + \frac{\partial p}{\partial x^\alpha} = 0 \tag{9.130}$$

With the help of the equation of continuity (9.120), Eq. (9.130) may be written in the following form:

$$A_m q^\beta \frac{\partial q^\alpha}{\partial x^\beta} = -\frac{\partial p}{\partial x^\alpha} - q^\alpha q^\beta \frac{\partial p}{\partial x^\beta} \tag{9.131}$$

The first three components of Eq. (9.131) are the Euler's equations of motion in relativistic fluid mechanics and the fourth component of Eq. (9.131) is

$$\frac{\partial S q^\alpha}{\partial x^\alpha} = 0 \tag{9.132}$$

which is the relativistic energy equation of an ideal fluid, S is the entropy per unit proper volume and the entropy S is given by the thermodynamic identity:

$$d\left(\frac{A_m}{n}\right) - \frac{dp}{n} = Td\left(\frac{S}{n}\right) \tag{9.133}$$

where T is the temperature.

If we linearize Eq. (9.130) by considering the sound wave in a uniform state, we have the following linearized equations:

$$\frac{\partial U'_m}{\partial t} = -A_m \frac{\partial q^i}{\partial x^i}, \quad \frac{A_m}{c^2} \frac{\partial q^i}{\partial t} = -\frac{\partial p'}{\partial x^i} \tag{9.134}$$

where prime refers to the perturbed quantities and q^i is the ith component of velocity in the three dimensional space.

For an adiabatic process, we have

$$U'_m = \left(\frac{\partial U_m}{\partial p}\right)_{ad} \cdot p' \tag{9.135}$$

By the help of Eq. (9.135), Eqs. (9.134) give

$$\frac{1}{a^2}\frac{\partial^2 q'}{\partial t^2}=\frac{\partial^2 p'}{\partial x_i^2} \tag{9.136}$$

where

$$a=\text{sound speed}=c\left(\frac{\partial p}{\partial U_m}\right)^{1/2}_{ad} \tag{9.137}$$

Eq. (9.136) is the sound wave equation in relativistic fluid mechanics. The only difference of Eq. (9.136) from that of ordinary gasdynamics is that the mass density is replaced by U_m/c^2.

We may generalize the momentum-energy tensor by including the disipative processes or by including the electromagnetic forces.[7, 15]

In the above analysis, we consider the relativistic fluid mechanics from the continuum theory point of view. We may also consider it from the microscopic point of view, i.e., the kinetic theory of a relativistic gas. Let us consider the Boltzmann equation for photons:

Instead of using the specific intensity I_ν as functions of frequency and direction cosine l^α, as we did in chapter VII, we describe the thermal radiation as being a gas of point like photons, each of which being characterized, at any time t, by a position vector x^α and a momentum p^α. It is easy to show that

$$p^\alpha=\frac{h\nu l^\alpha}{c}, \ |\ p\ |=\frac{h\nu}{c} \tag{9.138}$$

We denote the distribution function of photons by F_R which is defined by the relation that

$$F_R(t,\ x^\alpha,\ p^\alpha)dxdp \tag{9.139}$$

is the number of photons in $dx=dx^1dx^2dx^3$ at x^α and in $dp=dp^1dp^2dp^3$ at p^α and at time t.

The relation between the specific intensity I_ν and F_R is

$$I_\nu=\frac{h^4\nu^3}{c^2}F_R \tag{9.140}$$

The 4-momentum of a photon whose 3-momentum is p^i will be

p^α with

$$p^\alpha p_\alpha = 0, \quad p^4 = |p| \tag{9.141}$$

The equation which governs F_R can be derived from the radiative transfer equation (7.12) and relation (9.140), i.e.,

$$p^\alpha \frac{\partial F_R}{\partial x^\alpha} = \frac{|p|}{c}\left(\frac{\delta F_R}{\delta t}\right) \tag{9.142}$$

where

$$\frac{|p|}{c}\left(\frac{\delta F_R}{\delta t}\right) = \frac{ck_\nu}{h^3\nu^2}(J_\nu - I_\nu) \tag{9.143}$$

The term $\delta F_R/\delta t$ is the time variation of F_R due to collision of photons with material particles, i.e., due to emission and absorption processes. Eq. (9.142) may also be written as

$$p^\alpha \frac{\partial F_R}{\partial x^\alpha} = B_1 - B_2 F_R \tag{9.144}$$

where

$$B_1 = \frac{ck_\nu}{h^3\nu^2}j_\nu \text{ and } B_2 = \frac{b\nu k_\nu}{c}$$

Eq. (9.144) is the relativistic Boltzmann equation for photons.

From the distribution function F_R, we may define the macroscopic quantities such as radiation pressure, etc., in the same manner as those for material particles discussed in chapter VIII. The final results are the same as those obtained in chapter VII from the radiative transfer equation (7.12).[35]

7. Biofluid Mechanics

Biological science has been recently advanced from pure descriptive method to some analytical methods. Many analytical methods of physical science have been used successfully in the study of biological science. Hence we have a new science known as biophysics. Biofluid mechanics is a branch of biophysics in which we study the fluid flow problems associated with biological systems. In principle, biofluid mechanics does not involve any new development of the general principles of fluid mechanics but it does involve some new applications of the method of fluid mechanics. In some of the cases, the conditions for the fluid flow problems in the biological system are so unique that it has not been studied extensively in problems other than those associated with the biological system. The most common flow problem in biological system is the flow of blood. In 1840, French physician J. L. M. Poiseuille was interested in the blood flow and conducted the study of flow in capillaries. It is the well known Poiseuille flow in pipes of ordinary fluid dynamics. But we know that ordinary Poiseuille flow does not represent the actual blood flow in a cardiovascular system. Even though we may consider as a first approximation of blood flow by the Poiseuille pipe flow, many improvements have been made since the Poiseuille investigation. Many biofluid mechanics problems are concerned with the classical fluid mechanics but also its modern aspects such as chemical reactions, electrical effects, rheology, etc. Biofluid mechanics is still a newly developed science and it is not feasible to give a complete detailed survey here. What we are going to discuss are some general principles concerning how we may apply some well known methods of fluid mechanics to the biological system.

An excellent review of biomechanics, its scope, history, and some problems, has been given by Fung in reference [11]. It is interesting to notice that biomechanics had been a very active subject more than a hundred years ago but it was not very active during the last few decades. Famous scholars of fluid mechanics, such as Leonhard Euler (1707—1783), Jean Poiseuille (1799—1869), Hermann von Helmholtz (1821—1894), Adolf Fick (1829—1901), Horace Lamb (1849—1934), and many others made valuable contributions in biomechanics. Fung suggested that von Helmholtz may be regarded as the father of bioengineering. Von Helmholtz was professor of physiology and pathology at Konigsberg, professor of anatomy and physiology at Bonn, professor of physiology at Heidelberg, and finally professor of physics in Berlin (1871). His theorem of vorticity is well known in fluid mechanics, but his contributions in the mechanism of the eye and that of hearing and many other biophysical problems are well known in physiology.

We are going to discuss the following four items which may give us a general view of the scope of biofluid mechanics:

(1) Biological similarity.[37] In mathematical physiology, one assumes that various organisms and physiological mechanisms operate similarly. For instance, we may assume that all mammals have the same type of respiratory, renal or cardiovascular system. Hence we may use the dimensional analysis of chapter IV to study the geometrical similarity and dynamical similarity of these biological systems. Such similarity criteria have been extensively studied by research workers in the field of biofluid mechanics. At first, they tried to find some simple invariant relations and later when they have more informations of these systems, they apply the π-theorem of Eq. (4.42) to various problems. Let us give a few examples:

For mammals, the following approximate mass invariants for various nondimensional quantities have been found:

(a) Shape factor:

$$\frac{\text{tidal lung volume}}{\text{heart volume}} = 0.94\, M_a^{0.03} \qquad (9.145)$$

where M_a is the ratio of masses of two mammals. Since the power 0.03 is rather small, we may assume that the volume ratio of Eq. (9.145) is independent of the mass.

(b) Time ratio:

$$\frac{\text{breath time}}{\text{pulse time}} = 3.9\, M_a^{0.01} \qquad (9.146)$$

(c) Flow rate ratio:

$$\frac{\text{minute air volume}}{\text{minute oxygen volume}} = 32 M_a^{0.01} \qquad (9.147)$$

(d) Concentration ratio:

$$\frac{\text{urine sodium ion concentration}}{\text{potassium ion concentration}} = 3.0 M_a^{0.00} \qquad (9.148)$$

The above results are some first order approximate values. If we study the operation of a given biological system, we will not have such simple invarient relations. On the other hand, we may find that the nondimensional quantity which we are interested in is a function of various nondimensional parameters. For instance, if we study the gas flow velocity in bronchi, we may have the following nondimensional relation:

$$\frac{v}{Lf} = \phi\left(\frac{v_r}{v}, \frac{K_1}{f}, \frac{L}{D}, \frac{V_T}{L^3}, \frac{p}{\frac{1}{2}v^2\rho}, \frac{\mu L}{A\rho v}, S\right) \quad (9.149)$$

where v is the velocity of the air flow at corresponding point in pulmonary trees of two similar respiratory systems, L is a reference length such as lung radius and f is respiratory frequency. Hence the left-hand side of Eq. (9.149) is the Strouhal number which characterizes the gas flow velocity in bronchi. This Strouhal number is a function of various nondimensional parameters given in the right-hand side of Eq. (9.149). The first term is the ratio of the membrane transport velocity of oxygen v_r to the flow velocity v; the second term is the ratio of the metabolic turnover constant K_1 to the respiratory frequency f; the third term is the ratio of lung radius L to the diameter of bronchi D; the fourth term is the ratio of the tidal volume V_T to the volume represented by L^3; the fifth term is the ratio of the pleural driving pressure p to the dynamic pressure of the air where ρ is the density of the air; the sixth term is essentially the Reynolds number of the problem where A is the cross sectional area of the bronchi and μ is the coefficient of viscosity of the air and the seventh term is a shape factor S of the bronchi. The complete solution of this problem is rather complicated. But the relation such as Eq. (9.149) is useful to correlate the experimental data.

(2) Hydrodynamics of blood flow.[5] When we study the blood flow in a body of mammal, we have to know the mechanical aspects of the circulatory system. The complete system is too complicated to be studied as a whole and we have to study various parts separately. From fluid dynamics point of view, we may roughly divide the circulatory system into four major parts as follows:

(a) The heart. The heart acts as a pump which causes the circulation of blood in the body. In fact, the heart acts as two pumps: one is a high pressure system which causes the peripheral circulation and the other is a low pressure system which is associated with pulmonary. These two systems are closely connected. The study of such a pump, the artificial heart, is one of the most interesting problems in biofluid mechanics.

(b) Arteries. The arteries form a distribution system for the blood flow which connects the heart to various organs and back as well as the heart and pulmonary. The study of blood flow at present time is mainly concerned with the flow of blood in arteries. The simplest type of blood flow in arteries is the Poiseuille flow in pipes. However, when we study the actual situation a little more in detail, we will find that the actual situation is much more complicated than the well known steady laminar flow in a circular pipe. First, the flow is not steady but pulsating. The actual variation of the blood flow with time is very complicated. Hence we have to study unsteady flow in pipes. Second, the arteries are not rigid pipes but elastic pipes. The size and/or the shape of the arteries change according to the pressure of the blood flow. We have to study the interaction of the blood flow and the elastic deformation of the arteries. Such an interaction problem has been studied by engineers for external flow associated with high speed airplanes and high speed hydrofoils, and is known as aeroelasticity or hydroelasticity. Now we have to apply the principles of aeroelasticity or hydroelasticity for the internal flow in a pipe. Third, the blood may not be considered as a simple Newtonian fluid with Navier-Stokes stress-strain relations. We may need to study the non-Newtonian fluid flow in an elastic pipe with a variable cross section and under the pulsatile condition. It is a very complicated situation. The complete analysis has not been carried out yet. One of the most interesting current research problems concerning the blood flow is to determine the stress-strain relation of the blood flow.[11-13]

In literature on the hydrodynamics of blood flow, many simple assumptions are made. For instance, we assume that (i) the blood is an incompressible Newtonian fluid, (ii) the blood flow is laminar, (iii) the artery is an infinitely long circular pipe of uniform cross section, (iv) the physical properties of the pipe are linear, and (v) the wall of the pipe is thin and is made of isotropic and homogeneous material. Under these conditions, we may write down the fundamental equations of hydroelasticity in cylindrical coordinates for the pipe flow with axial symmetry, i.e., all the variables are independent of angular coordinate and there is no tangential velocity nor tangential displacement of the wall.

The equation of continuity for incompressible fluid of density ρ is

$$\frac{\partial u_r}{\partial r} + \frac{u_r}{r} + \frac{\partial u_z}{\partial z} = 0 \qquad (9.150)$$

where u_r and u_z are respectively the radial (r–) component and the axial (z–) component of velocity of the blood flow.

The equations of motion are

$$\rho\left(\frac{\partial u_r}{\partial t} + u_r\frac{\partial u_r}{\partial r} + u_z\frac{\partial u_r}{\partial z}\right) = -\frac{\partial p}{\partial r}$$
$$+ \mu\left(\frac{\partial^2 u_r}{\partial r^2} + \frac{1}{r}\frac{\partial u_r}{\partial r} + \frac{\partial^2 u_r}{\partial z^2} - \frac{u_r}{r^2}\right) \qquad (9.151)$$

$$\rho\left(\frac{\partial u_z}{\partial t} + u_r\frac{\partial u_z}{\partial r} + u_z\frac{\partial u_z}{\partial z}\right) = -\frac{\partial p}{\partial z}$$
$$+ \mu\left(\frac{\partial^2 u_z}{\partial r^2} + \frac{1}{r}\frac{\partial u_z}{\partial r} + \frac{\partial^2 u_z}{\partial z^2}\right) \qquad (9.152)$$

The equations of motion of the wall:

$$\rho_w H\frac{\partial^2 s_r}{\partial t^2} - p + \frac{E^*h}{1-\sigma^{*2}}\left(\frac{\sigma^*}{r}\frac{\partial s_z}{\partial z} + \frac{\partial s_r}{\partial r^2}\right) = 0 \qquad (9.153)$$

$$\rho_w H\frac{\partial^2 s_z}{\partial t^2} + Ks_z + \left(\frac{\partial u_z}{\partial r} + \frac{\partial u_r}{\partial z}\right)_{r=r_i}$$
$$- \frac{E^*h}{1-\sigma^{*2}}\left(\frac{\partial^2 s_z}{\partial z^2} + \frac{\sigma^*}{r}\frac{\partial s_r}{\partial z}\right) = 0 \qquad (9.154)$$

where s_r and s_z are respectively the r- and z-displacement of a small wall element, ρ_w is the density of the wall material, H is the weighted volume of the wall substance, μ is the coefficient of viscosity of the blood, p is the pressure, K is the spring constant per unit inner pipe wall area of the external constraint, h is the wall thickness, E^* is the complex elastic modulus, the real part of which is Young's modulus, and σ^* is the complex Poisson ratio.

For ordinary Poiseuille flow, the pressure gradient $\partial p/\partial z$ is a constant and u_z is a function of r only, and u_r, s_r, and s_z are all zero. Thus we have to solve Eq. (9.152) only. Now we have to solve the five equations (9.150) to (9.154) simultaneously. Furthermore, we have to study unsteady flow. Further simplifications should be made to give some information on this system of equations. For instance, Womersley solved for the case of sinusoidal pressure gradient with small displacement and rate of change of displacements and small velocities and that the wave length is much larger than the radius of the pipe.[5] Womersley found that as the frequency tends to zero, his solution tends to the Poiseuille solution as a limit.

(c) Other organs. The blood will flow through other organs or parts of the body, such as the brain, kidney, skin, muscle, etc. The flow of blood in each case should be treated as a special biofluid mechanics problem which will be briefly discussed in next section.

(d) Pulmonary. Here we have a two-phase flow problem, i.e., the interaction of the blood flow with the air flow.

(3) Artificial organs. One of the most important practical applications of biofluid mechanics or rather bioengineering is the design and construction of artificial organs. There are many artificial organs in use which have been used as practical clinical and research tools and which are the artificial kidney, oxygen-generator, the heart-lung machine, and some other devices. However, because of the delicate and complicated nature of the natural organs, it is not possible at the present time to construct an artificial organ which is both geometrically and dynamically similar to the natural one. The main purpose of the artificial organs at the present time is to construct a device which may replace the natural organ for its main functions. To develop these artificial organs, the method of dynamical similarity may be used. First we may develop a model, either mechanically or mathematically, which will do the main functions

of an organ. If we know the dependence of various functions of this organ on some important nondimensional parameters, we may construct a successful device to replace the organ. For instance, the Kolff twin coil artificial kidney has been in daily use in numerous hospitals. For the detailed discussions of these artificial organs, special treatises should be referred to (see some of the references given in Ref. [37]).

(4) Flagellar motion. One of the most interesting problems of biophysics is the study of the motility, the capacity of biological objects to exhibit gross movement or internal motion, which is really an important feature of life. One of the studies which has been extensively investigated in recent years is the flagellar movement. Experimental observation of flagellar motion shows that the achievement of a forward velocity is a second-order effect in the amplitude of the wave travelling in the flagellum. Sir Geoffrey Taylor gave the following ingenious theory for the flagellar motion:

Since the size of flagellum is extremely small, the Reynolds number of flagellar movement is much smaller than unity. As a result, we may apply the Stokes flow of a viscous fluid to investigate the flagellar movement. The fundamental equation is the Laplace equation for the pressure p.

A flagellum may be represented by a cylinder of radius r_0 in which waves of amplitude b, wave number $k = 2\pi/\lambda$, and angular frequency ω are travelling. We assume that the amplitude of the wave is much smaller than the radius of the cylinder. The surface of the cylinder may be written as

$$r = r_0 + b \cos \theta \sin (kz + \omega t) \qquad (9.155)$$

where r, θ, and z are the cylindrical coordinates. Taylor found the solution of the Stokes flow equations with the no-slip boundary condition on the surface of the cylinder given by Eq. (9.155) and with vanishing disturbances of the fluid at infinity.[32] The velocity components and the pressure can be expressed in terms of modified Bessel functions. From the velocity distribution, Taylor found that the forward velocity of a flagellum is

$$V = \pi \ \{[K_0(kr_0) - \tfrac{1}{2}]/[K_0(kr_0) + \tfrac{1}{2}]\} \frac{\omega b^2}{\lambda} \qquad (9.156)$$

and the flagellum has a rotation with the frequency:

$$f = \frac{\omega b^2}{r_0 \lambda} \left\{ \frac{K_0(kr_0) - \tfrac{1}{2}}{K_0(kr_0) + \tfrac{1}{2}} \right\} \qquad (9.157)$$

where K_0 is the modified Bessel function of zeroth order. The results of Eqs. (9.156) and (9.157) agree with the observed properties of flagellar motion in the following points:

(a) the forward velocity is independent of the viscosity of the medium;

(b) the forward velocity is proportional to the square of the amplitude of the wave b^2, and

(c) there is a rotation whose frequency is proportional to the forward velocity.

PROBLEMS

1. Find out the velocity distribution of the following fluids in a straight circular pipe far away from the entrance so that the velocity depends on the radial distance only: (i) Dilatant fluid with $n = \frac{1}{2}$, (ii) Pseudo-plastic with $n = 2$, and (iii) Rivlin-Ericksen fluid, and (iv) Oldroyd's fluid (reference [36]).

2. Repeat problem 1 for the case of a Couette flow between two concentric rotating cylinders.

3. Discuss the critical cavitation number for some simple bodies of revolution or disc (reference [10]).

4. Discuss briefly the supercavitating flows (reference [10]).

5. Calculate the Rankine-Hugoniot relations of a normal shock in an ideal gas with a constant amount of heat release.

6. By mean of Mangler-Dorodnitsyn transformation, transform the fundamental equations of ablation near a stagnation point of body of revolution and the boundary conditions in the form of similar solutions, i.e., Eqs. (9.11) in terms of similarity variable.

7. Discuss the concept of critical tractive force in the study of sedimentation (reference [2]).

8. Find the flow rate of an aqueous solution through a vertical packed tower saturated with the solution. The tower is 5 ft high with a cross section of 10 sq. ft and a permeability of 5 darcys. The solution has a viscosity of 1.1 centipoise and a density of 65 lb/ft^3. The fluid is injected into the top of the tower and a pressure-gauge level with the top of the tower reads 10 psig and one level with the bottom reads 5 psig.

9. What is the formula of Lockart-Martinelli for the horizontal flow of liquid and gas (reference [40])?

10. Derive a formula for the sound speed of the mixture of small solid particles in a gas based on Eq. (9.17).

11. Determine the effects of small solid particles on the normal shock structure based on Eq. (9.17).

12. Study the effects of small solid particles on the Blasius laminar boundary layer of an incompressible fluid on a flat plate based on Eq. (9.17).

13. Derive a generalized Ohm's law based on multifluid theory including the effects of the pressure gradient of electrons and the tensor effects of electrical conductivity, i.e., both Hall current and ion slip.

14. Derive the coefficients S_4, S_2, and S_0 of Eq. (9.91) and discuss the wave motion of these longitudinal-transverse waves in an ideal plasma for the cases (i) The electron plasma frequency ω_e is much larger than the cyclotron frequencies, i.e., very weak magnetic field and (ii) the electron plasma frequency is much smaller than the cyclotron frequencies, i.e., very strong magnetic field.

15. Derive the coefficient C_0 to C_4 of Eq. (9.93) and show that in an ideal plasma, the three undamped waves are the fast, slow, and transverse waves of classical theory of magneto-gasdynamics, and the fourth wave is damped when the frequency of the wave tends to be zero.

16. Study the one-dimensional nozzle flow of an ideal fully ionized plasma based on the two-fluid theory (ref. [26]).

17. Discuss the effect of external applied electric field on the plane Couette flow between two parallel plates on a slightly partially ionized argon (ref. [27]).
18. Derive the formulas of the two sound speeds, Eqs. (9.101) and (9.102) for liquid helium II.
19. A small temperature difference dT is maintained between the ends of a capillary containing liquid helium II. Determine the heat flux along the capillary (ref. [4]).
20. Find the solution of the equations of relativistic fluid mechanics for a one-dimensional unsteady simple wave, a generalization of Riemann invariant in the relativistic fluid dynamics.
21. Discuss the energy-momentum tensor for a dissipative process.
22. From the fundamental equations of relativistic fluid dynamics of an ideal gas, i.e., without dissipative process, derive the generalized Euler's equations (3.110) in relativistic fluid dynamics up to the order of $R_r^2 = q^2/c^2$ only where R_r is assumed to be a small quantity.
23 Derive the momentum-energy tensor in relativistic magneto-gasdynamics and the corresponding oblique shock relations (ref. [7]).
24. From the Boltzmann equation of photon, Eq. (9.142), derive the expressions of radiation stress tensor up to the order of R_r^2.
25. Discuss briefly the similarity equation for a cardiovascular system (ref. [37]).
26. Discuss the similarity criteria in the analysis of artificial kidneys (ref. [37]).
27. Discuss the similarity criteria in the analysis of heart-lung machines and respirators (ref. [37]).
28. Discuss briefly the derivation of the fundamental equations of blood flow, Eqs. (9.150) to (9.154).
29. Solve Eqs. (9.150) to (9.154) for a sinusoidal pressure gradient by assuming that the movements of the wall are very small, the imaginary parts of E^* and σ^* are much smaller than the real parts of these quantities, and the wave length is much larger than the radius of the cylinder.

30. Find the general solution of flagellar motion satisfying the boundary condition (9.155), and derive the formulas (9.153) and (9.157) (ref. [32]).

REFERENCES

[1] Allen, J. F. and Jones, H. *Nature*, Vol. 141, No. 3562, 1938, pp. 243—244.
[2] Anderson, A. G. "Sedimentation". Sec. 18 in Handbook of Fluid Dynamics. V. L. Streeter (ed.). McGraw-Hill, 1961.
[3] Andronikashvilli, E. L. *Jour. of Phys.* (Moscow), Vol. 10, 1946, p. 201.
[4] Atkins, K. R. Liquid Helium. Cambridge University Press, 1959.
[5] Attinger, E. O. "Hydrodynamics of blood flow", in Adv. in Hydrosci. (Vol. III). V. T. Chow (ed.). Academic Press, 1966.
[6] Cole, R. H. Underwater Explosion. Dover Publications, 1965.
[7] de Hoffman, F. and Teller, E. "Magneto-hydrodynamic shocks", *Phys. Rev.*, Vol. 80, No. 4, 1950, p. 692.
[8] Donnelly, R. J. Experimental Superfluidity. Chicago University Press, 1967.
[9] Eirich, F. R. Rheology, Theory and Applications (Vol. I). Academic Press, 1956.
[10] Eisenberg, P. and Tulin, M. P. "Cavitation", Sec. 12 in Handbook of Fluid Dynamics. V. L. Streeter (ed.). McGraw-Hill, 1961.
[11] Fung, Y. C. B. "Biomechanics". *Appl. Mech. Rev.*, Vol. 20, No. 1, 1968, pp. 1—20.
[12] Fung, Y. C. B. "Microcirculation dynamics", in Biomedical Sci. Instrum. (Vol. 4). Plenum Press, 1968, pp. 310—320.
[13] Fung, Y. C. B. and Tong, P. "Theory of the sphering of red blood cells", *Biophysical Jour.*, Vol. 8, No. 2, 1968, pp. 175—198.
[14] Kamerlingh Onnes, H. *Proc. Acad. Sci. Amt.*, Vol. 11, 1908, p. 168.
[15] Landau, L. D. and Lifshitz, E. M. Fluid Mechanics. Pergamon Press, 1959.
[16] Lees, L. "Ablation in hypersonic flows". Proc. 7th Anglo-American Aero. Conf., Inst, Aero. Sci., 1959, pp. 344—362.
[17] London, F. Superfluids (Vol. I and II). John Wiley, 1954.
[18] Marble, F. E. "Dynamics of a gas containing small solid particles", in Proc. 5th AGARD Combustion and Propulsion Coll. Pergamon Press, 1963, pp. 175—215.
[19] Marble, F. E. "Nozzle contours for minimum particle-lag loss", *AIAA Journal*, Vol. 1, No. 12, 1963, pp. 2793—2801.
[20] Merrington, A. G. "Flow of visco-elastic materials in capillaries", *Nature*, Vol. 152, No. 3866, 1943, p. 663.
[21] Metzner, A. B. "Flow of non-Newtonian Fluids". Sec. 7 in Handbook of Fluid Dynamics. V. L. Streeter (ed.). McGraw-Hill, 1961.

[22]   Murray, J. D. "On the mathematics of fluidization, 1. Fundamental equations and wave propagation", *J. Fluid Mech.*, Vol. 21, 1965, pp. 465—494.

[23]   O'Keefe, J. A. and Adams, E. W. "Tektite structure and lunar ash flows", *J. Geophy. Res.*, Vol. 70, No. 16, 1965, pp. 3819—3829.

[24]   Pai, S.-I. Magnetogasdynamics and Plasma Dynamics. Springer-Verlag, 1962.

[25]   Pai, S.-I. "Wave motion of small amplitude in a fully ionized plasma under applied magnetic field", *Phys. Fluids*, Vol. 5, No. 2, 1962, pp. 234—240.

[26]   Pai, S.-I. and Tsao, C. K. "Nozzle flow of a fully ionized plasma based on two fluid theory", *ZAMP*, Vol. 16, No. 3, 1965, pp. 360—370.

[27]   Pai, S.-I. and Powers, J. O. "Nonequilibrium effects on energy transfer in an ionized fluid flow", in Dynamics of Fluids and Plasmas. S. I. Pai (ed.). Academic Press, 1966, pp. 179—198.

[28]   Pai, S.-I. Radiation Gasdynamics. Springer-Verlag, 1966.

[29]   Prandtl, L. Essentials of Fluid Dynamics. Hafner Publ., 1952.

[30]   Reiner, M. Deformation, Strain and Flow: An Elementary Introduction in Rheology. H. K. Lewis, 1960.

[31]   Richardson, J. G. "Flow through porous media". Sec. 16 in Handbook of Fluid Dynamics. V. L. Streeter (ed.). McGraw-Hill, 1961.

[32]   Rikmenspoel, R. "Physical principles of flagellar motion", in Dynamics of Fluids and Plasmas. S. I. Pai (ed.). Academic Press, 1966, pp. 9—34.

[33]   Rivlin, R. S. "The fundamental equations of nonlinear continuum mechanics", in Dynamics of Fluids and Plasmas. S. I. Pai (ed.). Academic Press, 1966, pp. 83—126.

[34]   Scala, S. M. and Sutton, G. W. "The two phase hypersonic laminar boundary layer. — A study of surface melting", in Proc. 1958 Heat Transfer and Fluid Mech. Inst. Stanford Univ. Press, 1958, pp. 231—240.

[35]   Simon, R. "The conservation equations of a classical plasma in presence of radiation". A & ES Report 62—1, School of Aero. & Eng. Sci. Purdue Univ., Feb. 1962.

[36]   Skelland, A. H. Non-Newtonian Flow and Heat Transfer. John Wiley, 1967.

[37]   Stahl, W. R. "The analysis of biological similarity", in Adv. in Biological and Medical Physics (Vol. IX). Academic Press, 1963, pp. 355—464.

[38]   Synge, J. L. The Relativistic Gas. North-Holland Pub. Co., 1957.

[39]   Tanenbaum, B. S. and Mintzer, D. "Wave propagation in a partly ionized gas", *Phys. Fluids*, Vol. 5, No. 10, 1962, pp. 1226—1237.

[40]   Tek, M. R. "Two phase flow". Sec. 17 in Handbook of Fluid Dynamics. V. L. Streeter (ed.). McGraw-Hill, 1961.

[41]   Tolman, R. C. Relativity: Thermodynamics and Cosmology. Oxford University Press, 1934.

[42]   van Driest, E. R. "Problems of high speed hydrodynamics", *Jour. Eng. for Ind. ASME Trans.*, Vol. 91, Ser. B, No. 1, 1969, pp. 1—12.

[43] von Kármán, Th. From Low Speed Aerodynamics to Astronautics. Pergamon Press, 1963.
[44] Wagener, P. P. and Pouring, A. A. "Experiments on condensation of water vapor by homogenous nucleation in nozzle", *Phys. Fluids*, Vol. 7, No. 3, 1964, pp. 352—361.
[45] Weissenberg, K. "A continuum theory of rheological phenomena", *Nature*, Vol. 159, No. 4035, 1947, pp. 310—311.
[46] Pai, S.-I. Two-Phase Flows, Vieweg-Verlag, 1977.
[47] Cramer, K. R. and Pai, S.-I. Magnetofluid Dynamics for Engineers and Applied Physicists, McGraw-Hill, 1973.

APPENDIX I. Important Symbols Used in This Book

Since modern fluid mechanics covers many branches of classical physics, many well established symbols in one branch of physics may have entirely different meaning in other branches of physics. Furthermore since we cover many different subjects, it is unavoidable that the same symbols may mean different things in different parts of this book. Hence we list below most of the important symbols for reference. Those less important symbols which occur only in one chapter are not listed, but are explained in the text. In principle, we try to use the well established symbols as much as possible if there is no confusion. However, if confusion may occur, we shall put a subscript in one of them to distinguish the meanings of these symbols.

Symbols	Meanings
a	sound speed
A	area
a_e	acceleration
$a_R = 7.67 \times 10^{-15} \dfrac{\text{erg}}{\text{cm}^3 \cdot {}^\circ\text{K}^4}$	Stefan-Boltzmann constant defined in Eq. (7.1)
a_ν	hemispherical absorptivity coefficient of radiation [see Eq. (7.65)]
$\text{Å} = 10^{-8}\text{cm}$	Angstrom unit
$B(T) = \dfrac{\sigma}{\pi} T^4 (\text{or } B)$	Planck integrated radiation function
$B(B_x,\ B_y,\ B_z) = \mu_e H$	magnetic induction with magnitude B
$c = 2.9979 \times 10^8$ m/sec	speed of light
c_a (c_i or c_z, etc.)	random or peculiar velocity
c_f	friction coefficient, Eq. (6.101)
C_F	force coefficient, Eq. (4.52)
C_M	moment coefficient, Eq. (4.55)
c_p	specific heat at constant pressure
C_p	pressure coefficient, Eq. (4.56)
c_s (or c_r)	number concentration
c_{sp}	specific heat in general

Symbols	Meanings
c_v	specific heat at constant volume
D_{am_1}	first Damkohler number, Eq. (4.68)
D_{am_2}	second Damkohler number, Eq. (4.70)
D_R	Rosseland diffusion coefficient of radiation, Eq. (7.40)
D_{sr} (D_{12}, etc.)	diffusion coefficient between species s and r
$e = 1.602 \times 10^{-19}$ coulomb	absolute electric charge of an electron
$E(E_a, E_e,$ etc.)	energy
$E(E_x, E_y, E_z)$	electric field strength with magnitude E
E_R	radiation energy density, Eq. (7.18)
$E_t = \bar{e}_m$	total energy of a gas per unit mass
$E_u = E + q \times B$	electric field in moving coordinates
F	molecular distribution function, Eq. (8.9)
F (F_x, F_y, etc.)	force
F_e	electromagnetic force, Eq. (6.4)
F_r	Froude number, Eq. (4.34)
g (magnitude: g)	gravitational acceleration
G_r	Grashoff number, Eq. (4.65)
$h = 6.624 \times 10^{-27}$ erg/sec	Planck constant
H (or h)	enthalpy
$H(H_x,$ etc.; $h: h_x,$ etc.)	magnetic field strength
H_s (or H_0)	stagnation enthalpy
$i = (-1)^{\frac{1}{2}}$	unit imaginary number
i, j, k	unit vectors along x-, y-, and z-axis respectively
I	electric conduction current
I_ν	specific intensity of radiation, Eq. (7.4)
J	electric current density
j_ν	emission coefficient of radiation
J_ν	source function of radiation
$k = 1.379 \times 10^{-16}$ erg/$^\circ$K	Boltzmann constant
K	potential energy
K_D	electric Knudsen number, Eq. (3.6)
K_f	Knudsen number, Eq. (3.4)

Symbols	Meanings
K_L	magnetic Knudsen number, Eq. (3.9)
K_p	Planck mean absorption coefficient of radiation, Eq. (7.49)
K_r	Knudsen number of radiation, Eq. (4.75)
K_R	Rosseland mean absorption coefficient of radiation, Eq. (7.41)
$k_s(k_r)$	mass concentration, Eq. (1.67)
k_ν	mass absorption coefficient of radiation
k'_ν	reduced absorption coefficient of radiation, Eq. (7.15)
l, m, n	directional cosines
L	length
L_D	Debye length, Eq. (3.5)
L_e	Lewis-Semenov number, Eq. (4.80)
L_f	mean free path of a gas, Eq. (4.22)
L_L	Larmor radius, Eq. (4.27)
$L_{R\nu}$	mean free path of radiation, Eq. (4.25)
m	mass
M	Mach number, Eq. (4.60)
M_m	magnetic Mach number, Eq. (4.71)
m_w	molecular weight
n	number density
$n(n_i,$ etc.$)$	unit normal vector
$n_A = 6.025 \times 10^{23}$ molecule/mole	Avogadro's number
n_e	number density of electrons
n_s	number density of sth species in a mixture
N_u	Nusselt number, Eq. (4.59)
p	pressure
P_e	Peclet number, Eq. (4.78)
P_m	magnetic Prandtl number, Eq. (4.81)
p_R	radiation pressure, Eq. (7.20)
P_r	Prandtl number, Eq. (4.62)
$q(u_i$ or $q_i)$	flow velocity vector
Q	quantity of heat

Symbols	Meanings
$q_m(\xi, \eta, \zeta)$	molecular velocity vector, Eq. (8.10)
q_R	heat flux of radiation, Eq. (7.22)
$r(x, y, z)$	spatial coordinates
R	gas constant, Eq. (1.18)
R_e	Reynolds number, Eq. (4.61)
R_E	electric field number, Eq. (6.28)
R_f	frequency parameter, Eq. (4.63)
R_F	radiative flux number, Eq. (4.86)
R_h	Hartmann number, Eq. (4.83)
R_H	magnetic pressure number, Eq. (4.71)
R_m	magnetic number, Eq. (4.82)
R_p	radiation pressure number, Eq. (4.73)
R_r	relativistic parameter, Eq. (4.74)
R_t	time parameter, Eq. (4.63)
R_σ	magnetic Reynolds number, Eq. (4.72)
s	distance along a curve or a ray
S	entropy
S_c	Schmidt number, Eq. (4.67)
S_t	Stanton number, Eq. (4.79)
t	time
T	temperature (kinetic)
T_s	temperature of sth species in a mixture
T_v	vibrational temperature
u, v, w	x-, y-, and z-component of flow velocity q respectively
U	typical velocity
U_m	internal energy
V	volume
$V_H(V_x, V_y,$ etc.)	velocity of Alfven's wave, Eq. (4.71)
V_i	ionization potential
w_s	diffusion velocity, Eq. (1.71)
x, y, z	Cartesian coordinates
α	degree of ionization
$\gamma = c_p/c_v$	ratio of specific heats
δ_{ij} (or $\delta_{\alpha\beta}$)	$=0$ if $i \neq j$; $=1$ if $i=j$

	Symbols	Meanings
ϵ	inductive capacity, Eq. (6.9)
θ	angle
κ	coefficient of heat conductivity, Eq. (1.56)
κ_R	coefficient of heat conductivity by thermal radiation, Eq. (7.43)
λ	wave number, Eq. (7.77)
μ	coefficient of viscosity, Eq. (1.48)
μ_e	magnetic permeability, Eq. (6.8)
ν	frequency
ν_g	coefficient of kinematic viscosity
ν_H	magnetic diffusivity, Eq. (6.23)
ρ	density of a fluid
ρ_e	excess electric charge
$\sigma = 5.75 \times 10^{-5} \dfrac{\text{erg}}{\text{cm}^2 \cdot {}^{\circ}\text{K}^4 \cdot \text{sec}}$	Stefan-Boltzmann constant for radiative flux, Eq. (7.2)
σ_e	electric conductivity
σ_s	source function
τ^{ij}	ith component of a stress tensor
τ_R	radiation stress tensor
τ_v	viscous stress
τ_ν	optical thickness, Eq. (7.9)
ϕ	velocity potential
Φ	dissipation function of viscous stress Eq. (3.53)
ψ	streamfunction, Eq. (3.42)
ω	frequency or solid angle
ω_c	cyclotron frequency, Eq. (4.21)
ω_p	plasma frequency, Eq. (4.20)
$\nabla = i\dfrac{\partial}{\partial x} + j\dfrac{\partial}{\partial y} + k\dfrac{\partial}{\partial z}$	gradient operator
$\dfrac{D}{Dt} = \dfrac{\partial}{\partial t} + q \cdot \nabla$	total time derivative
Subscripts i, j, x, y, z	refer to components of a vector in the direction of i, j, etc.
Subscript s or r	refer to value of sth or rth species in a mixture

APPENDIX II. Physical Constants and Conversion of Units

(A) Physical Constants

The following are some important physical constants useful in the study of modern fluid mechanics:

Avogadro number	$n_A = 6.025 \times 10^{23}$ molecule/mole
Boltzmann constant	$k = 1.379 \times 10^{-16}$ erg/°K
	$= 1.379 \times 10^{-23}$ joule/°K
electron mass	$m_e = 9.106 \times 10^{-31}$ kilogram
electronic charge	$e = 1.602 \times 10^{-19}$ coulomb
	$= 4.801 \times 10^{-10}$ esu
energy associated with 1 ev (electron volt)	$= 1.602 \times 10^{-12}$ erg
energy equivalent of electron mass	$m_e c^2 = 0.51097$ Mev
frequency associated with 1 ev	$\nu = 2.41814 \times 10^{14}$ sec^{-1}
gas constant per mole	$R_0 = 8.314 \times 10^7$ erg/(mol·°K)
gravitational acceleration	$g = 9.807$ m/sec^2
gravitational constant	$G = 6.670 \times 10^{-8}$ dyne·cm^2/gr^2
inductive capacity in vacuum	$\epsilon = 8.854 \times 10^{-12}$ coulomb^2 sec^2/ (kg·m^3)
Loschmidt number	$= 2.687 \times 10^{19}$/cm^3
magnetic permeability in vacuum	$= 4\pi \times 10^{-7}$ kg·m/(coulomb)$^2 = \mu_e$
mass of a proton	$m_i = 1.672 \times 10^{-27}$ kg
mass of hydrogen atom	$m_H = 1.673 \times 10^{-27}$ kg
mechanical equivalent of heat	$= 4.185$ joule/calorie
Planck constant	$h = 6.624 \times 10^{-27}$ erg/sec
radius of first Bohr orbit	$= 0.52917 \times 10^{-8}$ cm
ratio: e/m_e	1.759×10^{11} coulomb/kg
ratio: m_i/m_e	1836.12
speed of light	$c = 2.9979 \times 10^8$ m/sec
Stefan-Boltzmann constant	$a_R = 7.67 \times 10^{-15}$ erg/(cm^3·°K^4)
Stefan-Boltzmann constant for radiative flux	$\sigma = 5.68 \times 10^{-5}$ erg/(cm^2·°K^4·sec)

(B)   Conversion of Units

Fluid mechanics is a subject of engineering and science.   Hence in papers and books on fluid mechanics both engineering units and science units have been used.   Some people prefer engineering units such as those used by engineers in U. S. A., i.e., feet, pounds, seconds, etc., while others prefer the scientific units such as meters, kilograms, etc.   Furthermore, there are many unit systems in electromagnetic variables which cause even more confusion in modern fluid mechanics, such as magnetogasdynamics.   It is convenient if we collect here some important conversion factors from one system to the other.   The following are some most common conversion factors between different unit systems.

As we have discussed in chapter IV, there are five basic units in modern fluid mechanics: time t, length L, mass m, temperature T, and electric unit such as quantity of electricity in coulomb.   We discuss various unit systems of these five basic units first.   In practical problems, our main interests are not in these basic units but in some combinations of these units such as velocity, pressure, work, etc. We shall also give some conversion factors of these quantities.

(i)   Time t

The time unit is only one kind which may be expressed in hours (hr), minutes (min) or seconds (sec).   We have

$$1 \text{ hr} = 60 \text{ min} = 3600 \text{ sec}$$

(ii)   Length L

In engineering units of U. S. A., the length may be expressed in inches (in), feet (ft), or miles.   There are some other measures of length such as yard, rod, hand, etc., which will not be discussed here.

The relations between inches, feet and miles are:

$$1 \text{ foot} = 12 \text{ inches, } 1 \text{ mile} = 5280 \text{ ft}$$

In oceanology, a nautical mile may be used which is

$$1 \text{ nautical mile} = 6080.20 \text{ feet}$$

In scientific units, the length may be measured in meters or some powers of 10 of a meter such as kilometer $= 10^3$ meters; centimeter(cm) $= 10^{-2}$ meter.

The conversion factors of these two systems of length may be derived from the following relations:

1 foot = 0.3048 meter, 1 meter = 3.281 ft

(iii) Mass *m*

The mass is defined by Newton's second law of motion, i.e.,

$$\text{mass } m = \frac{\text{force}}{\text{acceleration}} = \frac{F}{a_c}$$

Hence the units of mass may be derived from the units of force and the units of acceleration. In engineering units, if the force of 1 pound is applied to a body of a mass of 1 slug, the body will have an acceleration of one foot per sec per sec. There is a confusion in terminology in force and mass. The word "pound" (lb) may be used to express mass too. The body of a mass of one pound will be accelerated at an acceleration of 1 ft/sec^2, if it is applied by a force of one poundal. One slug is equal to 32.1739 (approximately 32.2) times pound mass. Or one pound force is equal to 32.1739 (approximately 32.2) poundals.

In scientific units, when the mass is one gram and the acceleration is 1 cm/sec^2, the corresponding force is one dyne. If the mass is one metric slug (one Newton) and the acceleration is one meter per sec per sec, the corresponding force is one kilogram. One gram force is equal to 980.665 dynes.

1 pound = 0.4536 kg, 1 kg = 2.205 lbs

(iv) Temperature *T*

There are two most popular units of temperature: one is the degree centigrade (°C) or degree Kelvin (°K) and the other is degree Fahrenheit (°F) or degree Rankine (°R). The relations between these degrees of temperature are as follows:

$$°C = \frac{5}{9}(°F - 32°), \quad °F = \frac{9}{5}°C + 32°$$

and

$$°K = °C + 273.16°, \quad °R = °F + 459.69°$$

In engineering units, degrees Fahrenheit are often used while in scientific units, degrees Kelvin are often used.

(v) Electrical unit

The most common systems of units used in electromagnetic theory are:

(a)   The absolute cgs systems of electrostatic units in which the basic unit is usually abbreviated as esu. In this unit system, the inductive capacity ϵ is taken as unity in free space. The unit of electrical quantity is one esu of charge.

(b)   The absolute cgs system of electromagnetic units in which the basic unit is usually abbreviated as emu. In this unit system, the magnetic permeability μ_e in free space is arbitrarily chosen as unity. The unit of electrical quantity is one emu of charge. The relation between esu and emu of charge is

$$1 \text{ emu of charge} = c \times 1 \text{ esu of charge}$$

where c is the speed of light in free space $= 2.99790 \times 10^{10}$ cm/sec.

In the above two unit systems, the length is in centimeters, the mass in grams, and the time, in seconds.

(c)   Gaussian or mixed units in which we use emu for magnetic field and magnetic induction and esu for all other quantities including current and charge.

(d)   Practical units. They are based on emu but differ from emu by some arbitrary power of 10. For instance, the following are some practical units:

Quantity	Practical Unit	emu
current	1 ampere	$= 10^{-1}$ emu of current
potential	1 volt	$= 10^8$ emu of potential
work/second	1 ampere $\times$ 1 volt	$= 1$ watt $= 10$ erg/sec
resistance	1 ohm	$= 10^9$ emu of resistance
magnetic induction	1 weber/cm	$= 10^8$ gausses

(e)   MKS or Giorgi system. This is a rationalized MKS unit system in which the meter is used as the unit length; the kilogram, as unit of mass; the second, as the unit of time and the coulomb, as the electrical unit. The values of magnetic permeability μ_e and inductive capacity ϵ are given respectively in Eq. (6.8) and (6.9).

All of the electrical units are more or less scientific units. But since there are already many scientific units, it is in a complicated state already!

(vi) Velocity

The dimensions of a velocity are L/t. Since we have engineering unit and scientific unit in length, we have the corresponding unit in velocity. In engineering units, the velocity may be expressed in feet per second (ft/sec) or in miles per hour (mph). 1 mph = 1.467 ft/sec, 1 ft/sec = 0.6818 mph. In scientific unit, the velocity may be expressed in meters per second (m/sec) or kilometers per hour (km/hr). 1 m/sec = 3.281 ft/sec = 2.237 mph = 3.6 km/hr, 1 km/hr = 0.6214 mph, and 1 mph = 1.609 km/hr = 0.4470 m/sec.

In nautical units, the velocity is expressed in knots. One knot is one nautical mile per hour. 1 knot = 1.15155 mph = 1.8532 km/hr.

(vii) Pressure p

The dimensions of a pressure are $m/(t^2L)$ or force per unit area. In engineering units, the pressure may be expressed in pounds per square inch (lb/in²) while in scientific units, the pressure may be expressed in kilograms per square centimeter (kg/cm²). In USA engineering practice, the pressure is sometimes expressed in terms of atmospheres. One atmosphere of pressure is equal to 14.7 lb/in² which corresponds to the pressure of 760 mm Hg (29.921 in Hg). One atmosphere is equal to 1.033 kg/cm². One kg/cm² = 14.22 lb/in² and 1 lb/in² = 0.0703 kg/cm².

(viii) Work W

The work is force times distance. Hence we may express the work in foot·pounds (ft·lb) or kilogram·meters (kg·m). There are some other common units of work as follows:

1 joule (absolute) = 10^7 ergs = 10^7 dyne·cm

1 kilowatt·hour (absolute) = 3.6×10^6 joules (ab.)

1 horsepower·hour = 0.7457 kilowatt·hours

1 kilogram·meter = 7.233 ft·lb, 1 ft·lb = 0.1383 kg·m

(ix) Power

Power is work per unit time. Hence the unit of power is closely related to that of work.

The most common units of power are:

1 horsepower (hp) is equal to 550 ft·lb per sec

1 watt = 1 joule/sec, 1 kilowatt = 10^3 watt = 1.341 hp

(x) Heat units

In principle, the heat unit is the same as that of work. However,

544   *Modern Fluid Mechanics*

in practice, some special heat units have been widely used instead of the units of work discussed above. One of the heat units is the British thermal unit (Btu) which is the heat required to raise the temperature of one pound of water one degree Fahrenheit. The other heat unit is the calorie (cal) which is the quantity of heat to raise the temperature of one gram of water one degree centigrade. The relation between these heat units and the ordinary work units are as follows:

$$1 \text{ Btu} = 252 \text{ calories} = 778.2 \text{ ft·lb} = 107.6 \text{ kg·m}$$
$$1 \text{ cal} = 3.968 \times 10^{-3} \text{ Btu} = 4.187 \text{ joules} = 3.088 \text{ ft·lb}$$

(xi) Electromagnetic units

For electromagnetic variables, we list the relations between those in MKS system, esu and emu as follows:

Quantity	MKS system	esu	ems
current	1 ampere	$= 3 \times 10^9$ esu current	$= 10^{-1}$ emu current
charge	1 coulomb	$= 3 \times 10^9$ esu charge	$= 10^{-1}$ emu charge
capacitance	1 farad	$= 9 \times 10^{11}$ esu cap.	$= 10^{-9}$ emu cap.
inductance	1 henry	$= (1/9) \times 10^{-11}$ esu ind.	$= 10^9$ emu ind.
energy	1 joule	$= 10^7$ esu energy	$= 10^7$ emu energy $= 10^7$ ergs
conductivity	1 ohm/meter	$= 9 \times 10^9$ esu cond.	$= 10^{-11}$ emu cond.
resistance	1 ohm	$= (1/9) \times 10^{-11}$ esu res.	$= 10^9$ emu res.
potential	1 volt	$= (1/3) \times 10^{-2}$ esu pot.	$= 10^8$ emu pot.
magnetic flux	1 weber	$= (1/3) \times 10^{-2}$ esu mag. flux	$= 10^8$ emu mag. flux

(xii) Viscosity

The coefficient of viscosity is expressed in poise (gr/cm·sec) in the scientific unit system, and in lbf·sec/ft² or slug/ft·sec in the engineering unit system. The relations between these units are

$$1 \text{ poise} = 0.002088 \text{ lbf·sec/ft}^2 = 0.06721 \text{ slug/ft·sec}$$
$$1 \text{ slug/ft·sec} = 14.88 \text{ poise} = 0.03107 \text{ lbf·sec/ft}^2$$

APPENDIX III. Properties of Some Typical Fluids

From the macroscopic point of view, many properties of a fluid, such as density, specific heats, viscosity, heat conductivity, and electrical conductivity, are assumed to be given functions of state variables, e.g., the temperature T and the pressure p of the fluid. Of course, from the microscopic point of view, these properties of a fluid may be expressed in terms of some basic physical constants listed in appendix II. Here we list the values of a few important properties of common fluids which may be useful in the calculations of fluid mechanics problems.

(A) Density ρ, Specific Weight w_s, Specific Volume V_s, and Specific Gravity g_s

Density is one of the main properties of a fluid. For a given fluid, the density is a function of its temperature and pressure by equation of state as we discussed in chapter I. Some typical values of density of various common fluids at certain given temperature and pressure are listed below:

Fluids (gases)	Air	Argon	Helium	Hydrogen	Nitrogen	Oxygen
Density	0.00251	0.00346	0.00035	0.00017	0.00243	0.00277

Fluids (liquids)	Benzene	Glycerine	Mercury	Water
Density	1.74364	2.44297	26.38777	1.94039

In the above table, the temperature T is 32°F, the pressure p is 14.7 lb/in², and the density is in terms of slug/ft³.

Density of a fluid is closely related to the following terms:

(i) Molecular weight m

For a perfect gas, the density of the gas is given by Eq. (1.18). Hence if we know the molecular weight m, the density of a perfect gas may be easily expressed in terms of pressure p and temperature T. The values of molecular weight m and gas constant R of a few common gases are given below:

Gases	Air	Argon	Helium	Hydrogen	Nitrogen	Oxygen
Molecular weight m	29.0	40.0	4.0	2.0	28.0	32.0
Gas constant R/g	53.3	38.7	386.3	766.8	55.16	48.3

where the gas constants are in terms of the foot·pound·second system of units. $Rm/g = k = 1545$ ft·lbf/lb·mole·°R.

(ii) Specific weight w_s

The density of a fluid is its mass per unit volume while the specific weight w_s of a fluid is its weight per unit volume. Hence we have

$$w_s = g\rho$$

where g is the gravitational acceleration.

(iii) Specific volume V_s

There are two ways to define a specific volume. In scientific system, the specific volume V_s is defined as $1/\rho$, while in American engineering practice, the specific volume is defined as $1/w_s$.

(iv) Specific gravity g_s

In engineering practice, it is customary to use specific gravity to express the density of a fluid. The specific gravity of a fluid is the ratio of the density of the fluid to the density of some standard fluid at standard temperature and pressure. For liquids, the standard fluid is pure water at standard atmospheric pressure and a standard temperature is 4°C, while in engineering practice, the standard temperature is taken as 60.0°F. For gases, there is no international agreement of the standard fluid. Sometimes, the density of a standard atmosphere at sea level may be used as the reference density.

For fluids near the gas-liquid line or for a mixture of gases in which chemical reactions may be occurred, we have to use very complicated formula for the equation of state. Usually, we have to use tables or charts to express the equation of state for such cases. Even though we may use a simple formula such as Eq. (1.22) for the equation of state, the function Z is a very complicated function of temperature and pressure which is usually given in tables or charts. Hence specific references should be used to find the values of Z.

(For instance: (1) Streeter, V. L., Handbook of Fluid Dynamics, Chapter I, McGraw-Hill Book Company, Inc., (2) Marks, L. S., Mechanical Engineers' Handbook, McGraw-Hill Book Co., or similar books.)

For a first approximation, van der Waals equation (1.21) may be used. The van der Waals constants for a few fluids are given below. We would like to write the van der Waals equation in the conventional form, i.e.,

$$(p+b_2\rho^2)\left(\frac{1}{\rho}-b_1 \right) = RT$$

For a fluid, there is a critical temperature T_c with a corresponding critical pressure p_c. Above the critical point (T_c, p_c), there will be no distinguishing boundary marking off the liquid and vapor. According to the approximation of van der Waals equation, this critical point of a fluid is determined by the conditions:

$$\left(\frac{\partial p}{\partial \rho}\right)_T =0 \text{ and } \left(\frac{\partial^2 p}{\partial \rho^2}\right)_T = 0$$

We also list the critical temperature T_c and critical pressure p_c below:

Gases	Argon	Helium	Hydrogen	Nitrogen	Oxygen	Water vapor
b_1	0.0322	0.0237	0.0267	0.0391	0.0318	0.0305
b_2	1.345	0.0341	0.244	1.39	1.36	5.46
T_c	150.7	5.3	33.3	126.1	154.4	647.4
p_c	48.7	2.26	12.8	33.5	49.7	218.3

where b_1 is in 1/mole, b_2 is in atm²/mole², T_c is in °K, and p_c is in atm.

(B) Specific Heats

The specific heat of a fluid is; in general, a function of its temperature. The variation of the specific heat with temperature for a fluid near its gas-liquid line, i.e., for a vapor, is in general very complicated. We have to use tables or charts to express this variation. Similarly, for a mixture of gases such as air, the variation of specific heat with temperature over a large range of temperature is also complicated because chemical reactions may occur under high temperature. Special tables and charts should be used. For many gases at room temperatures, their specific heats are practically constant. The following are specific heats c_p and c_v of some common gases at room temperatures.

Gases	c_p	c_v	$\gamma = c_p/c_v$
Air	0.241	0.1725	1.40
Argon	0.124	0.0743	1.67
Helium	1.25	0.754	1.66
Hydrogen	3.42	2.435	1.40
Nitrogen	0.247	0.1761	1.40
Oxygen	0.217	0.1549	1.40
Water vapor in air	0.48	0.36	1.33

where the specific heats are in Btu/lb·°F and the temperature is ordinary room temperature. Because the pressure of water vapor in air is small, the water vapor may be considered as a perfect gas.

The mean specific heats of several liquids between the temperature range from 32°F to 212°F in Btu/lb·°F are as follows:

Liquids	Alcohol	Gasoline	Glycerine	Mercury	Sea water
c_v	0.58	0.50	0.58	0.033	0.94

(C) Viscosity

The coefficient of viscosity of a fluid is, in general, a function of its temperature. For liquids, the coefficient of viscosity decreases with the increase of temperature, while for gases, the coefficient of viscosity increases with the increase of temperature. We list the variation of the coefficient of viscosity with temperature for some common fluids as follows:

Liquids	T°C	Centipoise	Liquids	T°C	Centipoise
Benzene	0	0.900	Mercury	20	1.524
	50	0.433	Water	0	1.79
	100	0.263		50	0.550
	150	0.170		100	0.282
Glycerine	0	5400.0		150	0.184
	50	175			

Gases	T°C	Centipoise	Gases	T°C	Centipoise
Air	−100	0.120	Hydrogen	−100	0.061
	0	0.171		0	0.085

	100	0.218	100	0.103
	200	0.259	Nitrogen −100	0.114
	300	0.296	0	0.165
Argon	0	0.209	100	0.208
Helium	−100	0.140	Oxygen −100	0.1325
	0	0.190	0	0.193
	100	0.234	100	0.247

(D) Heat Conductivity

The coefficient of heat conductivity or thermal conductivity of a fluid is also a function of its temperature. From the kinetic theory of gases, we know clearly that there is a definite relation between viscosity and thermal conductivity (chapter I). It is sometimes convenient to use the Prandtl number Eq. (1.58) to show the relation between viscosity and thermal conductivity.

The coefficient of thermal conductivity of various fluids are given below:

Liquids	$T°C$	Btu/ft·hr·°F	P_r
Benzene	20	0.0887	7.22
Ethylene Glycol	20	0.144	200
Mercury	20	5.4	0.023
Water	20	0.346	6.88
	100	0.394	1.71

Gases	$T°C$	Btu/ft·hr·°F	P_r
Air	0	0.0140	0.696
	93	0.0177	0.695
Argon	0	0.00915	
Helium	0	0.0818	
Hydrogen	0	0.09660	
Nitrogen	0	0.0140	
Oxygen	0	0.0142	

APPENDIX IV. Some Useful Formulas for Fluid Mechanics

The fundamental equations of fluid mechanics can be greatly simplified by using the vector notation. We are going to discuss vector notation in this appendix. In many problems of fluid mechanics, it is convenient to use curvilinear coordinate system rather than the simple Cartesian coordinate system because of special shape of the boundary of the flow field. We are going also to discuss some well known curvilinear coordinate systems in this appendix too.

(A) Vector Notation

A scalar quantity is one which is specified completely by its magnitude. For instance, temperature, density, volume, etc., are scalar quantities. A vector is one which is specified completely by its magnitude and one direction. For instance, velocity, temperature gradient, etc., are vector quantities. A tensor is a quantity which should be specified by its magnitude and more than one direction.

In the ordinary three-dimensional space, a vector has three components. Any vector q with component u, v, and w along the x-, y-, and z-axes may be expressed as follows:

$$q = iu + jv + kw \tag{A-1}$$

where i, j, and k are respectively the unit vector in the direction of x-, y-, and z-axis. The addition of two vectors q and q_1 gives

$$q + q_1 = (iu + jv + kw)(iu_1 + jv_1 + kw_1)$$
$$= i(u + u_1) + j(v + v_1) + k(w + w_1) \tag{A-2}$$

If s is a scalar quantity, the multiplication of q by s gives

$$s q = i su + j sv + k sw \tag{A-3}$$

The scalar or dot product of two vectors q and q_1 is defined as follows:

$$q \cdot q_1 = q q_1 \cos \theta = q_1 \cdot q \qquad \text{(A-4)}$$

where q and q_1 are the magnitude of q and q_1 respectively and θ is the angle between these two vectors.

From Eq. (A-4), it is easy to show that

$$i \cdot i = j \cdot j = k \cdot k = 1$$

$$i \cdot j = j \cdot i = j \cdot k = k \cdot j = k \cdot i = i \cdot k = 0$$

Using these results, Eq. (A-4) becomes

$$q \cdot q_1 = u u_1 + v v_1 + w w_1 \qquad \text{(A-4a)}$$

We also have $a \cdot (q + q_1) = a \cdot q + a \cdot q_1$ where a is a vector.

The vector or cross product of two vectors q and q_1 is defined as follows:

$$q \times q_1 = \epsilon q q_1 \sin \theta = -q_1 \times q \qquad \text{(A-5)}$$

where ϵ is a unit vector in the direction perpendicular to both q and q_1. From Eq. (A-4), we have

$$i \times i = j \times j = k \times k = 0$$

$$i \times j = -j \times i = k, \; j \times k = -k \times j = i, \; k \times i = -i \times k = j$$

Using these results, Eq. (A-5) may be written as follows:

$$q \times q_1 = \begin{vmatrix} i & j & k \\ u & v & w \\ u_1 & v_1 & w_1 \end{vmatrix} = i(vw_1 - v_1w) + j(wu_1 - w_1u) + k(uv_1 - vu_1) \qquad \text{(A-5a)}$$

We also have

$$a \times (q + q_1) = a \times q + a \times q_1$$

The scalar triple product of three vectors r, q, and q_1 is defined as

$$r \cdot (q \times q_1) = \begin{vmatrix} r_1 & r_2 & r_3 \\ u & v & w \\ u_1 & v_1 & w_1 \end{vmatrix} = (r \times q) \cdot q_1 = [r q q_1] \qquad \text{(A-6)}$$

where $r = i r_1 + j r_2 + k r_3$.

The following are some useful formulas for various products

of vectors:

$$A \times (B \times C) = (A \cdot C)B - (A \cdot B)C \tag{A-7}$$

$$(A \times B) \cdot (C \times D) = A \cdot [B \times (C \times D)] = (A \cdot C)(B \cdot D) - (A \cdot D)(B \cdot C) \tag{A-8}$$

$$(A \times B) \times (C \times D) = [(A \times B) \cdot D]C - [(A \times B) \cdot C]D \tag{A-9}$$

If the vector q is a continuous function of a scalar s, the derivative of q with respect to s is

$$\frac{dq}{ds} = \lim_{\Delta s \to 0} \frac{\Delta q}{\Delta s} = \lim_{\Delta s \to 0} \frac{q(s + \Delta s) - q(s)}{\Delta s} \tag{A-10}$$

If we write $q(s) = iu(s) + jv(s) + kw(s)$, then

$$\frac{dq}{ds} = i\,\frac{du}{ds} + j\,\frac{dv}{ds} + k\,\frac{dw}{ds} \tag{A-11}$$

$$\frac{d(q \cdot q_1)}{ds} = q \cdot \frac{dq_1}{ds} + \frac{dq}{ds} \cdot q_1 \tag{A-12}$$

$$\frac{d}{ds}(q \times q_1) = \frac{dq}{ds} \times q_1 + q \times \frac{dq_1}{ds} \tag{A-13}$$

If q is a function of more than one scalar, such as

$$q(x, y, z, t) = iu(x, y, z, t) + jv(x, y, z, t) + kw(x, y, z, t) \tag{A-13a}$$

then

$$\frac{\partial q}{\partial x} = i\,\frac{\partial u}{\partial x} + j\,\frac{\partial v}{\partial x} + k\,\frac{\partial w}{\partial x} \tag{A-11a}$$

$$\frac{\partial uq}{\partial x} = u\,\frac{\partial q}{\partial x} + q\,\frac{\partial u}{\partial x} \tag{A-14}$$

Vector integration of a vector $q(s)$ with respect to s gives

$$\int q(s)ds = Q(s) + \text{constant} \tag{A-15}$$

or

$$\int_a^b q(s)ds = Q(b) - Q(a) \tag{A-15a}$$

If A is a constant vector, then

$$\int A \cdot q(s) ds = A \cdot \int q(s) ds \qquad \text{(A-16)}$$

$$\int A \times q(s) ds = A \times \int q(s) ds \qquad \text{(A-17)}$$

If a curve c in space is specified by a radius vector $r(s)$, the line integral along this curve of any vector q is defined as

$$\Gamma = \int_c q \cdot dr = \int_c q \cdot t ds \qquad \text{(A-18)}$$

where t is the unit vector in the curve of the curve c, i.e., $r = ts$.

If the area of a surface S is specified by a vector $A = nA$, the surface integral of any vector q over the area S is defined as

$$Q = \iint_S q \cdot dA = \iint_S q \cdot n dA \qquad \text{(A-19)}$$

where n is the unit normal vector of the surface S.

Now we define an operator ∇(del) as follows:

$$\nabla = i \frac{\partial}{\partial x} + j \frac{\partial}{\partial y} + k \frac{\partial}{\partial z} \qquad \text{(A-20)}$$

where x, y, and z are the Cartesian coordinates. The radius vector r is defined as

$$r = ix + jy + kz \qquad \text{(A-21)}$$

We consider a scalar function $T(x, y, z)$. On the surface of $T(x, y, z) = $ constant, we have a point P represented by the radius vector r, and on a neighboring surface of $T + dt = $ constant, we have a point P' represented by the radial vector $r + dr$. We have then

$$dT = \frac{\partial T}{\partial x} dx + \frac{\partial T}{\partial y} dy + \frac{\partial T}{\partial z} dz$$

$$= (idx + jdy + kdz) \cdot \left(i \frac{\partial T}{\partial x} + j \frac{\partial T}{\partial y} + k \frac{\partial T}{\partial z} \right)$$

$$= dr \cdot \nabla T \qquad \text{(A-22)}$$

Let n be the unit normal vector to the surface $T=$ constant, then

$$dn = n \cdot dr \tag{A-23}$$

and

$$dT = \frac{dT}{dn} dn = dr \cdot n \frac{dT}{dn} \tag{A-24}$$

Comparing Eqs. (A-22) and (A-24), we have

$$\nabla T = n \frac{dT}{dn} = \text{gradient of } T = \text{grad } T \tag{A-25}$$

where dT/dn is the greatest spatial rate of change of T.

The divergence of a vector function q is defined as

$$\nabla \cdot q = \frac{\partial u}{\partial x} + \frac{\partial v}{\partial y} + \frac{\partial w}{\partial z} \tag{A-26}$$

The curl of a vector q is defined as

$$\nabla \times q = i\left(\frac{\partial w}{\partial y} - \frac{\partial v}{\partial z}\right) + j\left(\frac{\partial u}{\partial z} - \frac{\partial w}{\partial x}\right) + k\left(\frac{\partial v}{\partial x} - \frac{\partial u}{\partial y}\right) \tag{A-27}$$

If q is the gradient of a scalar function ϕ, i.e., $q = \nabla \phi$, then

$$\nabla \times q = \nabla \times \nabla \phi = 0 \tag{A-28}$$

If the vector q is the curl of a vector function A, then

$$\nabla \cdot q = \nabla \cdot \nabla \times A = 0 \tag{A-29}$$

The following are some other useful formulas:

$$\nabla \cdot (uq) = q \cdot \nabla u + u \nabla \cdot q \tag{A-30}$$

$$\nabla \times (uq) = u \nabla \times q + (\nabla u) \times q \tag{A-31}$$

$$\nabla \cdot (q \times q_1) = -q \cdot (\nabla \times q_1) + q_1 \cdot (\nabla \times q) \tag{A-32}$$

$$\nabla (q \cdot q) = 2(q \cdot \nabla)q + 2q \times (\nabla \times q) \tag{A-33}$$

$$\nabla \times (q \times q_1) = -(q \cdot \nabla)q_1 + q(\nabla \cdot q_1) + (q_1 \cdot \nabla)q - q_1(\nabla \cdot q) \tag{A-34}$$

(B) General Orthogonal Coordinates

In chapter III section 14, we discussed the general orthogonal coordinates and the expressions of the gradient of a scalar function, divergence of a vector function, and the curl of a vector function. In deriving the fundamental equations of fluid mechanics, we also need the expressions of the strain tensor in these coordinates which will be given here.

If the components of a velocity vector q in the direction of a_1, a_2, and a_3 be q_{a1}, q_{a2}, and q_{a3} respectively, the expression for the components of strain tensor are as follows:

$$\left.\begin{aligned}
e_{a_1a_1} &= \frac{1}{h_1}\frac{\partial q_{a1}}{\partial a_1} + \frac{q_{a2}}{h_1h_2}\frac{\partial h_1}{\partial a_2} + \frac{q_{a3}}{h_3h_1}\frac{\partial h_1}{\partial a_3} \\[6pt]
e_{a_2a_2} &= \frac{1}{h_2}\frac{\partial q_{a2}}{\partial a_2} + \frac{q_{a3}}{h_2h_3}\frac{\partial h_2}{\partial a_3} + \frac{q_{a1}}{h_1h_2}\frac{\partial h_2}{\partial a_1} \\[6pt]
e_{a_3a_3} &= \frac{1}{h_3}\frac{\partial q_{a3}}{\partial a_3} + \frac{q_{a1}}{h_3h_1}\frac{\partial h_3}{\partial a_1} + \frac{q_{a2}}{h_2h_3}\frac{\partial h_3}{\partial a_2} \\[6pt]
e_{a_2a_3} &= \frac{h_3}{h_2}\frac{\partial}{\partial a_2}\left(\frac{q_{a3}}{h_3}\right) + \frac{h_2}{h_3}\frac{\partial}{\partial a_3}\left(\frac{q_{a2}}{h_2}\right) \\[6pt]
e_{a_3a_1} &= \frac{h_1}{h_3}\frac{\partial}{\partial a_3}\left(\frac{q_{a1}}{h_1}\right) + \frac{h_3}{h_1}\frac{\partial}{\partial a_1}\left(\frac{q_{a3}}{h_3}\right) \\[6pt]
e_{a_1a_2} &= \frac{h_2}{h_1}\frac{\partial}{\partial a_1}\left(\frac{q_{a2}}{h_2}\right) + \frac{h_1}{h_2}\frac{\partial}{\partial a_2}\left(\frac{q_{a1}}{h_1}\right)
\end{aligned}\right\} \quad \text{(B-1)}$$

where the first subscript of the strain component refers to the direction of the normal of the surface considered while the second subscript refers to the direction of change of the velocity. Thus, $e_{a_1a_2}$ denotes the strain component along a_2 of the shearing strain in the surface perpendicular to a_1.

The stress components are still given by the formulas:

$$\left.\begin{aligned}
\sigma_{a_1a_1} &= -p + 2\mu e_{a_1a_1} + \lambda\nabla\cdot q, & \tau_{a_2a_3} &= \tau_{a_3a_2} = \mu e_{a_2a_3} \\
\sigma_{a_2a_2} &= -p + 2\mu e_{a_2a_2} + \lambda\nabla\cdot q, & \tau_{a_3a_1} &= \tau_{a_1a_3} = \mu e_{a_1a_3} \\
\sigma_{a_3a_3} &= -p + 2\mu e_{a_3a_3} + \lambda\nabla\cdot q, & \tau_{a_1a_2} &= \tau_{a_2a_1} = \mu e_{a_1a_2}
\end{aligned}\right\} \quad \text{(B-2)}$$

The dissipation function (3.53) in the curvilinear orthogonal coordinates a_1, a_2, and a_3 is

$$\Phi = \Phi_a = \mu[2(e_{a_1a_1}^2 + e_{a_2a_2}^2 + e_{a_3a_3}^2) + e_{a_2a_3}^2 + e_{a_3a_1}^2 + e_{a_1a_2}^2]$$
$$+ \lambda(e_{a_1a_1} + e_{a_2a_2} + e_{a_3a_3})^2 \qquad \text{(B-3)}$$

The above formulas can be applied to the special cases of cylindrical or spherical coordinates as follows:

(a) **Cylindrical coordinates.** In this case, if r, θ, z are taken as a_1, a_2, a_3 respectively, then

$$h_1 = 1, \quad h_2 = r, \quad h_3 = 1 \qquad \text{(B-4)}$$

because $x = r \cos \theta$, $y = r \sin \theta$, and $z = z$. The velocity vector q has components q_r, q_θ, and q_z. The strain components are:

$$\left.\begin{array}{l} e_{rr} = \dfrac{\partial q_r}{\partial r}, \quad e_{\theta\theta} = \dfrac{1}{r}\dfrac{\partial q_\theta}{\partial \theta} + \dfrac{q_r}{r}, \quad e_{zz} = \dfrac{\partial q_z}{\partial z} \\[3mm] e_{\theta z} = \dfrac{1}{r}\dfrac{\partial q_z}{\partial \theta} + \dfrac{\partial q_\theta}{\partial z}, \quad e_{rz} = \dfrac{\partial q_r}{\partial z} + \dfrac{\partial q_z}{\partial r} \\[3mm] e_{r\theta} = \dfrac{1}{r}\dfrac{\partial q_r}{\partial \theta} + \dfrac{\partial q_\theta}{\partial r} - \dfrac{q_\theta}{r} \end{array}\right\} \qquad \text{(B-5)}$$

and the dissipation function is

$$\Phi = \Phi_c = \mu[2(e_{rr}^2 + e_{\theta\theta}^2 + e_{zz}^2) + e_{\theta z}^2 + e_{zr}^2 + e_{r\theta}^2]$$
$$+ \lambda(e_{rr} + e_{\theta\theta} + e_{zz})^2 \qquad \text{(B-6)}$$

(b) **Spherical coordinates.** In this case we take r, θ, ϕ as a_1, a_2, a_3 respectively, then

$$h_1 = 1, \quad h_2 = r, \quad h_3 = r \sin \theta \qquad \text{(B-7)}$$

because $x = r \sin \theta \cos \phi$, $y = r \sin \theta \sin \phi$, and $z = r \cos\theta$. The velocity vector q has components q_r, q_θ, and q_ϕ. The strain components are:

$$e_{rr} = \dfrac{\partial q_r}{\partial r}, \quad e_{\theta\theta} = \dfrac{1}{r}\dfrac{\partial q_\theta}{\partial \theta} + \dfrac{q_r}{r}, \quad e_{\phi\phi} = \dfrac{1}{r \sin \theta}\dfrac{\partial q_\phi}{\partial \phi} + \dfrac{q_r}{r} + \dfrac{q_\theta \cot \theta}{r} \Big]$$

$$e_{\theta\phi} = \frac{\sin\theta}{r}\frac{\partial}{\partial\phi}\left(\frac{q_\phi}{\sin\theta}\right) + \frac{1}{r\sin\theta}\frac{\partial q_\theta}{\partial\phi}$$

$$e_{\phi r} = \frac{1}{r\sin\theta}\frac{\partial q_r}{\partial\phi} + r\frac{\partial}{\partial r}\left(\frac{q_\phi}{r}\right), \quad e_{\theta r} = r\frac{\partial}{\partial r}\left(\frac{q_\theta}{r}\right) + \frac{1}{r}\frac{\partial q_r}{\partial\theta} \qquad (\text{B-8})$$

and the dissipation function is

$$\Phi = \Phi_s = \mu[w(e_{rr}^2 + e_{\theta\theta}^2 + e_{\phi\phi}^2) + e_{\theta\phi}^2 + e_{\phi r}^2 + e_{r\theta}^2] + \lambda(e_{rr} + e_{\theta\theta} + e_{\phi\phi})^2 \qquad (\text{B-9})$$

Subject Index

EGD generator 315
Einstein formula 516
elastic collision 25
elastic modulus 526
E-layer 71
electric charge 42,535
electric conducting fluid 64
electric conduction current 42,265,491, 535
electric conductivity 42,266,317,494
electric convection current 42,265,491
electric current density 42,155,265,483, 535
electric efficiency 282
electric field 42,156,535
electric field number 270,537
electric Knudsen number 83,535
electric potential 65
electric units 541
electric wind 311
electrogasdynamic approximations 311
electrogasdynamics 263,310,330
electromagnetic force 264,486,535
electromagnetofluid dynamics 263,330
electron 1
electron excitation 23
electron mass 539
electron sound speed 500
electron volt 539
electronic charge 539
element of length 139
elliptical coordinates 141
EM-domain 329,452
emission coefficient 340,359,535
encounter 7
energy density of radiation 27
energy flux 416,434
energy level 18
energy source 488
enlargement of a pipe 109
enthalpy 21,24,208,535
entropy 126,221,537
equation of conservation of electric charge 266,485

equation of continuity 91,117,215, 264,417,483
equation of diffusion 62,91,215
equation of diffusion velocity 429,487
equation of electric current 265
equation of energy 62,114,117,215, 265,417,485
equation of heat flux 420
equation of motion 97,117,215,264, 417,485
equation of pressure stress 419
equation of radiative transfer 342
equation of state 14,62,117,215,264,483
equation of transfer 416
equilibrium condition of a gas 411
equilibrium constant 26,225
equilibrium flow 207,221,223,480
equilibrium sound speed 225
equipartition of energy 44
equipotential surface 50
ergodic theorem 136
Euler equation 105,151,163
Eulerian method 86
Euler's turbine theorem 111
evaporation 41,467
excess electric charge 42,65,155,483, 538
excitation energy 23
excited state 20
exosphere 71

Faraday MHD generator 281
fast shock 297
fast wave 294
film coefficient 172
film effect 508
finite mean free path of radiation 370
first law of thermodynamics 209
first-order chemical reaction 39,211
first-order collision regime 406
fixed bed flow 475
flagellar motion 527
flame front 236
flat plate 437
F-layer 71

Author Index